Biotheology

A New Synthesis of Science and Religion

Michael Cavanaugh

Foreword by
Loyal Rue

University Press of America, Inc.
Lanham • New York • London

4720 Boston Way
Lanham, Maryland 20706

3 Henrietta Street
London, WC2E 8LU England

Printed in the United States of America
British Cataloging in Publication Information Available

Library of Congress Cataloging-in-Publication Data

Cavanaugh, Michael.
Biotheology : a new synthesis of science and religion / Michael Cavanaugh ; foreword by Loyal Rue.
p. cm.
Includes bibliographical references and index.
1. Biology--Religious aspects. 2. Evolution--Religious aspects. 3. Ethics. 4. Religion and science. I. Title.
BL255.C38 1995 215 --dc20 95-33439 CIP

ISBN 0-7618-0104-9 (cloth : alk: ppr.)
ISBN 0-7618-0160-X (pbk: alk: ppr.)

™ The paper used in this publication meets the minimum requirements of American National Standard for information Sciences—Permanence of Paper for Printed Library Materials, ANSI Z39.48—1984

for Carolyn

With all my love

Contents

Part Two: Theology

Part Three: Biotheology in Individual Life

Part Four: Biotheology in Group Life

Foreword

Theodosius Dobzhansky once observed that nothing in biology makes any sense apart from the concept of evolution. E.O. Wilson extended the thesis with his suggestion that the social sciences can no longer proceed constructively apart from the paradigm of evolutionary biology. With a gentle and steady hand Michael Cavanaugh carries the insight the rest of the way: theology, too, is now implausible apart from the ultimate reference point of the evolutionary paradigm. Theology, henceforth, is biotheology.

Yet *BIOTHEOLOGY* does not offer us a flimsy reductionist thesis—it does not insist that concepts in theology are nothing but primitive attempts to speak about something else. God-talk is still talk about God. But while theology is not merely biology, neither is is extra-biological. Theology is an imaginative and constructive human enterprise, and humans are—don't forget it—biological beings. This basic fact should be enough to persuade us that biology can and should be a powerful resource for analyzing and reconstructing theological concepts. Herein lies the basic program of *BIOTHEOLOGY*.

Cavanaugh gives his reader a leg up to the program by providing a concise and selective overview of relevant evolutionary theory, culminating in a discussion of the emergence of human thought, language and culture. This lays the groundwork for a sustained examination of the ways in which theological concepts actually work, that is, how they emerge, how they enable us, and how they manage to change in response to new challenges. All of which amplifies the point of having a biotheological program: to equip us to respond theologically under the conditions of a world that is scientifically sophisticated and environmentally threatened.

The historical analysis offered in this book makes it clear how theological concepts have conferred adaptive payoffs in the past. But as any good evolutionist knows, the game is not well played by resting on laurels. Unyielding orthodox supernaturalism must be seen for what it is: an idolatrous recipe for fossilhood. In contrast, biotheology, i.e., Life-Theology, invites us to be venturesome and experimental in our attempts to image divine reality naturalistically. At the very least we are invited to reconstruct theological concepts that bear upon our personal and collective existence.

In the second half of the book the reader will find much to stimulate hope that our traditions are still viable, provided we have the courage and imagination to reconstruct their theological foundations. Here Cavanaugh considers the payoffs of a new religious naturalism for self-fulfillment as well as for social progress.

Cavanaugh undertakes to reconstruct our vision of human nature by integrating traditional doctrines of theology (e.g. free will, sin, grace, righteousness) with contemporary insights from biology. Finally, the author draws out the implications of a biotheological view of human reality for the domains of economics, politics, education and religion.

The result is both satisfying and compelling. The reader will discover that Cavanaugh has overlooked nothing in his attempt to synthesize traditional theology with contemporary biology. One comes away from the book with a sense that theology has been biotheology all along, and that we are only now coming to an explicit recognition of the fact. But this recognition is no trivial thing—it amounts to a sea change in self-understanding, an opportunity to recover our bearings. Given our present circumstances it is difficult to imagine anything we need more urgently.

Loyal Rue

Preface

Twentieth Century Theology is a highly creative enterprise, but its very creativity has produced a near hopeless state of fragmentation. Several dynamic developments converged to produce this profound pluralism, including the Protestant Reformation, freedom of the press, and human imagination itself. At a popular level these dynamics resulted in dozens and dozens of religious denominations. At an academic level they inspired differentiation and combinations of various theological and secular traditions to yield almost as many separate theologies as there are theologians. Even in the realm of naturalistic theology, where the emphasis on science offers hope for commonality, there is surprisingly little consensus as to either methodology or conclusions.

That is frustrating. It is frustrating for honest theologians working hard to construct a cohesive worldview, for intelligent laypersons who are looking for a plausible framework on which to hang their beliefs and behaviors, and for students who are trying to make sense of the many theological and philosophical currents in our culture.

One of the best strategies for overcoming frustration is to organize it, breaking it down into manageable categories. Fortunately, several outlines or "typologies" of modern theology have been published, which can indeed help us get a handle on the many complex issues addressed by twentieth-century theologians. None of the typologies is absolutely persuasive, and the authors are themselves uneasy about their own categories, largely because it is so difficult to place any given theology within one and only one category.[1] Yet despite these problems we can get some resolution, some clearer insight into modern theology and especially into naturalistic theology, by discussing three separate typologies and looking at how they intersect. At the least, a cross-sectional look at these typologies will locate biotheology within the landscape of contemporary ideas, and that is the primary goal of this preface.

The three typologies I will discuss are those of John Macquarrie, Hans Frei, and Arthur Peacocke. The first divides modern theology into 19 categories, the second lists 5 categories, and the third lists 8 categories.

Macquarrie's listing is not strictly typological so much as historical, in that it labels various traditions as they were formulated in the first half of the 20th Century (through 1960). Nonetheless, by setting out these 19 categories with some of their representatives, we will begin to get a flavor of the many thoughts and ideas that must somehow be incorporated in a modern synthesis. The categories are found in Macquarrie's *Twentieth Century Religions Thought*,[2] and they are as follows:

1. Absolute Idealism. (E. Caird, H. Jones, F.H. Bradley, B. Bosanquet, C.A. Campbell, J. Royce, J.R. Illingworth, R.J. Campbell, R.W. Trine)
2. Personal Idealism. (A.S. Pringle-Pattison, W.E. Hocking, G.H. Howison, J.M.E. McTaggart, H. Rashdall, C.C.J. Webb)
3. Philosophies of Spirit. (R. Eucken, B. Varisco, E. Boutroux, J. Ward, B.P. Bowne, E.S. Brightman, W.R. Sorley, A.E. Taylor, W.G. de Burgh, F.R. Tennant)
4. The Idea of Value in Philosophy and Theology. (W. Windelband, H. Cohen, H. Vailinger, H. Höffding, W. Herrman, T. Haering, J. Kaftan, A. Harnack, H.C. King)
5. Positism and Naturalism. (E. Mach, K. Pearson, E. Haeckel, E.B. Tylor, J.G. Frazer, S. Reinach, J.H. Leuba, S. Freud, C.G. Jung)
6. Philosophies of History and Culture. (W. Dilthey, O. Spengler, E. Carrirer, B. Croce, G. Gentile, R.G. Collingwood, A.J. Toynbee, W.M. Urban, C. Dawson)
7. Christianity, History and Culture. (O. Pfleiderer, E. Troeltsch, J. Weiss, A. Schweitzer, A. Drews, W.R. Inge, F. von Hügel, F. Heiler)
8. Sociological Interpretations of Religion. (E. Durkheim, M. Weber, A. Kalthoff, K. Kautsky, V.I. Lenin, W. Gladden, W. Rauschenbusch, S. Mathews)
9. Pragmatism and Allied Views. (H. Bergson, M. Blondel, C.S. Peirce, W. James, J. Dewey, A.F. Loisy, L. Laberthonniére, E. Le Roy, G. Tyrrell, H.N. Wieman, H.E. Fosdick)
10. Philosophies of Personal Being. (M. Buber, K. Heim, M. de Unamuno, J. Ortega y Gasset, N. Berdyaev, S. Bulgakov, J. Macmurray)
11. The Religious Consciousness and Phenomenology. (R.R. Marett, R. Otto, J.W. Oman, E. Husserl, M. Scheler, G. van der Leeuw, M. Eliade)
12. The New Realism. (F. Brentano, G.E. Moore, B.A.W. Russell, C.D. Broad, H.H. Price, R.B. Perry, G. Santayana)
13. The New Physics, Philosophy and Theology. (M. Planck, A. Einstein, W.K. Heisenberg, A.S. Eddington, B.H. Streeter, E.S. Barnes)
14. Realist Metaphysics and Theology. (C.L. Morgan, S. Alexander, A.N. Whitehead, C.E.M. Joad, N. Hartmann, W. Temple, L.S. Thornton, P.

Teilhard de Chardin, C. Hartshorne, A.C. Garnett)
15. Neo-Thomism and Roman Catholic Theology. (D.J. Mercier, M. de Wulf, P. Coffey, J. Maritain, E. Gilson, F.C. Copleston, A.M. Farrer, K. Adam, K. Rahner, H.U. von Balthasar, E. Pryzwara, J. Daniélou, F.J. Sheen)
16. Logical Empiricism. (L. Wiggenstein, R. Carnap, H. Reichenbach, K.R. Popper, A.J. Ayer, G. Ryle, R.B. Braithwaite, J. Wisdom)
17. The Theology of the Word. (K. Barth, E. Brunner, O. Cullmann, G. Aulén, A. Nygren, D. Bonhoeffer)
18. Post-Liberal Theology in the English-speaking Countries. (J. Baillie, D.M. Baillie, H.H. Farmer, Reinhold Niebuhr, H.R. Niebuhr)
19. Existentialism and Ontology. (M. Heidegger, K. Jaspers, J.P. Sartre, G. Marcel, L. Lavelle, R. Bultmann, F. Gogarten, F. Buri, P. Tillich)

If you carefully read the list above, you will recognize some non-theologians, who are quite properly included because of their contributions to theology. Even without the non-theologians, however, and even without considering developments since 1960 (such as environmental theologies, liberation theologies, feminist theologies, etc.), the number of Macquarrie's categories demonstrates the contemporary fragmentation and/or creativity of theology. Nineteen categories is just too many, and makes one yearn for a more efficient typology with fewer categories.

This need is answered by a typology that only has five categories, and I intend to discuss it next. But before doing that it may be helpful to locate Biotheology's ancestors in the listing above, as a first effort at placing it within the 20th Century context.

Unfortunately, that is not easy. Biotheology draws threads from several categories, and discusses or references many of these writers. It draws especially from the "realistic" categories (#'s 5, 12, 14, and 16), but it also has a strong value component (# 4), and it is highly indebted to studies in culture and sociology (#'s 6, 7, and 8). Its older heritage preceded even these categories, insofar as it participates in what has traditionally been called natural theology. It is tempting to discuss that ancient heritage further, but it would be impossible to do so in a reasonable number of pages, and other sources discuss it at length.[3]

Let us therefore turn to the typology of Hans W. Frei found in a 1992 book called *Types of Christian Theology*,[4] which was compiled by Frei's friends after his death from various of his papers and speeches and unfinished works. Here are Frei's categories, with representative writers again indicated in parentheses:

1. Theology as a Philosophical Discipline. (Kant, Kaufman, Fichte).
2. Theology as a Manifestation of the Spirit of the Age. (Bultmann, Pannenberg, David Tracy's early works, Karl Rahner, Carl Henry)

3. Theology as a Procedural Study. (Schleiermacher, Tillich, David Tracy's later works)
4. Theology as a Practical Discipline. (Karl Barth, John Henry Cardinal Newman, Jonathan Edwards)
5. Theology as Internal Self-Description. (D.Z. Phillips, later Wiggenstein, maybe Richard Rorty and Alasdair MacIntyre)

You might notice that Frei doesn't limit himself to the 20th Century, though most of his examples come from there. You might also notice how difficult it would be to harmonize this 5-category listing with Macquarrie's listing above. Perhaps some brands of idealism or phenomenology could be merged with # 1 or # 5 here, but it would seem quite strange to make bedfellows of Karl Barth and John Dewey, which would be necessary if categories # 9 and # 17 above were squeezed into Frei's category # 4, for example.

Despite its difficulties this typology is useful for our purposes, because it has one category that biotheology clearly fits within, and that is # 2. Although biotheology has some affinities with other categories,[5] Frei's description of the "spirit of the age" decides the placement. In the grand tradition of G.F. Hegel, Frei perceives each age as focusing on some concept or set of concepts that has a great influence on subsequent history, and he identifies certain theologies as participating in and reflecting the spirit of their particular ages. In our case the spirit of the age is science (though some theologians would dispute this), and since biotheology is a science-based theology it fits neatly in category # 2, with some significant bleedover into category # 1.

The inclusion of Pannenberg's name in Frei's category # 2 is certainly correct, but it gives a hint of some additional explanation that is needed before listing the third and final typology. Pannenberg is a widely-respected theologian who takes most of his cues from physics instead of biology. Based on the principles and uncertainties of modern physics, he is able to make a fairly persuasive case for the existence of a somewhat classical God. Thus Pannenberg and other similar theologians fit squarely into the natural theology tradition mentioned earlier, whereas biotheology and other biology-based theologies slightly depart from that tradition. Instead of focusing on theology and seeing science and reason as resources for proving the truth of traditional doctrines, biology-based theologies tend to focus on science first, seeing theology as a way of describing nature and determining our ethical stance within it. That is why I said earlier that Biotheology has some of its roots in cultural and sociological approaches to theology.

But I am getting ahead of myself. The third typology was actually published eleven years before Frei's typology, but it represents an elaboration of his category # 2 (in which I have located biotheology), and therefore it serves as a good follow-up to Frei's typology even though it was published earlier. Unfortunately this third typology was never fully developed, and is merely a rough listing in an introduction to a collection of symposium papers. However, its author is one of the foremost articulators of the spirit of our age—science—in theological terms, and his typology thus turns out to be very helpful.

The author is Arthur Peacocke, who has himself written several books and articles closely related to biotheology such as *God and The New Biology* and *Creation and the World of Science*.[6] But it is as editor of *The Sciences and Theology in the Twentieth Century* that he constructs the typology discussed below.[7] In such a sketchy listing Peacocke gives very few examples of contemporary writers who fit in each category, but I have taken the liberty to add some. Peacocke says the relation between science and theology can be conceptualized in the following 8 ways:

> 1. *Science and theology are concerned with two distinct realms.* Despite the popularity of this dualistic approach from Descartes forward, it is losing ground today, at least among naturalistic theologians. (Karl Heim)
> 2. *Science and theology are interacting approaches to the same reality.* (Peacocke himself)
> 3. *Science and theology are two distinct non-interacting approaches to the same reality.* In popular parlance, this means science asks "how" questions and theology asks "why" questions. (Paul Tillich, Peter Winch, Michael Bradie[8])
> 4. *Science and theology constitute two different language systems.* (late Wiggenstein)
> 5. *Science and theology are generated by quite different attitudes in their practitioners—science that of objectivity and logical neutrality, theology that of personal involvement and commitment.* (Karl Barth, Paul Tillich)
> 6. *Science and theology are both subservient to their objects and can only be defined in relation to them.* One has either faith in science or faith in God. (T.F. Torrance, H. Richard Niebuhr)
> 7. *Science and theology may be integrated.* (Moltman, Pannenberg, Bowker, Ian Barbour, Arthur Peacocke, Gordon Kaufman, Phil Hefner)
> 8. *Science generates a metaphysics in terms of which theology is them formulated.* (Whitehead, Polanyi, Torrance again, Ian Barbour)

Again it is impossible to make real writers conform exactly to these categories, as the inclusion of some of them in two categories reveals. Nonetheless it is safe to say that biotheology fits quite closely in category # 7. It has some affinity with other categories (especially # 2 and 4) and it discusses three others favorably (#'s 3, 5, and 8). As you will see in the book itself, it rejects only # 1 and # 6, both of which depend on a strong dualism, the belief that there are two completely different kinds of reality in the universe.

So far I have pulled some threads from various schools of thought in the first half of the 20th Century, with an allusion and references to the ancient tradition of natural theology, and we have located biotheology within Frei's category of theologies that articulate the spirit of their ages. In our age that is science, and therefore we further isolated biotheology by locating it within Peacocke's category # 7. Of course the book itself bears the burden of locating biotheology more specifically than that, but it will be helpful to end this preface by describing several recent books that are closely related to biotheology—first cousins if not siblings—in order to further orient the reader as to biotheology's location within the family of modern theology.

In addition to Peacocke's own *God and The New Biology*, mentioned above,[9] three books will represent the specific category in which biotheology fits, which may be identified as a sub-set of Peacocke's category # 7 above, and labeled "biology-based theologies." The books I will discuss are Gordon D. Kaufman's *In Face of Mystery*,[10] Philip Hefner's *The Human Factor*,[11] and Loyal D. Rue's *Amythia*.[12]

The full title of Kaufman's book is *In Face of Mystery: A Constructive Theology*. One of his reviewers says it should be called a "Reconstructive" theology because it reaffirms so many traditional doctrines.[13] That is not entirely true, however, because Kaufman ultimately challenges many of those doctrines, at least insofar as they depend on supernatural or dualistic concepts.

In many ways this book is a culmination of Kaufman's career. He is a long-time professor of theology at Harvard Divinity school, and has written several other influential books, including *The Theological Imagination*, which will be discussed further in Chapter 5. This latest book is more comprehensive than the earlier ones, however, and clearly locates Kaufman as a biology-based theologian. That is because he begins by constructing a reliable doctrine of humanity, which he accomplishes by showing how our historical experience grew out of and built upon our evolutionary origin. After discussing some implications of this biohistorical foundation (such as the obligations that grow out of our social nature), he begins to establish a doctrine of the world.

This gets Kaufman into complexities which he deems mysterious, and he concludes that evolved mental processes require us to resolve these mysteries with "small steps of faith." Here Kaufman is trying to distinguish his argument from a Kierkegaardian "leap" of faith, and yet he avoids the accusation of reductionism by acknowledging our ignorance on many points. From my perspective this is a bit of a dodge, because Kaufman's small steps of faith are not much different from a normal conception of thinking and decision-making, in which we use what we know to decide how to approach what we don't know, relying on knowledge as far as we can and then making judgments based on that knowledge. Nonetheless, Kaufman's approach does allow him to fit a modern body of realistic ideas into traditional theological garments, without too much squeezing.

One of the most important of Kaufman's small steps of faith is his well-known insight into imagination. Human brains always create symbols, which summarize subconscious musings into conscious if still abstract forms. Chief among those abstract symbols is the concept of God, which once constructed becomes a source of orientation for individuals as they navigate within society, and for cultures as they navigate within the world.

Incidentally, Kaufman undergirds his project not only with "small steps of faith," but also with a concept of serendipity. He acknowledges that the universe is blind, distinguishing himself from those who see some grand and supernatural purpose in the universe. This does not lead him into nihilism (the idea that nothing really matters), but to a celebration of how well our nature corresponds to the realities of the world. Yet by no means does he lapse into syrupy pollyannia; indeed he soberly admonishes us to emphasize those parts of our social nature which give hope of solving the serious problems now facing humanity, and he acknowledges our power—even our tendency—to make erroneous decisions. He is ultimately optimistic, however, because he believes we will always construct and re-construct our theological symbols so as to render them useful in attacking the very kinds of problems we now face. He employs the concept of "trajectory" to argue that we are on a good path if we can just stay there.

Hefner's book is much to the same effect, but with some differences. For one thing he spends an entire chapter making sure his theory-building is considered sound by modern philosophical standards. This prompts one reviewer to shout "Stop fixing your uniform and play ball!," but in the current intellectual climate Hefner is probably wise to spend so much time on this issue.[14]

More important, Hefner grapples with the problem of purpose. He too acknowledges the blindness of the universe, but he relies on the concept of "teleonomy" to construct a theory of ordinary purpose, as opposed to the widely discounted "ultimate" purpose that once supported theological constructs. Yet for humans teleonomy is more than the ordinary purposes of daily life, however meaningful they might be; Hefner's teleonomy is based on the biological concept of survival, which he sees as a problem that must be addressed communally rather than individually.

Building on the ideas of Ralph Wendell Burhoe, Hefner gives special attention to cultural processes, but his most interesting concept describes human beings as "created co-creators." We were originally created by "what really is," a deliberately vague concept of God which is meant to be entirely consistent with biological evolution and other natural processes. Once we emerged, our own qualities and choices became critical to life on earth, not only for our own lives but also for all other species. At the present time we hold in our hands the power to make life either wonderful or horrible (or even non-existent), depending on how we exercise our role as co-creators.

As in Kaufman, myth-making is a valuable part of co-creation. Of course myth does not refer to untrue stories, but to any poetic or narrative structure that summarizes our experience and communicates it in powerful ways. There can be bad myths, as some of our technological ideology and potential for destruction makes clear, but the goal is to create psychologically sound myths, which organize and encourage us toward healthy cooperation in using the resources of this planet.

Finally, I want to return to the idea of fragmentation that began this preface, by discussing the work of Loyal Rue. In his *Amythia* (which means "the lack of myth"), Rue documents and decries the fragmentation in western culture, but he avoids the reactionary impotence of glorifying an imagined golden age from the past. Instead he traces a history of cycles, featuring relatively long periods of stable social life, punctuated by occasional periods of amythia, during which society wallows in indecisiveness while struggling to find an acceptable new worldview. His Chapter 4 traces specific paradigm shifts in the judeo-christian culture, which Rue sees as "modes of piety" connected only by loose social structures and by the powerful idea of covenant. One is reminded of the family axe, which has had three heads and eleven handles, and yet is somehow the same axe; it is the ongoing *usage and relationship* that give continuity, and not a rigid adherence to a changeless myth.

Interestingly, Rue's diagnosis is widely affirmed today by otherwise divergent groups. Fundamentalists and liberals, democrats and republicans, academics and laypersons, scientists and humanities scholars, all bemoan the lack of cultural unity and the fragmentation of religious and civil life. The liberal vision includes more diversity than the conservative vision does, but liberal theologians like Rue are clearly in agreement with more conservative citizens in the desire for a basic level of unity.[15] It even seems fair to suggest that this is one place where C.P. Snow's "two cultures" can stand on the same soil; almost everyone yearns for more safety and integrity that we seem to have, and almost everyone agrees that we need enough basic agreement about our moral and philosophical underpinnings to avoid the worst consequences of disagreement.[16]

Rue's cure will not be as widely accepted as his diagnosis is, but it firmly places him among the biology-based theologies. He calls for a new myth characterized by the twin requirements of 1) a plausible cosmology and 2) a persuasive morality based on that cosmology. Specifically, Rue suggests evolution as the "root metaphor" out of which any proposed myth would have to emerge. Above all, Rue interprets both biology and theology as mandating *community*, the one because community is inherent in our nature and the other because community is critical to mythic cohesiveness.[17]

Rue's book was published four years before the two books discussed earlier, and those books respond to Rue's call for a new myth based on biology. The book in your hand is intended to answer Rue's call even more precisely, however, by proposing specific doctrines and mechanisms for transposing classical supernatural concepts into plausible natural ones. It takes a more gentle approach than most theology books do, by starting at a very basic level and building its theology step by step. It does take a scholarly approach with lots of notes,[18] but it aims to set out its concepts so clearly that they can easily be understood by persons not already well-grounded in both biology and theology. Accordingly, its audience may be divided into three main groups, namely 1) intelligent laypersons who are struggling to find a worldview that accommodates both modern science and traditional theology; 2) students who are trying to get a handle on the many complex issues presented in courses with names like "Science & Theology" or "Philosophy of Science," which are becoming increasingly popular in both secular and religious universities; and 3) scholars who are already familiar with the data and ideas presented here, but who need them arranged in a more compact and useful way.

The book is divided into four parts, each with four chapters. Part One deals with basic biological questions, placing thought within a biological construct which actively incorporates culture. Part Two deals with basic theological questions, tracing both the history and the conceptual status of imagination, scripture, faith, and God. Part Three begins to establish the logical and emotional payoff for those concepts, by updating certain traditional doctrines (free will, sin, salvation, and abundant life) and showing how they intertwine with individual life. Part Four expands the potential payoff to include larger social issues.

Notes

[1] In defense of the typologists, one must admit the problem is compounded by the irritating tendency of theologians to improve their work with adjustments and expansions, so that it no longer fits whatever category it almost fit before. For example "realistic metaphysics" has been re-moulded into process theology.

[2] New York: Harper & Row, 1963. Incidentally, Macquarrie included three "Comments by the way" chapters that sought common themes in each group of categories, but those chapters do not amount to a typology.

[3] See for example Andrew D. White's *A History of the Warfare of Science with Theology in Christendom* (New York: D. Appleton and Company, 1936); Charles E. Raven's *Natural Religion and Christian Theology* (London: Cambridge at the University Press, 1953); and Chapter 1 of Loyal Rue's *By the Grace of Guile* (New York: Oxford University Press, 1994). From 1888 to the present the best original source is the Gifford lectures, most of which have been reduced to book form; an anthology of selections from 1888 to 1968 is *Earnest Enquirers After Truth*, ed. Bernard E. Jones (London: George Allen & Unwin Ltd., 1970). Beginning in 1966 an even more representative original source is *Zygon: Journal of Religion and Science*.

[4] New Haven, Connecticut: Yale University Press, ed. George Hunsinger and William C. Placher.

[5] Indeed, I struggled between categories 1 and 2 as biotheology's proper locus. Since Kaufman's later work places biotheology squarely in his camp (see below) it is tempting to place it there at the outset, especially since Frei is vague on whether science or some other construct expresses the spirit of our age. In any case Frei sees categories 1 and 2 as closely related, so my ambivalence is apparently one he would share.

[6] The first book listed was published by Dent (London, 1986) and the second by Oxford University Press (New York, 1979).

[7] Notre Dame, Indiana: University of Notre Dame Press, 1981. The typology is found at pages xiii-xv.

[8] See his *The Secret Chain: Evolution and Ethics* (State University of New York Press, 1994) and "What Does Evolutionary Biology Tell Us about Philosophy of Religion?" in *Zygon: Journal of Science & Religion* Vol. 29 No. 1 (March 1994).

[9] It is a real injustice not to discuss this book further here, and were not the preface already too long I would indulge in the luxury of doing so. More than anyone else Peacocke deals with the accusation of reductionism against biotheology, focusing on its emergent qualities instead; and he also deals effectively with the interplay between chance and lawfulness.

[10] Cambridge: Harvard University Press, 1993.

[11] Minneapolis: Fortress Press, 1993.

[12] Tuscaloosa: The University of Alabama Press, 1989. Other writers that I would place in this category include Lindon Eaves and Karl Peters.

[13] See Frederick Ferré's "Unfazed by Mystery," pp. 363-70 in *Zygon: Journal of Religion & Science* Vol. 29 No. 3 (September 1994).

[14] See Eugene D'Aquili's "The Summa Hefneriana: Myth, Megamyth, and Metamyth," pp. 371-381 in *Zygon: Journal of Religion & Science*, Vol. 29 No. 3 (September 1994).

[15] See for example the article by Lindon Eaves and Lora Gross entitled "Exploring the Concept of Spirit as a Model for the God-World Relationship in the Age of Genetics" at pp. 261-285, Vol. 27 (September 1992) of *Zygon: Journal of Religion and Science*; Langdon Gilkey's "Nature, Reality, and the Sacred: A Meditation in Science" at pp. 283-289 Vol. 24 (September 1989) of the same journal; and Mary Midgley's "Toward a new Understanding of Human Nature: The Limits of Individualism," Ch. 18 (pp. 517-533) in *How Humans Adapt: a Biocultural Odyssey*, ed. Donald J. Ortner (Washington, D.C.: Smithsonian Institution Press, 1983).

[16] Admittedly, some radical deconstructionists might not agree on the need for basic commonality, but even they seem to say that such a unity is desirable if we can persuade one another to cooperate to that extent. See pp. 260-261 of Paul Feyerabend's *Against Method* (London: Verso, 1988) and Richard Rorty's "The Priority of Democracy to Philosophy," pp. 175-196 in *Objectivity, Relativism and Truth* (New York: Cambridge University Press, 1991).

[17] I want to distinguish myself (and biotheology) from Rue in one respect, because of three objectionable pages in his new book *By the Grace of Guile* (New York: Oxford University Press, 1994). At pages 272-274 Rue embraces the postmodern critique in what seems an unnecessary (though provocative) extension to his argument. As you will see, I ultimately reject the postmodern critique, though I think it underscores the need for humility. In every other important respect, however, Rue is a prime mover in the establishment of the principles of biotheology.

[18] Footnotes (and endnotes) are important for many reasons other than the technical demands of scholarship or to squeeze in side comments like this one. They let readers second-guess authors, which tends to keep authors honest. They remind readers of the broad-based foundation upon which any worthwhile work is built. Finally, they permit the expression of an author's gratitude to other researchers and writers, and thus supplement the formal acknowledgments in an absolutely necessary way. Incidentally, my endnotes will usually give a complete citation at the first mention of any book or article, and thereafter enough information will be given to permit one to find the reference in the bibliography.

Acknowledgments

One could not possibly know everything one needs to know to write a book of this scope, without relying greatly on others. As I mentioned above, each endnote is a kind of acknowledgment, but there were so many additional persons with whom I had conversations about the subjects in this book, and so many authors who do not appear here, that it seems impossible to adequately acknowledge the debt. Because of both the acknowledged and the anonymous debt, I always felt a profound sense of humility as I wrote and rewrote, and a profound sense of gratitude as I read and researched.

I would like to specifically acknowledge and thank a group of twenty or so persons who gave feedback as I wrote. Because of them I made a thousand minor changes and many major ones, but all too often I stubbornly resisted their good advice. They should get much of the credit for whatever good is in this book, but the errors of fact and judgment are mine. The readers who stuck with me through all 16 chapters were especially helpful, because they were able to point out inconsistencies between earlier and later chapters. Those faithful readers were Susann Dorman, Bill Stone, Celia Girard, Pamela Davidson, Arthur Wang, Carolyn Cavanaugh, and Vaughn Crombie. Others who read a chapter or more were Bob Ardoin, Brad Buckhalter, Kevin Johnson, Richard Pevey, Daniel Salter, and Grant Smith.

After the book was accepted for publication but before it was put in final form, I benefited immensely from comments by 10 scholars in particular areas. They were Bill Irons at Northwestern, Nancey Murphy at Fuller Theological Seminary, Paul Kurtz at S.U.N.Y., Larry Greenfield at The Park Ridge Center in Chicago, Gerd Theissen at the Wissenschaftlich-Theologisches Seminar in Heidelberg, Robert Glassman at Lake Forest College, Timothy Earle at U.C.L.A., Holmes Rolston at Colorado State University, Linda Jones at Grace Lutheran Church in LaGrange, Illinois, and the Egyptologist Jane Sellers. Since each of these scholars read an isolated chapter, they cannot be held accountable for how I used their insights, but they were very helpful. I am also grateful to Loyal Rue (at Luther College) for his kind remarks in the Foreword, to Dora Jordan, who prepared the manuscript, and to Robert Segal (now at Lancaster University), who offered generous insights into the publishing process.

Which brings me to the publisher. What a positive experience! Michelle Harris at UPA was not only patient with my plodding ways, but she was always cheerful and uplifting. I do not believe the public has any idea how important good solid presses represented by encouraging people like Michelle are to our society. Without them it would be too burdensome to write even a simple book, much less a complex one. Every suggestion was useful, and some of them were profound. When it came time for production, even my quirks were respected by Helen Hudson (for example, she let me keep the "brief outlines" which open each chapter). In short I wish for every writer a publisher like University Press of America.

Finally, I want to express my special thanks to my wife Carolyn. Not only did she serve as a faithful reader and competent critic, but she gave support in a hundred other ways, from the emotional to the practical. She offered sensitive encouragement for the several years I was pre-occupied with this project, and she backed up that encouragement with concrete help and many thoughtful ideas. How blessed I feel to have such a partner.

Part One: Evolution

Chapter 1: Four and a Half Billion Years

Chapter 2: The Evolution of Thought and Language

Chapter 3: Culture, a Kind of Learning

Chapter 4: Concept-Formation: Thought, Language, and Culture Together

Chapter One

Four and a Half Billion Years

An Outline of the Chapter

A. Evolution Synopsis
B. Four Observations about Evolution
   1. Variety and Commonality
   2. Jury-Rigging
   3. Trade-Offs
   4. Niches

Biotheology's goal is to make theology plausible again, by placing it on the strong foundation provided by modern biology. It is especially interested in the human brain, because current concepts in brain biology allow us to build a new and stronger structure of moral and theological doctrine than has ever existed before. Such a structure is possible, however, only because excellent researchers and writers have clarified both biological and theological concepts in the last few years, and their work thus supplies the building blocks of biotheology. My task is largely to fit those building blocks together in a logical and pleasing order.

Let us begin by discussing the fundamental concept that underlies all of biology, namely evolution. This chapter will not try to convince the reader of evolution, but it does present findings which support evolution, not only from biology but also from several other disciplines.[1] And although certain creationists claim that lots of scientists disbelieve evolution, there are actually only a very few such scientists, at least if we limit the field to those speaking within their area of expertise.[2]

Evolution Synopsis

The earth was formed as a spinoff from the sun about 4.5 billion years ago. As it cooled, the oceans and atmosphere slowly formed, producing a particular mixture of chemicals in a "primordial soup" which was constantly charged with energy from such sources as lightning, sunshine, and volcanoes. That energy changed some of the chemicals into molecular compounds structurally inclined to replicate and develop, and those modified molecules turned out to be the precursors of life. After about another billion years of additional changes, those replicating compounds became simple lifeforms.

This chemical nature of life is easy to take for granted, but it profoundly influences every moment of our lives. Not only is medicine built on it, but we are beginning to understand how spiritual life depends on it too, as we will see.

The first simple organisms were more complex than the precursor compounds only by small incremental degrees. They evolved over vast time[3] until they became the myriad forms we know today, each tremendously complex, but each a variation on a simpler form that preceded it. As the underlying structures evolved, so did the behaviors that grew out of them, with results that seem electrifyingly dramatic only if you weren't there to see—or cannot now imagine—the slow changes and intermediate forms: Eyes evolved from light-sensitive cells, anger evolved in many species, care of the young evolved, symbolic communication evolved, sex evolved, kindness evolved, hunger-pains evolved. Wow and double-wow! These glorious things (and many others) evolved at different rates and in many different species, and often evolved in different ways through independent paths.

From the point at which evolution began, it took a long time for higher creatures like fish to evolve, and even longer for a recognizable mammal like the small tree shrew to appear. Indeed, of the 4.50 billion years since Earth formed, 4.44 billion passed before the shrew emerged. In the 60 million years since then, we have evolved from her (and from *him*, sex having evolved long before the shrew did[4]); thus the shrew is important to our history, and we need to understand what happened after its debut.

Sixty million years is only a bit over 1% of the 4.5 billion total, but it is still long enough to boggle our minds. To unboggle them, let me divide 60 million years into smaller parts, calculating how long the shrew's little leg had to grow each year to become my long one. If its leg was say 1/2 inch long and mine is say 30 1/2 inches long, the leg grew 30 inches in 60,000,000 years of evolution. That works out to 1/2,000,000th of an inch per year, which is also 1/16th of an inch every 125,000 years. That small yearly growth is easily believable, especially considering that we average a full inch taller than our grandparents.

Not only have we grown physically; Lorenz and others have brilliantly shown how *behavior* evolved in the same small degrees,[5] and Darwin himself wrote the first explanation of the evolution of emotions in 1872.[6] The evolution of behavior and emotions depended on changes in physical structures like hormones and neurons, which reminds us again how thoroughly physics and chemistry interlink with evolution, and impact on us today.

Four Specific Observations About Evolution

Such a brief synopsis should make it clear that a complete treatment of evolution has been left to the endnotes and bibliography, but because four principles will come up often in the following chapters, they should be discussed here at the outset, so the remaining chapters can flow more efficiently. Those four principles are 1) Variety and Commonality, 2) Jury-rigging, 3) Trade-offs, and 4) Niches.

1. Variety and Commonality

Variety is one of the things that caused evolution in the first place and one of the things that keeps it going; biologists say it is one of evolution's *engines*.[7] Barnyard variety helped start Charles Darwin thinking about variety in nature, and ultimately led to his exciting new theory.

For example, think about the differences between two breeds of horses. We started with similar horses, but by breeding to emphasize naturally occurring differences, we produced animals with very different appearances; thus one breed became the huge Clydesdales and one became the tiny Shetland ponies. Observing such manmade selection made Darwin wonder if the process could happen naturally, and whether it would ultimately lead to different species altogether.

If we put two breeds on different continents, and if we continued to select for completely different characteristics, we now know they *would* eventually become different species, and would not interbreed if put back on the same continent again. If the in-between breeds died out in the meantime, and the two remaining species changed enough, the differences might be so great as to hide the fact that they had a common ancestor. Darwin finally realized that this kind of thing happens naturally, over and over again, and is responsible for all the species that exist.

To get differences that result in separate species, nature begins with differences inside a species. We always knew individuals were different, but molecular studies now demonstrate that a surprisingly high number of individuals differ from their parents genetically. In other words, genetic material often reproduces itself inexactly. That is what starts each new phase of evolution. Geographic isolation and other natural processes merely magnify those differences to produce new species.

These comments about variety are standard fare in evolutionary literature,[8] but they have profound effects on philosophy too. Primarily they lead to an understanding of our commonality, its origins, and its limits.

Commonality is what binds us together as a species. Even though we differ, we are enough alike to expect basic similarities. We need similar clothes in similar climates (with due allowance for body build); we need similar foods to stay healthy (with due allowance for genetic differences such as lactose intolerance); we need similar sexual mores (perhaps with due allowance for genetic differences).

You can see by the parentheses above, however, that commonality bumps up against variety at every turn. Even apart from the variety caused by individual experience and culture (more on that later), commonality depends on body chemicals which evolved, but which evolved so as to produce variety in the gene pool of any population.[9] For example aggressiveness and risk-taking appear to depend on brain chemicals which vary from person to person[10] in the same ways eye color varies across our species.

This doesn't change the fact that our commonality is profound—a species is defined by hereditary similarities. But these observations also show how profound inherent differences are. A modern theology must account for both the commonality and the variety.

That will make it hard. To say something that applies to everybody will be very difficult. I will often remind you that there are exceptions to what I say. On the other hand, our commonality gives hope of identifying some general truths, and I will point out such truths when we seem to be in their presence.[11]

2. Jury-Rigging

A second principle of evolution will also affect our later understanding, and that is its hodge-podge aspect. I don't mean to imply that evolution leads to inelegant or awkward solutions to life's problems (though at times it does). I mean rather that evolution builds on what was previously there. Let me explain further by discussing three similar but confusing phrases:

- Jury-fixing is an illegal activity wherein one tries to persuade a civil jury to vote differently than it otherwise might, usually by threats or bribes.

Of course this has nothing much to do with either evolution or theology, at least not in any direct way.

- Jerry-rigging is an ethnic slur growing out of the last days of World War II, when lack of time or supplies forced the Germans (called "Jerries") to put together products and weapons poorly. So to say something is jerry-rigged, means it is sloppily constructed.

Because of evolution's trial-and-error nature, genetic changes can be pretty sloppy. But environmental and competitive pressures usually eliminate unworkable changes from the genetic pool over time. So jerry-rigging is a poor metaphor for describing evolutionary products.

- Jury-rigging is a nautical phrase meaning "rigged for temporary service." Solving problems at sea with materials at hand can be a fine art; the goal is not necessarily mechanical precision or professional design, but simple elegance and practical workability.

Of the three concepts, evolution works most like jury-rigging. At any point in time, the nature and structure of a species is "given," and further evolution must grow out of that material, as it exists then. This results in solutions that are obviously not engineered "from the ground up," but rather are built on what was there before.

Jury-rigging can be elegant, but it would be a mistake to think evolution makes perfect fits between an organism and its environment, or between members of a species in their social life, or between the parts of an individual organism. Instead, evolution works like this: A species exists, and variety-producing mutations occur at microscopic genetic levels in some individuals of that species. When those individuals are full grown, or somewhere along the way, these genetic changes produce bodily or behavioral changes, some of them big and some small.

Each change makes an individual either better adapted to its environment than other members of its species, or worse adapted, or else it makes no real difference in adaptation. Another possibility is for the environment itself to change, in which case the modified individual will fit the new environment better or worse or about the same. In all probability the individual organism has no idea whatever that it carries the genetic change, much less whether the change helps it adapt better or worse to the challenges of its life. But in fact the change does cause it and its descendants to adapt better or worse, or else makes no difference.

Over a very long time, small accumulated changes in genetic makeup or in the external environment can make a big difference (and occasionally a single change can). The sub-groups that carry the change can become extinct because it was a non-adaptive change. Or they may be very successful and displace other species or sub-species, either closely related ones or entirely different ones, because the change was advantageous for them.

In fact, this process has produced many non-adaptive features in living species, humans included.[12] Some such features are downright disadvantageous, though not so disadvantageous as to cause extinction, so far anyway. Some are not particularly adaptive, but not particularly harmful. For example some birds have showy feathers that don't help them secure territory or attract mates, and which might even attract certain predators if they lived nearby, but as it is no harm is done.

In none of these cases does the species (much less the individual) plan ahead.[13] It simply happens that some random changes make the fit better, and some don't. Every living species is the result of a series of good fits, but that does not translate into perfect fits, and certainly not into any conscious planning.[14] Unfortunately, some evolution literature makes it sound as if organisms and entire species actively plan strategy for the future. Such language confuses the issue even though it is meant to explain, because the metaphorical nature of such words as "plan" and especially "strategy" are not made clear.

In humans this jury-rigged construction results in useful but cranky organs like the appendix and gall-bladder. It also results in thinking patterns and behavioral tendencies which are effective but problematical. That is not to say evolution produced a shoddy product, but it means we should not expect to find perfection in our natures. On the contrary, we should expect to find problems related to our jury-rigged construction.

There are many ways this jury-rigged aspect of evolution must influence our views of human nature and our concepts of God, which will surface as we go along. My theme will be that theology must account for the problematic aspects of our nature, both those so problematic as to be pathological, and those that are problematic though essentially healthy.

3. Trade-offs

The third evolutionary dynamic to be discussed is trade-offs. Medical pathology presents many examples, and I will report on one of them, namely sickle-cell anemia, to represent the hundreds of others in our nature.

Sickle-cell is a poignant part of a doctor's life. It is an inherited disorder affecting mainly black and Mediterranean people, causing sickle-shaped cells that can't easily pass through the capillaries. These cells cause excruciating pain and early death in persons who inherit the gene from both parents.

Medical detectives have discovered that possessors of the gene have an advantage most of us don't have, namely a resistance to malaria, a disease transmitted by mosquitoes. Nowadays this might not seem like a big advantage, but malaria was feared in America only two generations ago.[15] Overall it was worth some health problems to survive it—in fact the sickle cell could save the species from extinction under certain conditions.

So there is a trade-off between the sickle cell's advantages and disadvantages. Today we have other ways of getting the advantages (using mosquito-control poisons, for example), so we deal with the disadvantages through treatment and genetic counseling. Yet we cannot completely avoid the trade-off, since poisons can pollute the environment. Besides, it is probably impossible to get rid of mosquitoes; this could lead us to eliminate the sickle cell just as malaria comes charging back, and we'll have lost evolution's offer of that particular trade-off.[16]

4. Niches

Although evolution produces some non-adaptive features and some unusual features like the sickle-cell, its more usual course is to retain genetic changes that are advantageous. We can see this more clearly by discussing niches.

As a species expands, near-identical individuals get spread over quite a large area. Perhaps one group migrates to a lush jungle while another remains in a nearby savannah. Assuming some barrier between them (say a river or a mountain range), the variations that naturally appear in the two groups will have different impacts. In the jungle, variations will die out that might have been useful in the savannah, and variations will flourish which might have been useless there. And the opposite will happen in the savannah. Over time you can see that two different species will result. (If the two groups get back together before becoming separate species, the gene pool will have lots of variety in it, which can influence further evolution, perhaps conferring additional advantages and/or disadvantages in the competition with other species.)

Once the two groups are different enough to prohibit interbreeding, they will continue to evolve separately even if they get back together geographically. In that case, they probably won't compete for the same parts of the environment. Desert-evolved organisms will eat different foods than jungle-evolved ones, or will prefer different kinds of shelter, or will prefer to stay on the ground instead of the trees, or will have thicker skin. In this way hundreds of species can eventually live in the same area, but will occupy such different niches that they may as well live in different areas. Yet each one becomes part of the other's environment, either as prey or predator or competitor, or as mere background. The resulting interaction creates a tapestry of life that is profoundly dynamic. Although competition tends to weed out weaker organisms, it also opens up additional niches, including some very subtle ones. For example we will see how humans evolved as a cooperative species that competes partly by squelching competition, and partly by encouraging it in controlled ways.

The phrase "niche" sounds like each species is fitted for a very narrow piece of the environment, but humans and a few other species claimed our niches as generalists rather than specialists, which allows us to cultivate a broader slice of available resources. This led to an emphasis on cleverness in such species as the rat and the crow (who are also generalists), and in humans it was part of the development of reason. Again, we didn't claim this broader niche through planning or strategy, but through the normal random changes discussed in the jury-rigging section above. At this point it may be possible, as some have suggested, for us to determine our own future evolution,[17] but there are so many unforeseen consequences of any action that I doubt it. While rational planning has enormous potential for us as individuals and temporal groups, it would seem somewhat arrogant, at this stage of our development, to think we could consciously control our own evolution to a very large extent.

Incidentally, one writer has challenged the niche concept,[18] because it ignores the fact that most species modify the environment to create their niche. The concept thus leads to the artificial idea that specific physical niches lie waiting for specific species to come live in them. Of course the criticism is correct if we define niche so concretely, but the discussion above illustrates how dynamic the process is, so that "niche" describes an ongoing interaction between environments and species, rather than a set of static and pre-defined places.

* * *

From time to time in this book we will consider how these four dynamics of evolution (and others) affect us. In exploring culture, for example, I will develop the discussion about trade-offs between competition and cooperation; and in discussing morality I will consider how these dynamics might have produced both sexual fidelity and sexual promiscuity, side by side. We will also explore many other results of evolution which impact on our daily lives, both those we can do something about, and those we may not be able to change. Even in the latter case, however, increased understanding will offer us significant practical advantages and deeper spiritual meaning.

This concludes our overview of evolution. As already acknowledged, these broad strokes omit many important principles and mechanisms. But the chapter should provide enough background to let us think more clearly about the evolution of thought, language, and culture. We will discuss those topics in the next three chapters, and then Part Two will begin to concentrate on the theology which grows out of evolutionary understanding.

Notes

[1] For example see Theodosius Dobzhanzky's essay "Nothing in Biology Makes Sense Except in the Light of Evolution," in *Evolution versus Creationism: The Public Education Controversy*, ed. J. Peter Zetterberg (Phoenix: Oryx Press, 1983). For an especially readable primer on evolution see *An Introduction to Evolution*, 3rd edition, by Paul Amos Moody (New York: Harper & Row, 1970). And although time has naturally required some changes in the basic Darwinian theory, Charles Darwin's *On the Origin of Species* is still as profound and readable as ever, available not only in libraries but also in bookstores.

[2] For an interesting report by a scientist who was erroneously reported as having rejected evolution, see Richard Lewontin's introduction to *Scientists Confront Creationism*, ed. Laurie Godfrey (New York: W.W. Norton & Company, 1983). See also *Did the Devil Make Darwin Do It?*, ed. David Wilson (Ames: The Iowa State University Press, 1983); *The Creation Controversy*, by Dorothy Nelkin (New York: W. W. Norton & Company, 1982); *Is God a Creationist? The Religious Case Against Creation Science*, ed. Roland Mushat Frye (New York: Charles Scribner's Sons, 1983); and *McLean v. The Arkansas Board of Education*, 529 F. Supp 1255 (1982). For a contrary opinion as to the number of scientists who reject evolution, see Langdon Gilkey's *Creationism on Trial: Evolution and God at Little Rock* (Minneapolis: Winston Press, 1985).

[3] Evidence for an ancient earth comes from several independent dating techniques, compellingly presented in *The Earth's Age and Geochronology*, by Derek York and Ronald Farquhar (New York: Pergamon Press, 1973). A more technical but recent source is *The Chronology of the Geological Record*, ed. N.J. Snelling (London: Blackwell, 1985).

[4] See *Why Males Exist: an inquiry into the evolution of sex*, by Fred Hapgood (New York: Morow, 1979). For more on the long journey from the shrew to man (or more correctly, from a shrew-like mammal to man), see Wilfrid Edward LeGros Clark's *Early Forerunners of Man: a morphological study of the evolutionary origin of the primates* (New York: AMS Press, 1979).

[5] In his 1973 Nobel acceptance speech, Konrad Lorenz said that he best accomplished this in his paper comparing behaviors of various species of ducks (the anatinae). That and other remarkable Lorenzian ideas are most accessible to the English reader in his *Studies in Animal and Human Behavior*, 2 volumes (Cambridge, Massachusetts: Harvard University Press, 1971).

[6] *The Expression of the Emotions in Man and Animals* (Chicago: University of Chicago Press, <1965>).

[7] See "Patterns of Inheritance" by Hampton L. Carson in *Science as a Way of Knowing: III—Genetics.* At page 800 Dr. Carson explains how one trillion (1,000,000,000,000) different individuals could theoretically be produced from the recombinant genetics of two parents in a sexually producing species, even without considering the effect of genetic mutations.

[8] The recent argument about punctuated equilibrium (Is evolution slow and gradual, or is it instead highly stable, punctuated occasionally by rapid change?) does not modify evolution's basic observation about variety and change, even though some creationists have made a mountain of this academic molehill.

[9] In a slightly different but related vein, see Monrow W. Strickberger's "The Structure and Organization of Genetic Material," published as pp. 769-780 in *Science as a Way of Knowing: III—Genetics.* (The American Society of Zoologists, 1985). It is an excellent essay showing how constancy and variability operate in dynamic tension within the context of genetics.

[10] *Aggressive Behavior: Genetic and Neural Approaches*, ed. Edward C. Simmell, Martin E. Hahn, and James K. Walters, pp. 98-99 (Hillsdale, New Jersey: Lawrence Erlbaum Associates, 1983). See also p. 187 of Peter D. Kramer's *Listening to Prozac* (New York: Penguin Books, 1993).

[11]Incidentally, commonalities and differences aren't neatly packaged into different racial groups within the human family. Any small population of humans contains approximately 80% of the entire library of human genes; see p. 123 of Richard Lewontin's *Human Diversity* (New York: Scientific American Books, 1982). Thus the variety within any group can be intense, as anyone knows who thinks about the dynamics of an individual family. We are likely to encounter both the pleasures of communion, and the dynamics of differences, wherever we go.

[12] For several examples of non-adaptive responses see *Not in Our Genes*, by R.C. Lewontin, Steven Rose, and Leon J. Kamin (New York: Pantheon, 1984).

[13] This view, along with its important implications, is compellingly presented in several of the selections in *Learning, Development, and Culture*, ed. H.C. Plotkin (New York: John Wiley & Sons, 1982).

[14] It may be true that we are on the verge of being able to plan our own evolution, but that certainly has not been true in the past, and it isn't true for any other species. See Michael Ruse's "Genesis Revisited: Can We Do Better than God?" at pp. 297-316 in Vol. 19 No. 3 (September 1984) of *Zygon.* See note 17 below.

[15] Malaria came close to preventing my appearance on the earth, in fact, since my Dad almost died with it before I was conceived.

[16] I don't suggest we keep the sickle cell; our most humane guess is that its pains outweigh its potential gains. My point is that our most humane guess might be wrong.

[17] See Theodosius Dobzhansky and Ernest Boesiger, *Human Culture: A Moment in Evolution* (New York: Columbia University Press, 1983), p. 64.

[18] Lewontin, R.C. in "Organism and environment," Chapter 9 of *Learning, Development, and Culture*, ed. H.C. Plotkin.

Chapter Two

The Evolution of Thought and Language

<u>An Outline of the Chapter</u>

A. Introduction: The Chicken and the Egg
B. The Physical Brain as a Psychophysical Substratum
C. Memory Pathways, the Key to Biotheology
D. Two General Thought Processes and Two Specific Ones
   1. Rational and Non-Rational Processes
   2. Imagination and Learning
E. Language

It is hard to imagine how the great human capacities for thought and language might have originated. In trying to imagine that, one wonders 1) how language could have developed without a large brain; 2) how a large brain would be needed in the first place if we couldn't already think powerfully; and 3) how we could think powerfully without language. It is like the old question "Which came first, the chicken or the egg?"

Actually, evolution's answer to that ancient riddle provides an excellent analogy to show how thinking, talking, and the brain evolved, so let's take the question seriously for a change:

Biologists know the answer to the riddle—the egg came first. Egg-laying is so ancient it preceded even fish, the first animals with backbones. When amphibians evolved from fish and climbed out of the water, they kept laying eggs. Certain amphibians evolved into thick-skinned reptiles that could roam far from water, but they kept laying eggs. Then when one branch of reptiles slowly changed into gliding reptiles and then into flying reptiles, and finally into birds, it still laid eggs. Only much later did the specific bird called "chicken" evolve.

But since chicken eggs and reptile eggs differ slightly, maybe the question means "Which came first, the chicken or the chicken egg?" Again biologists know the answer, but it is more subtle. Think back to that flying reptile with her leathery eggs, and remember how evolution works a little at a time, with each new species beginning as an offshoot of an existing species. No doubt the last flying reptile with fringed scales had similar eggs to its first offspring with recognizable feathers—the first official bird. One line of that bird's descendants and its eggs continued to evolve in the same small increments already discussed, away from their flying reptile looks and toward chickenhood. Toward chicken and chicken egg they tumbled one after the other, and which finally came first depends on how one defines "chicken" and "chicken egg." Since we would probably define "chicken egg" as an egg laid by a chicken, the chicken had to come first. So the egg came before the chicken, but the chicken came before the chicken-egg.

Notice, however, that a chicken egg could be defined as an egg that gives birth to a chicken, rather than as one laid by a chicken. That would mean the chicken egg came before the chicken. But this depends on when a new species is deemed to come into existence, which is defined somewhat arbitrarily, and might be the moment an egg contains genetic material that will produce a grown chicken. Present definitions, however, declare that new species exist when adults from one group can successfully breed with each other but not with adults from closely related groups.[1] This leads to the same conclusion as before, namely that the chicken came before the chicken egg. But a different definition would produce a different answer, and the real import of the story is to show how completely integrated and unified the actual evolution of the chicken and its egg was.

When we apply this analogy to the evolution of the brain, thinking, and language, we see that they evolved together, like the chicken and its egg. It would be conceptually simpler if the brain had finished evolving first, followed by thinking and shortly afterward by language. But the evidence doesn't support such a history. Instead, our ancestors added a few brain cells to what their ancestors already had, giving them slightly more thinking power. Then came a grunt or two, then more thinking power, then simple words, then more brain cells and structures, then some rudimentary grammar, then more thinking power, then complex sentences, and so on until language and thought worked like they do today.[2]

Even though this evolution was not at all neat and measured, we can nonetheless analyze thinking and language separately, just as we can analyze chemicals separately which are not found free in nature. This separate analysis must not hide the fact, however, that thinking and language and the brain[3] evolved in normal jury-rigged fashion, operating in intimate interaction during every stage of the forward-tumbling process. That should be kept constantly in mind in these next sections, as we separately discuss the brain, memory, thought, and language.

The Physical Brain as a Psychophysical Substratum

The gradual buildup and grouping of nerve cells into that organ called the brain is one of evolution's most interesting stories,[4] if only because its details are not well understood like say, the horse's hoof is.[5] Once the brain is completely mapped perhaps our interest will turn elsewhere, but at this stage in intellectual history, nothing is more fascinating.[6]

In this mapping process we stand where Columbus stood when he began mapping the western hemisphere.[7] We have only touched the shore, more or less literally, since brain-probe experiments are limited to the outer cortex except where injury and disease give unfortunate opportunities for deeper research.[8] Solid analogies from animal research have gone deeper than that perhaps, but nobody denies that we are really just beginning.

Even with that small start, much is known. Some brain geography is well understood,[9] hemisphere differences are known to some extent,[10] hormone influences are being unravelled,[11] aphasia research gives hints as to how the brain processes language,[12] and we can even induce thoughts of certain kinds.

All this research shows that what we call "thinking" involves many different but related processes. Several of them, including some aspects of memory, have finally been linked to specific neural structures,[13] and biologists are now confident that all thinking processes have a physical basis. These physical structures, and the electrochemical processes depending on them, constitute what one writer calls the "psychophysical substratum of thinking."[14] This is in contrast to an earlier belief which held thought as something fundamentally different, perhaps dependent on a "vital force" that is non-physical in nature.[15]

Of course each part of the brain has its own evolutionary history. Most parts evolved in tandem with one or more of the others, and each is tied to the others by neural transmission lines of various kinds. Even the rational process, an evolutionary latecomer situated mostly in the outer cortex, has ties to older structures transmitting feeling, intuition, and basic life processes.[16]

It would be easier to see the relation between thought and its psychophysical substratum if thought pathways could be traced precisely, but that is not yet possible, and the effort to do so is part of the three-dimensional mapping effort mentioned above. One researcher (whose work already traces the neural pathways of concrete actions like kicking a ball,[17]) says we may soon be able to follow the path of an abstract thought through the neurons[18]. When the mapping is complete, we should finally understand what kinds of thinking originate at what sites in the brain, as well as the pathways used by various kinds of thought. We are of course far from finishing that task.

Yet the relation of thought to the psychophysical substratum is already becoming clearer, not only from research perspectives but also from philosophical ones. New metaphors to describe the relationship between thought and the brain are being fashioned, which express the purely physical nature of the relationship, and yet show how it is experienced at such a seemingly "mystical" level. Although some biophilosophers claim that thinking and the brain's electrochemical firings are exactly the same thing, more satisfactory metaphors divide the two, at least conceptually.[19] For example, one might say that thinking is to the brain what rolling is to a wheel[20], or what warmth is to an electric blanket, or what a screen is to a computer. In each metaphor the activity or sensation is conceptually separate from the physical structure which produces it, and we experience the activity as a free-flowing sensation, though of course a physicist would understand the rolling or the warmth as a physical phenomenon and not a magical one. Even non-physicists understand the computer monitor as a useful adjunct to the main business of the computer's calculations.[21]

Memory Pathways, the Key to Biotheology

The "psychophysical substratum" is roughly equivalent to the brain and nervous system, but that phrase highlight the fact that layers of memory structures (neurons, neurotransmitters, etc.) were gradually added during evolution, and did not suddenly appear as part of a fully developed brain. Slow evolutionary processes also led to particular relationships between various physical components, to produce particular mental processes. In the case of memory, for example, a structure called the hippocampus is particularly important, and definitely interfaces with other structures, yet it did not emerge until the mammals evolved.[22]

Dozens of researchers have worked hundreds of thousands of hours to understand memory, and only recently have they begun to achieve a consensus of how it works. A combination of several dynamic mechanisms is involved, which explains both why the search for memory's physical basis was so elusive, and why memory can be so powerful. Some of those dynamics are:

- The brain's neurons, or nerve cells, each send out shoots called dendrites to link with other neurons. The more enriched our experience is, the more dendrites are formed. Memory resides partly in the creation of new dendrites, and in the nerve-pathways they constitute.[23]

- Certain chemicals called neurotransmitters have evolved which facilitate the transmission of nerve impulses. When certain of these are present, memory is enhanced, and memory can actually be improved in some cases by administering them.[24] This and other evidence shows the role of brain chemistry at the point of contact (a synapse) between two neurons.
- A different set of chemicals also effects memory, namely the hormones. Although hormones are usually thought of in relation to such processes as sexual arousal, they actually have many functions in the body,[25] and one of them is carrying messages (thus hormones are also neurotransmitters). It has also been shown that memory works better when learning occurs in the presence of emotion, apparently because of the role of hormones.
- Nerve impulses actually travel along the outside sheath of nerve cells, and physical structures on the sheath may speed or slow the passage of a memory. Further, cell contents contribute by setting up an electrical potential, making the sheath more or less conductive. An electrical phenomenon called "long-term potential" seems especially important to memory. Naturally, any pathology in the cell or its operation can affect memory and other thinking processes dependent on it.

These technical aspects of memory are important to biotheology, because they demonstrate the physical nature of brain processes. For most purposes, however, less technical ways of understanding memory are needed, which do not depart too far from accuracy. Researchers use several non-technical words for our benefit, and I will list a few of them here which will be used again later. For example, "memory trace" conveys the idea that a pathway can be a slight one at first, until the memory is firmly embedded. "Memory pathway" implies that each embedded memory has a separate electrochemical circuit, though it now appears that strong memories may have alternate circuits, so that interfering with the main one may not extinguish the memory.

It would be accurate (though perhaps not precise) to say that memories sometimes operate in a deep rut. The idea is that a memory pathway becomes so well established that it can hardly be overcome. Likewise an experience may become "etched" in our memory, and scientists now understand why certain memories—like those experienced in traumatic circumstances—become etched in our minds even though they only happened in a brief moment. Other non-technical words can also describe how memory pathways work—the memory pathways may be "greased" by neurotransmitters, or we become "set in our ways." I hope such non-technical words will remind you of the vast research in memory and its neurochemistry, while avoiding the confusion a more technical explanation might cause. Where non-technical words might themselves cause confusion, I will use technical ones, but that will not happen often.

I call memory the "key to biotheology" because it, more than anything else, explains how we evolved with the capacity to reduce experience into electrochemical phenomena, and carry it around in our heads. Chapter Three will show how memory even incorporates broader group experience, through the learning of culture. Chapter Four will show how the brain uses language to organize experience into categories called concepts, so that the memory can work even more efficiently. Another kind of memory is the so-called genetic memory: in a real sense the most significant experiences of our ancestors are encoded in inherited neurons and molecules upon which individual memory builds.[26] Later chapters will show how pathological memories can arise from all three kinds of experience—genetic, cultural, or individual—and we will explore ways to correct those pathologies.

Two General Thought Processes and Two Specific Ones

Rational and Non-Rational Processes

The complexity of the brain is staggering. Its 15 billion nerve cells may each have up to 10,000 dendrites and synapses. That permits an incredibly large number of pathways, theoretically allowing for more of them in a single brain than there are atoms in the universe. All possible pathways don't actually connect, but a huge number do, so many that it is difficult even to name and classify the resulting processes confidently.[27] Therefore, when this section analyzes two kinds of processes and says they are representative of two general categories of thought, namely rational and non-rational, that categorization is primarily one of convenience, and it is an exceedingly humble one. It may turn out that the classic division between rational and non-rational was too simple, and should be replaced by some other organizational division. Since some thinking processes are faster than others, it is tempting to say that the ancient non-rational processes are the fast ones (as when somebody has a "quick temper"), and that later-evolved rationality is slower ("cool-headed"), but that isn't known yet.

Nonetheless, generations of thinkers have considered the rational/non-rational division valid. It is used routinely in the humanities and sciences, and may well turn out to be correct. To be sure, rational processes are located in a different part of the brain than non-rational processes are.[28] But the line between those two categories is hazy, and I use them mainly as traditional hat-racks (or heuristic devices, philosophers would say) on which to hang more specific processes. On the hat-rack of non-rational processes hang imagination and emotions and curiosity and instincts, providing a baseline response to life's challenges.[29] On the rational hat-rack hang such mental processes as logic, learning, problem-solving, trial and error, and contemplation.[30]

Fortunately, researchers have removed several thought-categories from these general hat-racks and studied them in more detail. In fact some categories have been so well described that we confidently rely on them for medical, psychological, and philosophical analysis. That is especially true of the two thought processes discussed below, namely imagination and learning. Imagination fits generally into the classical category of non-rational processes, and learning fits generally into the classical category of rational processes, so the two processes can be examined while maintaining the classical division.

After imagination and learning are introduced, my task will be to show how they interact with each other, with language, and with memory to produce results that strongly influence our theological and practical understanding.

Imagination and Learning

Non-rational thinking processes are very old,[31] and we use them every moment of our lives. Long before rational thought evolved, these processes served us well. Generally, they reside in older parts of the brain, namely the brainstem and mid-brain. Imagination is one of these non-rational processes, and it evolved from even earlier processes, as follows:

In the very earliest stages of evolution, mobile organisms began to exhibit "orienting" behavior.[32] Rudimentary sense organs like chemical receptors or light-sensitive cells allowed living things to locate themselves within the environment, and staying oriented is still a primary concern of both simple and complex organisms.

If we study the orienting process in small animals like mice, we see that they respond much as we ourselves do. When first placed into a new environment a mouse stays very still, testing for danger with eyes and ears and nose. If no danger appears, it starts moving about slowly, orienting itself in a larger space. Soon it has established escape routes,[33] and the orientation is complete until a new situation or a new stimulus[34] reactivates the orienting impulse. Humans behave similarly; for example we walk into a crowded room and involuntarily "check out" the situation, quickly scanning the room with eyes and ears.

Long before humans evolved, orienting behavior produced an off-shoot, a mental process that is like orienting behavior, but more complex.

Its common name is curiosity, and it is also called "exploratory drive." Its roots lie in the alertness to danger, but it evolved into a more active interaction with the environment. Rather than merely avoiding danger, curiosity seeks positive opportunities. Its advantages secured its place among evolved mental processes long before humans and apes inherited them from a common ancestor.[35] Thus we experience and understand curiosity even when we see it working in distant animal cousins.

(Other mental processes also grew out of this foundation of orientation and curiosity, several of which were evolving at the same time. Rationality—the new kid on the block—was starting its evolution, for instance, and has some roots in curiosity. We will return to that story in a moment.)

From curiosity it was a small step to imagination.[36] Notice how the two are related by listening to little Johnny exclaim "I want to go see what is in that room!" (expressing curiosity), and then "I bet it's a big hairy monster!" (expressing imagination). This example demonstrates that, whereas curiosity normally involves more active exploration than passive thinking, imagination is slightly different and may include elaborate flights of fancy, without any active exploration whatever.

I do not mean to imply that imagination and curiosity are entirely separate; indeed they are closely related. But if one accepts the idea that curiosity evolved from orienting behavior, it is easy to see how additional changes in the psychophysical substratum permitted the rise of imagination. In terms of time this small step probably took ages, and paralleled the evolution of humanity out of an apehood which is still capable of curiosity, but only dimly capable of imagination.[37] An advocate for the apes might dispute this, and after watching a young orangutan playing with a boat recently, I see their point. Yet as imagination is formally defined, it must be distinguished from play—which usually involves interaction with immediate objects—even though play and imagination are also closely related.[38]

Formally defined, *imagination is the ability to form conscious ideas or mental images of things never before perceived by the imaginer.*[39] This definition pinpoints the differences between curiosity and imagination, between play and imagination, and between humans and our nearest relatives. Our greater imagination allows us to see the world and its components in new ways, without the physical dangers of active exploration. In concrete application it begins the process of technological invention, where imagined solutions often work. In abstract realms we can imagine new models of government, or new kinds of interpersonal relationships.

As in evolution generally, imagination's benefits come with trade-offs. An obvious disadvantage is that it can fool us. A moth flies into fire thinking it is flying toward light, and we can be burned by our fantasies too.

The good that begins in imagination includes not only technology and relationships, but economic structures, theological understanding, and much more. On the other hand imagination also contributes to superstition and inquisitions and wars, and it helped produce the atomic bomb, which might yet extinguish us. At a minimum, it can waste time and energy. One goal of our inquiry will be to learn to recognize both good and bad products of imagination, and to use that knowledge effectively.

Among the several components of non-rational thought, I have chosen to highlight imagination for two reasons. First, it interacts with language and rational thought to produce *concepts*, explored further in Chapter 4. Second, it leads directly to a consideration of the "theological" imagination, which will be our point of entry into theology, in Part 2 of this book, which will begin with Chapter 5.

Now let us turn to a discussion of rational processes. They evolved much later than non-rational ones, and seem to act as filters through which non-rational impulses pass; thus the complete thinking process contains a mixture of non-rational and rational components. The part of the brain that makes most rational thinking possible is the outer cortex, which is far more developed in us than in our nearest animal relatives.

Yet the temptation to think of humans as the only creatures that can think rationally should be avoided. Other species, including some extinct cousins of ours, had basic rationality; nor are we modern humans the final word in rationality, since many further changes in that direction may await us in the distant future. Yet there is one rational process which modern humans seem particularly adept at, at this point in time, and that is learning.

A good way to demonstrate the slow changes that led to human learning is to look at animal research. Starting far down the scale of life, and working upward toward humans, learning ability increases, coinciding both with increased brain size and with the complexity of memory pathways.[40] In each of the following cases higher animals can usually do what lower animals can, in addition to the tasks listed for them.[41]

A. Among *invertebrates*, even worms can be taught to turn right or left, grasping the idea of direction in a primitive way. Octopuses and other mollusks can be taught to recognize several patterns, and in certain snails the exact neuro-circuitry of such learning is almost worked out. Bees have a pin-head brain, but they have 850,000 neurons in that brain. They can be trained to recognize colors, black and white patterns, and scents.[42]

B. *Fish* can be taught to recognize patterns too, and can remember them for months, even compensating for alterations in the patterns. Goldfish can be taught to swim toward or away from an object. The fish eye has evolved like ours to respond to color, and many color responses are pre-wired in fish, but not so rigidly that they cannot be overridden by training. Fish can be taught to prefer colors, even colors they do not "naturally" prefer. They can also be taught to make "relative" choices between one or the other of lighter and darker, finer and coarser, smaller and larger, less and more. Their emotion-displays (puffed body, etc.) activate when they are given an impossible choice, so evidently the choice is one they are "thinking" about, at some level.[43]

C. *Birds* respond to danger calls, and some have separate calls for "danger from above" and "danger from below." A crow responds to danger calls of other species. Birdcalls are not strictly instinctual, since some species have to learn the call, and since different populations can have different dialects.[44] Crows and mynahs can say a few words appropriately, and seem to understand the idea behind the words, like "hello," or "I want food." An African Gray parrot has apparently eclipsed even the lesser apes in its ability to learn language.[45] After long training some birds can count up to seven, and use the numbers appropriately. Though admittedly the bird's ability to manipulate numbers is limited, counting is usually considered a sophisticated kind of learning. Incidentally, birds not only learn new things themselves, but can teach their fellows: In England titmice learned to remove cardboard milktops to get at milk solids underneath; the practice slowly spread through the titmouse population until humans redesigned the milktops.[46]

D. Among *mammals* below primates, cats have been trained to recognize which of several choices was "unlike" the others, and without further training can select other unlike choices such as differences in shape, number, or arrangement. A dog easily learns its name. How deeply a dog thinks about its name is not known, but at the least name-knowing is an impressive kind of associative learning. (There is also some question about how deeply humans think about their names.[47]) A dog also recognizes its master by sight and smell, and responds to a wide variety of verbal cues—dogs for the handicapped are taught almost 200 different commands. Many mammals express joy, shame, and other emotions that seem as appropriate as human responses under similar circumstances.

E. Of the *primates*, monkeys have been taught to solve the oddity problem: presented with an even number of objects and then with an uneven number of the same objects, they can be trained to pick either even or odd. Presented with a group of things made up of all similar objects save one, they can be trained to select the discrepant item. The behavior persists even with more complex experiments, and the monkeys seem to understand both difference and similarity. In a famous set of observations, Imo the wild genius macaque spontaneously developed the concept of winnowing, dropping grain and sand together into the water, and then scooping the grain off the top.[48] She also learned to wash sweet potatoes, and she was probably the one who first swam to a nearby island, which was unusual behavior for her species. She taught these behaviors to her young peers, but older macaques refused to learn. Chimpanzees spontaneously use sticks to extend the reach, or use a box to reach an item suspended from the ceiling. They do this after looking at the problem and the solution, and appear to have an "Aha!" sensation just before getting the box or the stick. Such discoveries are subsequently used routinely. Chimpanzees can also be taught to work mazes. They sit thoughtfully planning strategy for a few minutes before moving the gamepiece,[49] and contemplation time is proportional to the difficulty of the maze. These experiments show that many animals, especially apes, grasp the concept of learning to solve a problem. Gorillas and chimps can be taught sign language, grasping a large number of concepts that can be symbolically expressed, including some abstract concepts.[50]

This survey of sophisticated animal accomplishments gives insight into how learning probably evolved in humans, beginning with quite primitive ancestors. The process is a basic one depending on physical memory structures modified through experience. Whether learning is thought of as creating a "rut," or more positively as creating "well-oiled" neural pathways, it is clearly related to memory, and shares its evolution.

But nature went further than the examples above, combining learning with other processes to produce something new. Learning combined only with imagination would produce quite a remarkable creature, but another ability, namely language, dramatically extended human abilities. Thus we can skip over the two-way combination of imagination and learning, and consider a three-way mixture of imagination, learning, and language. As we will see in Chapters 3 and 4, that mixture led simultaneously to culture, and to the amazing mental ability called concept-formation.

Language

In the course of human evolution, there came a time when we began to stand upright. Environmental features doubtless favored the change; perhaps standing upright allowed our ancestors to see over the tall grasses of a savannah niche, for example. Standing triggered other physical changes, including the migration of vocal chords deeper into our throats. That permitted us to make more hollow sounds,[51] and to shape those sounds with tongue and teeth and lips, emphasizing the facial expressions which had already evolved.[52] (Even today people communicate cross-culturally with facial expressions, and when body language as a whole is considered, 60% of our communication is non-verbal.[53] Indeed, shared backgrounds even permit us to communicate across species lines to some extent with body language[54].)

The ability to make new sounds allowed us to mimic nature. By associating nature's sounds with noises we could make, we learned to communicate natural phenomena through such sounds as the hiss of a snake or the waah-waah of a baby. New significance was attached to other vocalizations, and at some point we realized that verbal symbols could represent concrete objects. This ability to create a "representation" of our experiences through efficient word-symbols retained in memory, is language's most significant aspect. It permits us to share experiences with great precision and detail, without having to haul out actual objects or relive actual experiences.[55]

These word-symbols did something else besides enhancing communication. Retaining them in memory enhanced thought, allowing us to produce mental representations of experience inside our heads without vocalizing it aloud. Animals can do this to some extent, but not with the efficiency and clarity real language-users attain. On the other hand, people with certain disorders of the psychophysical substratum cannot properly do it.

Of course bodily changes such as migrating vocal chords were supplemented by physical changes in our brains, so that language, thinking, and the brain continuously evolved from early grunts forward.[56] Genesis telescopes the "naming" part of this development into a single human lifetime, by telling how Adam named each animal.[57] (This mention of Genesis reminds us of later language developments, namely writing and scripture, but they will not be pursued until the second part of this book.)

After maybe a hundred thousand years of symbolizing concrete experience in words like "rock" or "water" or the names of animals, language further increased its power by beginning to symbolize more abstract experience in words like "justice" and "love." However long that took, and however lacking in any neat straight-line development, it is safe to say our evolution has now brought us to a point where such symbolizing is necessary to existence as we know it. It would be hard to imagine humans without imagining the use of both concrete symbols like "rock" and abstract ones like "love," not to mention specialized symbols like our own names and the words for our relationships. Using such symbols is an inextricable part of being human,[58] like building dams is part of being a beaver. One could imagine a human without language, or one could imagine a beaver without a dam, but in both cases it would be nearly certain that nature had been interfered with in some pathological way.

Although this description of language's evolution has been brief, the changes leading to abstract language paralleled (and may be the same as) the emerging human ability to make and understand concepts. Concepts are a more complex part of the evolution of language and thinking, and understanding them is an important goal of these first four chapters. But concepts cannot be properly understood without considering their relation to both thinking and language, so it was necessary to discuss those faculties first.

In fact we are still not quite ready to discuss concepts directly. There is one additional aspect of our evolution which must be discussed before concepts can be adequately explained, and that is culture. So we will discuss culture next, and then end Part One with a comprehensive look at how language, thought, and culture combine to produce concepts.

Notes

[1] "Successfully" means (in the case of sexually reproducing animals) a pair can produce fertile offspring, not merely that they can physically copulate. Actually, members of different but closely related species can sometimes interbreed, and can even produce offspring, but not fertile offspring. A mule, for example, is the non-fertile offspring of a donkey and a horse.

[2] See Dobzhansky, Theodosius Grigorievich. *Mankind Evolving: The evolution of the human species* (New Haven: Yale University Press, 1962).

[3] "Brain" is too simplistic, since the entire nervous system evolved together, along with associated substances or structures like hormones and glands. I will use the general term "brain" except where necessary to explain further, and trust you to know I mean the whole system. See *Primate Brain Evolution*, ed. Este Armstrong and Dean Falk (New York: Plenum Press, 1982).

[4] See Paul D. MacLean's "On the Origin and Progressive Evolution of the Triune Brain," pp. 291-316 in *Primate Brain Evolution*, ed. Este Armstrong and Dean Falk.

[5] Simpson, George Gaylord. *Horses; The Story of the horse family in the modern world and through sixty million years of history*, pp. 190-204 (Garden City, New York: Doubleday, 1961).

[6] The human genome might be more fascinating and more basic, but by beginning with the most interesting organ resulting from genetic evolution, I hope to gain in efficiency what is lost in technical correctness. For further technical reading see *The Brain Code*, by Norman D. Cook (London: Methuen & Co., 1986).

[7] Shirley, Hunter B. *Mapping the Mind* (Chicago: Nelsen-Hall, 1983).

[8] *Electrical Stimulation Techniques*, ed. Michael Patterson and Raymond Kesner (New York: Academic Press, 1981). Note that researchers are beginning to use other non-invasive techniques like CAT scans.

[9] *Regions of the Mind*, by Susan Leigh Star (Stanford, California: Stanford University Press, 1989).

[10] *The Evolution of Lateral Asymmetries, Language, Tool Use and Intellect*, by John L. Bradshaw and Lesley J. Rogers (New York: Academic Press Inc., 1993). Also see *Left Brain, Right Brain*, by Sally P. Springer and Georg Deutsch (New York: W.H. Freeman and Company, 1993, Fourth Edition).

[11] For a good overview see the introduction in *Hormones and Evolution* Vol. 1, ed. Ernest James William Barrington (Princeton, N.J.: Academic Press, 1979).

[12] See *Neural Models of Language Processes*, ed. by Michael Arbib, David Caplan, and John Marshall (New York: Academic Press, 1982).

[13] *Synapses, Circuits, and the Beginnings of Memory*, by Gary Lynch (Cambridge, Mass: The MIT Press, 1986).

[14] See Chapter 6C of Bernhard Rensch's *Biophilosophy* (New York: Columbia University Press, 1971, translated by C.A.M. Sym).

[15] Joad, Cyril Edwin Mitchinson. *The Future of Life, a Theory of Vitalism* (London and New York: G.P. Putnam's Sons, 1928).

[16] See Franz Seitelberger's "Neurobiological Aspects of Intelligence" in *Concepts and Approaches in Evolutionary Epistemology*, ed. by F.M. Wuketits (Dordrecht and Boston: D. Reidel Publishing Company, 1984).

[17] Diamond, Marian C. "Love Affair with the Brain." *Psychology Today*, November 1984. See also *Neural Coding of Motor Performance*, ed. J. Massion, J. Paillord, W. Schultz and M. Wiesendanger (New York: Springer-Verlag, 1993).

[18] Personal communication dated December 7, 1984 from Dr. Marian C. Diamond, Professor of Anatomy, University of California, Berkeley. For a technical reference on tracing brain pathways see *Neuroanatomical Tract-Tracing Methods*, ed. Lennart Heimer and Martine J. Robards (New York: Plenum Press, 1981).

[19] For a discussion of identity theory see pp. 114-118 of *Concepts and Approaches in Evolutionary Epistemology*, ed. F.M. Wuketits, cited above. For a poetic and yet highly competent discussion of the relation of neurophysiology to consciousness see William Calvin's *The Cerebral Symphony* (New York: Bantam Books, 1990).

[20] See Robert L. Klee's interesting criticism and extension of Sperry's use of this metaphor at pp. 57-62 of "Micro-determinism and Concepts of Emergence," in *Philosophy of Science* 51:44-63 (March 1984).

[21] When the computer program allows interaction from the monitor, it becomes even more useful as a metaphor for human consciousness, because the feedback function is similar to that between our senses and thinking. See the later discussion of the gestalt (Chapter 7) for more detail.

[22] Its earlier origins are apparent, however, from the existence of similar structures in amphibians, reptiles, and birds. See p. 240 of *Learning and Memory:*

A Biological View, ed. Joe L. Martinez Jr. and Raymond P. Kesner (New York: Academic Press, Inc. 1986).

[23] Incidentally, it now appears that Alzheimer's disease may be a result of malfunction in the connection of dendrites. See "Beta-amyloid and the pathogenesis of Alzheimer's disease," by Bruce A. Yankner and M. Marcel Mesulam, in the *New England Journal of Medicine*, December 26, 1991 (v. 325 n. 26), p. 1849 (9).

[24] Deyo, Richard A.; Straube, Karen T.; and Disterhoff, John F. "Nimodipine Facilitates Associative Learning in Aging Rabbits." *Science*, vol 241, p. 809 (February 10, 1989).

[25] To get a flavor of how dramatic a threshhold we stand on in endocrine research, see Jane Kaltenbach's paper "Endocrine Aspects of Homeostasis" in *Science as a Way of Knowing Volume 5 (Form and Function)*. American Society of Zoologists, University of California, Riverside, 1988.

[26] Of course no specific facts are remembered through the genetic memory. But the experience of our ancestors resulted in the survival of some genes and the elimination of others. The retained genes often produce specific behaviors and abilities, and make the emergence of other behaviors and abilities more likely. For example we don't each individually learn how to learn, but we often learn the same things because "how to learn" is a function of neural pathways which are relatively similar from person to person.

[27] These pathways involve up to 50 **trillion** synapses. See "Neurobiological Aspects of Intelligence," Franz Seitelberger's contribution to *Concepts and Approaches in Evolutionary Epistemology*. The specific facts cited are on p. 129, but the entire paper underscores the relation of thought to the psychophysical substratum.

[28] The hemisphere research also supports the division between right-lobe processes like imagination, and left-lobe processes like language. That may or may not be important, since the bundle of nerves called the *corpus collosum* conducts impulses from one lobe to the other frequently. But at least it helps us organize our observations, and search for effects that flow from this physical division.

[29] They even originate stimulation if the environment provides none. For further discussion of this provocative idea see Konrad Lorenz, *Studies in Animal and Human Behavior*, Volume II, pp. 202-209 (in a paper entitled "Psychology and Phylogeny.")

[30] For more on the reason/emotion balance see Ian Barbour's "Science, Religion and the Counterculture" pp. 380-397 in Vol. 10 No.4 (December 1975) of *Zygon: Journal of Religion & Science* and Michael Gazzaniga's "Gut Thinking" at pp. 68-71 in the February 1995 issue of *Natural History* (Vol. 104 no. 2). For a useful discussion of "Two Types of Reason," variously described as intellectual/rational (Karl Jaspers), calculative/meditative (Martin Heidegger), and instrumental/meaning-oriented (George Grant), see pp. 388-402 of Douglas John Hall's *Thinking the Faith: Christian Theology in a North American Context* (Minneapolis, Minnesota: Augsburg Fortress Publishers, 1991).

[31] McClary, Robert A. and Moore, Robert Y. *Subcortical Mechanisms of Behavior* (New York: Basic Books, 1965).

[32] This is Lorenz's term. Actually he discusses instinctual "taxis" as a precursor to orientation, which he sees as the most primitive behavior distinguishable from pure instinct. Orientation is, in his view, the basis for all appetitive behavior. See pages 272-275 of his paper "The Establishment of the Instinct Concept" in Volume I of *Studies in Animal and Human Behavior.*

[33] Hewitt, J.K., Fulker, D.W., and Broadhurst, A.L., "Genetic architecture of escape-avoidance conditioning in laboratory and wild populations of rats: A biometrical approach," *Behavior Genetics*, 1981, 11, 533-544.

[34] For an illustration in primates, see the report of a gorilla that was moved to a new cage, at p. A-1 of the Atlanta *Constitution* for May 14, 1988.

[35] See p. 55 of Bernard Campbell's *Human Evolution* (Aldine Publishing Company, New York, 1985).

[36] See Chapter IX of Hubert Alexander's *Language and Thinking* (Princeton: Van Nostrand Company, 1967).

[37] This statement is supported even by evidence of imagination in primates, though I agree it is a difference in degree rather than in kind. See pp. 40-41 of Jane Goodall's *The Chimpanzees of Gombe: Patterns of Behavior* (Cambridge, Massachusetts: The Belknap Press of Harvard University Press, 1986).

[38] For more on the orientation-curiosity-imagination sequence, see p. 46 of *Curiosity, Imagination, and Play*, ed. Dietmar Görlitz and Joachim F. Wohlwill (London: Lawrence Erlbaum Associates, 1987).

[39] *Webster's Third New International Dictionary*, unabridged. G. & C. Merriam Co., Chicago 1976.

[40] The list below is not a chain of actual evolution from invertebrates to humans, but a summary of research suggestive of the course evolution probably took in our ancestors. See Gary G. Tunnell's *Culture and Biology: Basic Concepts in Anthropology*, pp. 15-16 (Minneapolis: Burgess Publishing Company, 1973).

[41] By "higher" and "lower," I am referring only to brain substance and organization, and not to inherent worth. For a discussion of "higher" and "lower" in this sense see Chapter 2 of *Adaptation and Natural Selection: A Critique of Current Evolutionary Thought*, by G.C. Williams (Princeton, New Jersey: Princeton University Press, 1966).

[42] They also do an amazing bit of symbolizing: they do a dance in the hive which communicates the direction and distance to a nectar source, and the time of day when the flowers open. See K. Von Frisch and M. Lindauer's "The 'Language' and Orientation of the Honey Bee," Chapter 32 (pp. 347-358) of *Readings in Animal Behavior*, ed. Thomas E. McGill (New York: Holt, Rhinehart and Winston, 1965). See also pp. 422-4 of *Animal Behavior*, by David McFarland (Menlo Park: The Benjamin/Cummings Publishing Company, Inc. 1985).

[43] As to whether we are justified in saying an animal is having internal frustrations or other mental experiences, see Konrad Lorenz's excellent paper "Do

animals undergo subjective experience?," recorded in his 1971 book *Studies in Animal and Human Behavior*, Volume II, pp. 323-337.

[44] See Chapter 28 of Jerram Brown's *The Evolution of Behavior* (New York: Norton, 1975).

[45] Kaufman, Kenn. "The Subject is Alex." *Audubon*, September-October 1991 [v. 93 n. 5 p. 52 (7)].

[46] In "Social Learning Without Imitation: More about milk-bottle opening by birds" (at p. 987 of *Animal Behavior* for November 1990), David F. Sherry and Bennett G. Galef Jr. dispute the earlier findings, but they actually confirm them, and even raise the new issue of whether bird "tutors" communicate details to the pupil.

[47] But see the report of a tribe that debates endlessly over who should possess a name, in *Stealing People's Names: History and Politics in a Sepik River Community*, by Simon J. Harrison (Cambridge University Press, 1990).

[48] Kawai, M. 1965. "Newly acquired precultural behavior of the troop of Japanese monkeys on Koshima islet." *Primates* 6:1-30. Also see *Animals* 5:450-455. These periodicals are not widely available, but the study is widely reported, and may be read about at p. 326 of *Human Evolution*, by Bernard Campbell (New York: Aldine Publishing Company, 1985). Perhaps the best description I have found so far is at p. 846 of Luigi L. Cavalli-Sforza's essay "Cultural Evolution," in *Science as a Way of Knowing: III—Genetics* (The American Society of Zoologists, 1985).

[49] Rensch, Bernard. *Homo Sapiens: From man to demigod* (New York: Columbia University Press, 1972).

[50] Chimps can learn over 100 words. For communication using plastic symbols see A.J. and D.D. Premack's "Teaching language to an ape" *Scientific American* 227 (4):92-99 (1972). For communication using sign language see B.T. and A.R. Gardner's "Teaching sign language to a chimpanzee" *Science* 165:664-672 (1969).

[51] See "Phonetic ability and related anatomy of the newborn and adult human, Neanderthal man, and the chimpanzee" by P. Lieberman, E.S. Crelin, and D Klatt (in *American Anthropologist* 74:287-307, 1972). Also J.G. Hill's "On the evolutionary foundations of language" in *American Anthropologist* 74:309, 1972. For a lay-oriented description of the evolution of language, see pp. 347-355 of *Human Evolution*, by Bernard Campbell.

[52] See Philip Lieberman's *The Biology and Evolution of Language* (Cambridge, Mass: Harvard University Press, 1984).

[53] If we limit the discussion to the communication of emotional content, it is reliably estimated that 93% is non-verbal. See p. 77 of Alfred Moravian's *Silent Messages: Implicit Communication of Emotions and Attitudes*, 2d ed. (Belmont, California: Wadsworth Publishing Company, 1981).

[54] For a discussion of the innateness of facial expressions see Sackett, G.P. 1966. "Monkeys reared in isolation with pictures as visual input." *Science* 154:1468-1473.

[55] An excellent treatment of the evolution of language is given in *Philosophy, Evolution, and Human Nature*, by Florian von Schilcher and Neil Tennant (London: Routledge & Kegan Paul, 1984).

[56] The brain developed new masses of cells in the course of this evolution, and fossil skulls show the increasingly indented ridge that held those parts of the brain. See *Biological Perspectives on Language*, ed. David Caplan, Andre' Lecours, and Alan Smith (Cambridge, Mass: The MIT Press, 1984).

[57] Genesis 2:19-20.

[58] Eric Lenneberg gives 5 reasons to think language is inevitable, and "built into" the human brain. For example, one of them is the fixed onset of language's sequences in the developing child. "A biological perspective on language." In *Evolutionary Anthropology*, ed. H. Bleibtreu (Boston: Allyn and Bacon, 1969).

Chapter Three

Culture, A Kind of Learning

An Outline of the Chapter

A. The Evolution of Culture
B. The Functions of Culture
C. How Culture Enters An Individual Brain
D. The Mechanisms of Culture

There are many definitions of culture; in fact Kroeber and Kluckhohn listed 164 different ones over 40 years ago,[1] and many others have been proposed since then. For our purposes these definitions largely agree in one important respect; they distinguish culture from two other kinds of behavior and/or knowledge. Culture is not "innate" knowledge or behavior that is passed down genetically. Neither is it learned by individual experience in the absence of group influence. It is instead behavior and/or knowledge that is learned from the group.

Of these three categories—the innate, the individually learned, and the cultural—the latter two are both learned by individual brains. The distinction between them is important, however, because even though humans can learn independently, much of what we learn is rooted in culture. One definition efficiently communicates this idea by calling culture "the totality of learned socially transmitted behavior."[2]

Nobody knows exactly when culture began,[3] although that largely depends on how it is defined. If culture means only the use of tools or the simplest passed-down knowledge, it certainly preceded the first humans. But if one defines it more broadly to include sophisticated behaviors like art and invention, its onset was much later. In any case culture has continually expanded from its beginnings until now, and some experts say it has multiplied geometrically rather than merely arithmetically.[4] For our purposes it is not so important to pinpoint the exact time of its emergence, as to establish 1) how culture evolved, 2) its functions and mechanisms, and 3) how it enters our individual lives.

The Evolution of Culture

Although this section will show that evolution pre-disposed us for culture, there is one sense in which culture did not evolve, in the way chickens did. If each cultural development has to be tied to a specific mutation before we say it evolved, then culture did not evolve. That is evident because humans had modern brains and language long before any true culture emerged.[5] Therefore it might seem that the development of culture required something in addition to evolution.

But it is not necessary to tie changes so directly to their results in order to establish the product's evolutionary nature. Neither is it necessary for the result to take place the moment it could theoretically take place. For example a spider might evolve the ability to catch a particular insect, and yet it may take a hundred thousand years before the spider actually starts catching the insect, perhaps because the spider has to first expand its range to a place where the insect exists. If the spider were intelligent, it might take a long time before it noticed how it could slightly modify its web to catch the insect. If the intelligence itself had evolved, it would certainly be reasonable to say the spider had evolved the ability to catch the insect, even though other things had to happen first.

With this broader understanding of evolution, three different lines of research show that culture did indeed result from trial-and-error evolution, just like chickens did. Let us review that research to see how elegant culture is, and yet how similar it is in principle to the rest of our jury-rigged nature.

The first line of research focuses on changes in the brain itself, which we already looked at when discussing the evolution of thought. Slowly, over the eons, brain mass and function changed, leading to new abilities and behaviors, some of which worked and some of which didn't, ultimately permitting the set of behaviors and abilities that produced culture.

The second line of research traces culture's evolution through various animal and pre-human ancestors. These studies show that evolution produced several species in pre-cultural niches, but ultimately produced only one species in a thoroughly cultural niche—us.[6] And just as it is no longer reasonable to speak of a "missing link" in our physical evolution, it is all but impossible to deny the slow changes that led to culture.

Both of these lines of research give insight into culture's evolution, but a third line is more interesting for our purposes, because it analyzes the *components* or *ingredients* of culture, and traces the evolution of each. Not only does it show how culture emerged from a mixture of ingredients, but it shows that culture was a foregone conclusion, given the evolution of those ingredients and given enough time.[7] In other words, the research leads to the conclusion that culture is innate in humans.[8]

Several such components have been identified, but eight appear on most lists. Chapter 2 discussed four of those, namely 1) memory; 2) standing upright; 3) language and 4) imagination. This chapter will emphasize a fifth one—sociality—and it will introduce another one which will be developed in Chapter Five, namely 6) systems of explanation. The other necessary ingredients usually listed include 7) the thumb and 8) our stereoscopic eyesight; these two are surely important, but because they are not directly relevant to biotheological conclusions, they will be mentioned only in passing.

Each component originated in normal evolutionary precedents, themselves the result of ongoing changes in specific parts of the body. By sorting out how these separate developments intertwined and looped back on one another to take mutual advantage of their respective features, we can begin to grasp a fuller history of culture than is usually presented.

All the factors are interrelated, so intimately that several of them evolved within a relatively short time, one right after the other, causing culture to emerge somewhat suddenly. Standing upright, you remember, helped move the vocal chords lower and thereby contributed to the emergence of language. Standing upright also freed our hands, which did two things. First it permitted gestures,[9] which served as a transition to verbal language. Second it permitted tool-making, in association with the opposable thumb and stereoscopic eyesight[10]. Tool-making then combined with imagination and learning to accelerate invention. Once invention began, culture was well on its way. Of course there was no straight-line development, with one component finishing its evolution before the next one began.[11] The ingredients emerged and developed together in the awkward leapfrog way already discussed for other evolutionary products.

The component of culture called sociality is particularly important, and if I were discussing the contributors to culture in the strict order of their evolution[12] instead of in an order that efficiently presents a conceptual framework, I would have discussed sociality sooner, because it evolved very early, before any other listed factor except memory. Not only is sociality important in our lives today, but it was important to our ancestors right from the start.[13] It had been part of primate makeup long before humans started using language or imagination, and it was separately established for hundreds of thousands of years before it intertwined with the later developments to produce culture. It was useful in many practical ways, most of which centered around the advantages of cooperation.

Nowadays we are realizing how profound cooperation is, not just for humans, but for all species which developed lifestyles arising out of it. A new discipline, called Sociobiology, is dedicated to a thorough exploration of sociality's common implications for the species which possess it, even if they have little else in common.[14] Sociobiology highlights our evolutionary theme by showing how the brains of each social species evolved mental organizations that either predetermine or predispose them to live in groups. Honeybees are absolutely determined by their wiring to operate in groups, whereas humans, with our more open architecture, may choose not to live in groups, though we are born with a strong predisposition to do so.

Being prewired with a predisposition toward togetherness meant early humans approached most aspects of their lives cooperatively. We hunted and gathered together,[15] rather than separately like some species do.[16] Indeed sociality is typical of all apes, though male orangutans are an exception. We expressed sociality from our earliest beginnings by sharing objects and information and concepts; that sharing in turn produced tools, taboos, medicines, hunting techniques, and gods, which became the outward substance of culture.

Not only is sociality a crucial component of culture; it also illustrates a general principle of human evolution not yet mentioned. Certain early writers after Darwin called evolution "red in tooth and claw," by which they meant to emphasize the struggle for survival that drives evolution forward. Unfortunately, the public has seized on this image and largely defines evolution in its terms. I say unfortunately, because raw competition is only a part of evolution. Certainly competition between individuals has affected our evolution, but not nearly so much as have the struggle against the environment and competition with other species. For us the premium has not been so much on individual fitness, but rather on group cooperation.[17] Thus sociality and cooperation are not contrary to evolution's notion of struggle, but are rather an adaptive response to a major part of the struggle.[18]

Again, this does not eliminate individual competition from our existence, and besides, there is some middle ground between competition and cooperation which will be discussed in a moment. But by first acknowledging cooperation's more significant role in our lives, competition is placed in a subordinate position, which is an accurate placement from both biological and theological perspectives.

We should also pause to note that being pre-wired for sociality has an important implication for the chapter's opening definition of culture. That definition depended on a distinction between innate behavior and learned behavior, especially the kind of learned behavior that is transmitted across generations. In light of our discussion of sociality, the delineation between "the innate" and "the learned" may now be made more explicit, and even though it will make the line between them seem quite fine indeed, it will leave the earlier definition intact. I say the line between learned and innate behavior is a fine one, because *learning is itself innate*. Humans come prewired with both the ability, and the predisposition, to learn. This is true both of learning that comes from individual experience, and learning that comes from our group.

With that principle in mind, the traditional distinction between learned and innate must be seen as expressing a subtle and important difference between flexible and rigid behaviors, and not as defining an absolute boundary. The responses learned through culture are likely to be the flexible ones, whereas instinctive responses are more rigid. Other labels besides "learned" and "innate" might actually be more satisfactory, such as those Chomsky uses when talking about language; he calls those aspects of language found in all cultures "deep structures," and those which vary from culture to culture "surface structures."[19] Both arise from innate responses interacting with the cultural and physical environment, but some are more malleable or "softer" than others. In this sense we could call symbolic communication (language) a deep structure, but writing was a fortunate extension of it that did not appear until five thousand years ago. Searching for food is inherent in our makeup, but horticulture is only ten thousand years old, and did not have to happen.

But even this adoption of Chomsky's language must be immediately modified. Since learning has a large "trial and error" component, it can persuasively be argued that our deep structures would eventually lead to such general responses as writing and horticulture, though not necessarily to such cultural specifics as golf balls or land-grant colleges. In other words, there are "mid-depth" structures that might not develop in every culture, but which would eventually develop in some because they arise naturally and logically out of deep structures. These mid-depth structures have arisen independently in several different cultures, and would probably arise again if they were lost.

Mid-depth structures can be very powerful, and like the deep structures themselves, can have good and bad results. Horticulture has had much to do with our success, but unless its dangers have been noticed in time to correct them, it may lead to a lethal stripping of forests from our planet. Writing augments the transmission of culture, but it too has unforseen consequences that can cause great suffering, as we will see.

These observations on sociality support the chapter's earlier contention that culture is the direct result of the interaction of components which themselves evolved. Thus culture exists today as a manifestation of our biology. While it may be studied as a separate discipline, it can only be fully understood if biology is the foundation of that discipline. Like most aspects of our evolution, it evolved because of its survival value for our species, and because it performs functions which aid in that survival. Now let us look at those functions in more detail.

The Functions of Culture

To say that culture evolved because it contributed to our survival means, at a minimum, that it increases our solidarity against the environment and against other species. But culture performs this basic function through subsidiary functions which are not obvious,[20] and this section will discuss three of them, in the following order:

1) Culture mediates conflicts between cooperation and competition.
2) Culture lessens the dangers of curiosity and imagination.
3) Culture provides options for lifestyle and belief, through its inevitable variety.

1) ***Cooperation and Competition***. As I said above, individual competition is real, despite our evolved sociality. We are complex enough, and have enough neural circuits, to allow both competition and cooperation, side by side. This allows us the advantages of both behaviors, and (in good trade-off fashion) brings important disadvantages into our lives. For example we can kill one another in disputes over food or mates, or cruelly exclude one another from group life.

The tension between competition and cooperation is often great. Some way is needed to try to get the advantages of cooperation without sacrificing the advantages of competition. One of culture's functions is to help do that, by using language to construct systems of rules. While that cannot remove the innate drive in each individual to improve his or her competitive position, it moderates it and helps prevent its most abrasive consequences.

Even more dynamically, culture creates a middle ground between cooperation and competition by encouraging competition up to a point, and discouraging it beyond that point. The resulting system for "competition within the rules" imposes cooperative boundaries around our competitive urges, and yet allows them some room in which to operate. Game-theory research shows how thoroughly this function permeates our lives,[21] not only in sports, but also in daily competitions for mates or business or academic success.

As the jury-rigging discussion showed, it would be a mistake to think culture performs its functions perfectly. For example, one way it resolves tension between competition and cooperation is to focus on the local group, calling for a relatively large degree of cooperation within that group, but for competition with outsiders.[22] This "competition between groups" worked well when humans were a small-group species. But because our success has led to overcrowding, inter-group competition now causes great harm, and it has the potential for civil unrest and world war.[23] Understanding this evolutionary heritage, and dealing with it through evolved mental abilities, is critical to answering the challenge of our future. Later chapters will develop this observation further, asking whether culture and rationality can somehow bridle the potential fury of intergroup competition.

2) ***Curiosity and Imagination***. Not only must competition between individuals and between groups be kept under control, but so must the inbuilt competition between the group and its own members. Curiosity and imagination have an inherent tendency toward a dangerous innovativeness, and culture functions to reduce that danger without squelching imagination altogether. This function may be called the conservative function in culture. In this sub-section I will explain its basic operation, the ways it is moderated by institutionalizing imagination, and the special dangers that result when conservativism and innovation get out of balance.

Many species face problems arising from curiosity and/or imagination. For species that learn by experience, curiosity can be fatal—it can kill not only cats but dogs or monkeys or people. For example, a wolf-pup playing with a porcupine is likely to get hurt, which is good because the pup learns from the experience. But puppies may also play with rattlesnakes, and since evolution's strategy (in this case a parent's watchful eye) cannot be so wasteful as to give 100% vigilance, sometimes the pup gets killed. Without such brakes as the usually watchful parent and its timely nip, or culture's discouraging stare, such dangers might kill more than an occasional juvenile, and could even extinguish an entire species.

Actually, nature compensates for curiosity's dangers in several ways. Various physiological checks act in a conservative way to overrule curiosity, for example taste buds that make us spit out obnoxious substances. Culture evolved more subtly but just as surely, to check other behaviors in analogous ways.

The tensions between curiosity and conservatism are further demonstrated by the Imo story referred to earlier. The old macaques (a kind of monkey) would not learn the new behavior invented by Imo—figuratively speaking, they spat it out. If that behavior had turned out to be dangerous, their reluctance would have saved the troop. As it was, Imo's imagination expanded the group's repetoire of behavior in positive ways.[24] Similarly, human young tend to have more curiosity and imagination, and old people exercise more caution, though no absolute generalization can be made for our species, because the complexity of our brains and the capacity for individual differences often results in great imagination in older individuals, or unusual caution in young ones. Culture, however, is almost invariably conservative, and thereby helps us avoid the excesses of curiosity and imagination.

Even if that leads to some bad effects, overall it is a good trade-off, and thus conservatism has its good side. The flip side of the coin is that new ideas take a long time to gain a foothold in human life, and some good ideas may never get a foothold.

Although this second function of culture emphasizes its conservative aspect, culture has found a way to permit innovation while keeping it under some semblance of control. That is by the establishment of culturally approved institutions. For example, universities slow change by establishing reviewing committees, publishers submit stories to panels of readers, and scientists criticize each other's conclusions. Again it would be naïve to believe such mechanisms work smoothly; they can easily go too far in one direction or another, and are tuned only roughly. Yet in the long run this rough tuning is quite efficient, allowing adequate "play" in the cultural engine to accommodate imagination.

On the short run, however, the interaction can be very inefficient, delaying or preventing much-needed change. Besides, institutionalizing a new idea does not *always* slow it down. Sometimes the implied approval of institutionalization gives the innovation an enhanced capacity to challenge culture. This raises the question of whether innovation can become too energized, and overcome culture's healthy braking function. Could a youth-oriented society, for example, embrace innovation to such an extent that it over-penalized conservatism, and thereby lost conservatism's contribution to group life?

The answer is clearly yes. Although both imagination and cultural conservativism are too deeply embedded in our makeup to be permanently squelched, temporary squelching can certainly happen; indeed it happens often. One recent example is China's Cultural Revolution of the 1970's, when young iconoclasts destroyed centuries of culture with the encouragement of Chairman Mao. On the other hand, historical events like the Inquisition show that the opposite also happens; in fact culture's conservative forces probably win out more frequently than innovative forces do, though not necessarily more powerfully.

These two historical examples show two extreme cases, one where innovation used undue violence, and one where conservatism used undue violence. But other historical events, such as the American Revolution, show another interesting development, which may have become the dominant one in recent centuries. Though superficially similar to the Cultural Revolution, the American event adhered to a more conservative protocol, one which allows conflict between groups, but only within certain cultural rules—in this case the rules of war. The colonists gave warnings, sent graceful emissaries to the British parliament, and then formally declared war, all in rough accord with the emerging ideals of international law. International law has continued to develop since then, with widely accepted rules for diplomatic relations and the imposition of sanctions. It is not yet clear whether international law will "take," but if it does, culture will have provided another successful adaptation for permitting and yet slowing change, this time in large-group interactions.

These examples show that it is possible for conservativism and innovation to find mutual accommodation within culture's framework, just as competition and cooperation can. But it is also clear that both kinds of balance can be upset by revolutionary or reactionary violence. Let us now look at how the balance can be disturbed more peacefully, though just as profoundly, by the dynamics of ordinary variety.

3) ***Variety and Interaction***. A final function of culture offsets its conservative elements in another way, by presenting new options for individuals and sub-groups within the larger group. I call this function the variety function because it recalls our earlier discussion of variety in evolution, and modifies it slightly.

As you remember, some individual differences originate from the inexact duplication of genetic material. These differences introduce a variety into human life which makes philosophy difficult, at least insofar as philosophy tries to make common statements valid for everyone. The emphasis on commonality is important, but philosophy must learn to account for variety by stating its truths with appropriate reservations and exceptions.

In addition to the variety produced by genetics, variety comes into our lives because of different personal experiences, and Chapter 7 will discuss some implications of that source of variety. Yet another source of variety, however, is the interaction of cultures. In geographic areas where cultures mix readily, the available variety can be intense. Large countries such as the United States experience this cross-cultural variety not only from interaction with foreign cultures, but also from interactions among sub-cultures within their own boundaries. This interaction brings advantages and disadvantages—it provides many new ideas, but it can also produce confusion and dissipation.

Chapter 7 will also discuss how one evaluates exciting new ideas, as a part of faith-development. But long before serious evaluation begins, culture typically reacts. The larger group feels threatened; it clings tightly to its traditions and symbols, protecting itself from cross-cultural stimulation, and trying to discourage its members from flirting with danger.

Nonetheless, some people are inevitably attracted to new opportunities, and this tendency ensures that variety will eventually have an effect, either overriding or modifying dominant traditions. In a few pages we will see how cultural intermingling destabilized the Hebrew tribes, and how cultural cross-currents profoundly influenced the development of theology.

Culture functions in other ways than by coping with the dynamics of competition versus cooperation, conservatism versus innovation, and variety. But the stage is now set for talking about the central question of this chapter, which is "How does cultural knowledge get into individual heads?" After discussing that question we will look at specific cultural mechanisms which impact on ethics and morality, philosophy and doctrine, and on the specific doctrines of biotheology.

How Culture Enters an Individual Brain

When we say culture is based in biology, it is important to acknowledge that "biology" has two slightly different meanings. We may think of the word with a capital "B," which refers to broad biological principles; or we may think of it with a small "b," which refers to daily biological operations. We have mostly been talking about the capital "B" so far in this chapter, showing how culture resulted from broad-based principles of evolution, especially the evolution of the brain.

As part of this broader biology, different species evolved different forms of memory and brain structures. In most of the apes, brain pathways evolved that did not move their possessors toward culture. The owners of those brains became modern species occupying a non-cultural niche (such as gorillas or chimpanzees), or else they perished for some reason, which may or may not have been related to their lack of culture. Other species evolved rudimentary culture and then died out, perhaps even because of competition from humans.[25] *Homo sapiens* and its culture flourished, however, both because the components listed above emerged, and because brain structures evolved to integrate and coordinate those components. Increases in brain size supported such changes, and studies already cited give further help in understanding how brain evolution predisposed us for culture.[26]

But culture also requires more mundane biological processes, in which brain chemicals and structures are modified by the things we learn each day. This "biology with a small b" operates to get culture into *individual* heads, and not just into our species. How that happens is explored by Developmental Psychology, a field which traces mental and emotional development over the entire lifespan, beginning with the newborn child;

you will remember that Chapter 2 mentioned the field when discussing learning. The founder of Developmental Psychology—Jean Piaget—was a student of brain biology. He taught that we must understand the brain in order to understand *any* psychology, but especially developmental psychology. Piaget inspired a large literature, which describes and documents the stages of growth at which we are most likely to learn various aspects of culture.

This literature shows that *what* we learn comes largely from culture, but the *way we learn* is determined by biology. It turns out that humans are particularly sensitive to learning from our group, and we are also programmed to teach our young. Children are not born with cultural knowledge inside their heads, but they are born with an inbuilt "schoolmarm" which prompts them to ask questions, accept answers, modify those answers, and teach their young in turn. The details are flexible, but the tendency is invariable. Again, a reasonable analysis of the various brain circuits involved—language, imagination, explanation-systems, sociality, etc.—shows that culture is always at the end of the complete set of circuits,[27] and that teaching culture to our young is part of our evolved sociality. Thus culture is something we learn, and as such fits squarely into our earlier discussion of learning. That is why this chapter is entitled "Culture, a Kind of Learning."

In the growth of each healthy child, the neural pathways are pre-programmed to "expect" to learn cultural knowledge[28] and to take advantage of its various components.[29] Indeed, the brain contains more dendrites during the juvenile period than in adulthood. These extra dendrites permit social and other learning, often at precise times of life by means of the powerful mechanism called imprinting. Unused dendrites wither after the period of early learning, though some new ones continue to be manufactured throughout life.[30]

This usage of some dendrites and the withering of others implies that children can learn different things, depending on which dendrites are connected. And of course they do learn different things. Each person's brain is therefore structured in slightly different ways, because different learning experiences cause differences in their dendritic connections. Differences in learning are further emphasized by small genetic differences in the hard-wiring of our brain structures (or by hard-wiring malfunctions), but people turn out differently mainly because of differences in learning experiences after birth, using essentially the same mental machinery.

If neural pathways develop differently in response to different experiences, it is clear that our brain structure permits and even encourages the emergence of different approaches to life. When a group of people has experiences similar enough to produce a relatively common set of behaviors and objects and beliefs, that set is called a culture. Each culture then circles back to implant its ideas into the brains of each individual within it,[31] through mechanisms of formal and informal education.

Culture is thus embedded in individual lives through normal learning processes. First a memory trace is created, which becomes a memory pathway if reinforced. Alternate circuits may develop for important pathways, and the path may wear down so deeply as to become a rut. Unfortunately, both unhealthy and healthy ruts develop, and culturally induced ruts can run deep indeed. When such stable pathways serve us well (which is most of the time), we can be grateful for the process that produced them. But even if they cause suffering, such pathways can be very stable, and thus hard to modify, just as it is hard to modify high blood pressure or chronic bronchitis, no matter how much suffering they cause. We can sometimes modify brain pathways, however, just as we can modify high blood pressure, and later chapters will show how.

Once culture is stored in our brains, it functions in the ways already described: What we learn helps us survive, and also helps us regulate our behavior in society. Although we think of culture in terms of groups and institutions, it is first embedded in individual brains through the kinds of learning mandated by our biology, and it is expressed by individuals if it is expressed at all.

The Mechanisms of Culture

Even though culture is expressed directly from individual brains, the fact that culture lasts longer than individual lives implies the existence of mechanisms by which it is maintained and transmitted to each new set of individuals. A typical list of these mechanisms would include education, literature, law and custom, religion, social organization, economic systems, science, and art. These mechanisms are each so important as to constitute separate disciplines, and each separate discipline has uncovered enough principles to require division into sub-disciplines. A cultural "mechanism" is therefore not a simple mechanical device for transmitting or maintaining culture, but a rather complex societal operation.

To underline the complexity, notice that some aspects of each mechanism are formal and some are informal, and most are mixed. Education, for example, takes place in both formal and informal ways. Further, some aspects of each mechanism arise out of "deep structures" and some come from "surface structures," and again one finds mixtures of the two. For example it is highly likely that incest taboos are innate,[32] but they are also institutionalized into law, so that our individual knowledge of them comes from learning formal rules as well as from heeding the deeper warnings of our neuronal and hormonal structures.

Moreover (and like other components of our nature), cultural mechanisms come with both advantages and disadvantages. We can only seize the one and avoid the other through application of our rational and imaginative processes. Just as one may inherit weak eyes which are compensated for rationally and mechanically, culture permits us to incorporate or reject various options from our mental pathways. Later chapters will address such choices more specifically; here our intent is only to notice the special possibilities inherent in culture, and to become aware that its advantages and disadvantages begin in particular mechanisms.

One mechanism, namely literature, will be discussed in Chapter Six. There culture's threads picked from this chapter, and theology's threads plucked from Chapter Five, will be woven into a discussion of writing and scripture. I will be trying to show how scripture acts as an agent of culture, of language, of learning, and of imagination, to capture and symbolize matters of utmost importance for us, offering stories and beliefs which we can incorporate into our brain pathways and thus into our lives.

We will also return to cultural mechanisms in Chapter 10's discussion of sin, and Chapter 15 will discuss education in more depth. For present purposes, however, these basic observations about culture and its evolution will enable us to finally resume the discussion of language and thought, and see how they interact with each other and with culture to produce concepts.

Notes

[1] In "Culture, a critical review of the concepts and definitions," *Papers Peabody Mus.Am.Archeol.Ethnol.* 47(1):1-223 [33-38], 1952.

[2] Keesing, F.M. *Cultural Anthropology*, New York: Rinehart 1958.

[3] Estimates begin at about 2 million years ago. See *The Evolution of Human Societies: From Foraging Group to Agrarian State*, by Allen W. Johnson and Timothy Earle (Stanford, California: Stanford University Press, 1987).

[4] Arithmetic progression proceeds as in 2+2+2, whereas geometric progression proceeds as in 2x2x2. Already after three steps in the progression the first yields 6 and the second yields 8, and from then on the difference becomes dramatic. An arithmetic progression of 2 taken to 10 steps yields only 20, for example, whereas a geometric progression of 2 taken to 10 steps yields 1024. See Tunnell, pp. 38-40, cited above.

[5] This has been disputed. Richard G. Klein speculates that the first skeletally modern people may not have had fully modern neurology, in his 1989 book *The Human Career* (University of Chicago Press). If Klein is correct, or course the argument of this chapter is even stronger.

[6] See *The Evolution of Culture in Animals*, by John Tyler Bonner (Princeton, New Jersey: Princeton University Press, 1980).

[7] Bielicki, T. "Deviation-amplifying cybernetic systems and hominid evolution." *Materialy i Prace Antrhopologicyne* 77:57-60. 1969: 57-60. Cited in Tunnell (above). Bielicki considers the interaction of 8 factors in organic evolution which underlie cultural development. Over time these factors amplified each other by feedback relationships, and ultimately produced culture.

[8] The culture-acquisition devices built into us are discussed at p. 13 of *The Imperial Animal*, by L. Tiger and R. Fox (New York: Holt, Rinehart and Winston, 1971). The inevitability of culture because of "spinoffs" is argued at pp. 32-39 of *The Concept of Culture*, by Leslie White with Beth Dillingham (Minneapolis: Burgess Publishing Co, 1973).

[9] For the innateness of hand-body signals and their contribution to language and culture, see Irenaus Eible-Eibesfeldt's *Ethology: The biology of behavior*, especially the chapter on flirting (New York: Holt, Rinehart and Winston, 1970).

[10] The opposable thumb and bifocal eyes apparently developed in response to trees. Obviously the thumb makes it easier to grasp limbs. Set-apart eyes allow separate foci and thus better 3-dimensional vision; that is also very helpful in grasping limbs.

[11] For example Australopithecus used tools and had linguistic abilities, so some of the developments traced here clearly preceded humanity. See *Lucy's Child: The Discovery of a Human Ancestor*, by Donald Johanson and James Shreeve (New York: Avon Books, 1990).

[12] For an excellent treatment that does take this approach see *Human Evolution* by Bernard Campbell.

[13] The consensus is that sociality arose from the need for taking care of the young, though other influences may have also played a part—See Bayliss, Jeffrey R. and Halpin, Zuleyma Tang. "Behavioural antecedents of sociality," Chapter 13 in *Learning, Development, and Culture*, ed. H.C. Plotkin.

[14] *Biological Foundations and Human Nature* is a good general source on sociobiology. It has several interesting essays, including one by Edward O. Wilson. It was edited by Miriam Balaban and published by Academic Press: New York 1983. I will discuss sociobiology further in Chapter 10.

[15] In fact, even Pithecanthropos (*homo erectus*) apparently conducted organized big game hunts. Pp. 85-100 of *Early Man*, by F. C. Howell (New York: Time-Life Books, 1965).

[16] It is interesting that virtually all pre-agricultural bands have 14-40 individuals (p. 200 of Robert J. Wenke's *Patterns in Prehistory* (Cambridge: Oxford University Press, 1984). Even with primitive agriculture the number is only about 500 individuals (Pp. 375-9 of *The Emergence of Man*, by J.E. Pfeiffer (New York: Harper and Row, 1972). Thus we might operate "more naturally" in small groups, a speculation we will return to later.

[17] See p. 323 of *Human Evolution*, by Bernard Campbell. See also Mary Midgley's "Toward a New Understanding of Human Nature: the Limits of

Individualism," Chapter 18 (pp. 517-533) in *How Humans Adapt: a Biocultural Odyssey*, ed. Donald J. Ortner (Washington, D.C.: Smithsonian Institution Press, 1983).

[18] J.B.S. Haldane recognized this subtlety in *The Causes of Evolution* (London: Longmans, Green; 1932).

[19] Chomsky, Noam. *Language and Mind.* New York: Harcourt, Brace and World, 1967; "Review of Skinner's verbal behavior." *Language* XXXV:26-58 (1959).

[20] William Irons has pointed out (personal communication, April 25, 1995) that I use culture as an autonomous entity in this section, as if it could do things by itself. I agree with him and other behavioral ecologists that only individuals do things, and thus culture must be seen as a metaphor of individual actions. Better yet, I prefer to say culture is a *summary* of actions taken by individuals, and may be expressed in either small or large groups. I hope any confusion this section introduces on this point will be cleared up in the next section, dealing with the question of how culture enters individual brains.

[21] For a basic explanation of game-theory in the context of social life see Henry Hamburger's *Games as Models of Social Phenomena* (San Francisco: W.H. Freeman and Company, 1979). See also John Maynard Smith's *Evolution and The Theory of Games* (Cambridge University Press, 1982); and Robert Axelrod's *The Evolution of Cooperation* (New York: Basic Books, Inc., 1984).

[22] For a critique of Richard D. Alexander's theory of group-group competition see William Irons's "How Did Morality Evolve?" Vol. 26 No. 1 (March 1991) of *Zygon: Journal of Religion & Science*, pp. 49-89.

[23] Konrad Lorenz thinks we have an innate aversion to drawing blood, which works to end combat in hand-to-hand struggles. The aversion may be triggered by the smell of blood, but weapons which work at a distance keep such innate mechanisms from operating. See his *On Aggression* (New York: Harcourt, Brace & World, 1963).

[24] I said above that non-human primates don't have imagination. Usually they don't, but Imo certainly seems to stand as an exception. For a citation to the Imo story, see Note 48 of Chapter 2.

[25] Cro-Magnon man, for example, may have had a larger brain than ours, with brain pathways that we might envy in some respects. Yet something about Cro-Magnon's evolution or fortunes fitted him less well to life than our ancestor's evolution did; he was Able and we were Cain. Or perhaps we and Cro-Magnon are one, and we displaced Neanderthal. See *Cro-Magnon Man*, by Tom Prideaux (New York: Time-Life Books, 1978).

[26] Gary Tunnell's book *Culture and Biology: Basic Concepts in Anthropology* (one of the most powerful little books I have ever read page for page), gives further explanation of the evolution of the brain and its resulting role in culture. It explains, for example, that culture is impossible without learning and memory and symbolic communication (p. 23). Tunnell even goes so far as to suggest (p. 35) that sociality is an innate attitude that relates directly to the idea of the holy. That

ties in neatly to Gordon Kaufman's idea of theological imagination presented in Chapter 5. The one quibble I have with Tunnell is that he implies that culture and biology are not related in a causal way (p.3). I thinks he means to make a specific point, namely that *race* is not directly related to culture in a causal way, and of course that is correct. But in the sense of our being prewired for culture, there is a direct causal relation between culture and biology, as Tunnell and several other books cited in this chapter make clear.

[27] See Henry L. Shapiro's paper entitled "Impact of Culture on Genetic Mechanisms" in *The Nature and Transmission of the Genetic and Cultural Characteristics of Human Populations*. Milbank Memorial Fund, 1957.

[28] James E. Black and William T. Greenough, "Developmental Approaches to the Memory Process," in *Learning and Memory*, ed. Joe L. Martinez, Jr. and Raymond P. Kesner.

[29] Language, for example. Although it is almost inevitable that healthy humans will learn some language, brain architecture is open enough for culture to determine which language is learned. This apparently happens through a kind of imprinting, which insures rigidity in the basic ability, but allows flexibility of content. Tunnell, pp. 31-32.

[30] There are other theories of the reason for these extra dendrites, but all of them relate to the special needs of childhood.

[31] See p. 325 of *Human Evolution*, by Bernard Campbell (New York: Aldine Publishing Company, 1985).

[32] See J. Maynard Smith in "The Concepts of Sociobiology," pp. 29-30 in *Morality as a Biological Phenomenon*, ed. Gunther S. Stent (Berkeley: University of California Press, 1980).

Chapter Four

Concept-Formation: Thought, Language, and Culture Together

An Outline of the Chapter

A. Pre-concepts
B. Language and Concepts
C. Modifying Concepts
D. Imagination and Concepts
E. A Summary and Model of Part One

In purely practical ways and even without culture, *concrete* words expanded thinking and communication far beyond body language and expressive grunts. But as language's *abstract* symbols began interacting with thinking and culture, the mix became even more powerful. This chapter will look at one of the major ways that mix operates, by discussing conceptualization.

A concept is defined as an idea or set of ideas, extracted from the general range of experience into a category, and identified by a descriptive word or phrase. For example, "medicine" is a concept that took a long time to formulate, but once separated from other experience, it stimulated a creative process of healing and learning that still continues today.

Concept-formation is fundamental to human life, because our repertoire of categories largely determines our approach to the world. We take the process for granted because we start forming concepts so early in life that we are not even aware they are forming. But since concepts influence our every decision, it behooves us to understand them better.

Fortunately, a great deal of work has been done toward understanding concept-formation.[1] Psychologists know when children become capable of forming concepts, and how they think before that. I will survey that research in terms of one child's life, whom I will call Mary. Then we will look at how concept-formation relates to language, to learning, and to imagination. Our evolutionary orientation will be maintained as we go, by pointing out similarities and differences between us and animals, to show how humans came to conceptualize in our particular way.

Pre-concepts

Young Mary begins to learn as soon as she is born,[2] but even without learning she has certain predictable responses. She starts smiling randomly at a predictable point; apparently this is an instinctive behavior which encourages her parents to smile back and bond to her.[3] A bit later she starts learning to smile appropriately. By that time she is already beginning to grasp concepts, but it would be more accurate to say she is at a pre-conceptual stage.[4] Grasping the smile's significance is probably not a cognitive act, but she easily grasps the idea that crying brings attention, and she seems to cry quite purposefully.

During these early months Mary's nervous system continues to develop, as if she were still in the womb. One reason some behaviors are predictable is that they are produced by normal developments in the psychophysical substratum. Such developments result from the genetic construction of neural pathways rather than from learning, and they are relatively similar from child to child. Individual experience may modify those pathways early, or embed them more deeply, but at the beginning they depend almost entirely on genetic inheritance.

Soon Mary learns to distinguish herself from the rest of the world, grasping at a rudimentary level the idea of a separate self. Not long afterwards, other basic ideas are grasped instinctively, like "the continued existence of objects no longer present," or the idea of "mine," even before she knows the words for such concepts. Mary is now about eight or nine months old.

Up to this time, Mary's ideas may still be called pre-concepts, or *averbal* concepts, as one writer calls them, to distinguish them from the powerful verbal concepts about to appear.[5] Some writers also refer to the mental images in animal brains as averbal concepts.

Language and Concepts

Earlier theorists routinely claimed that animals could not conceptualize, but pre-concepts or averbal concepts force us away from such a firm conclusion. Chapter 2 listed some learning accomplishments by lower animals, and in higher mammals that learning is accompanied by very concept-like images. Biologists remind us that our concepts seem so different from an ape's pre-concepts or averbal concepts, only because we weren't there to see the intermediate forms between us and the ancestors we share with the apes. Those intermediate species must have formed keener and keener pre-concepts in the course of human evolution.

Yet even heeding that reminder, it is necessary to acknowledge human evolutionary advances in brain structure and mass. To acknowledge those advances, it seems fair to limit the word "concepts" to ideas which can be contained within verbal boundaries by the holder of the ideas. Concepts as thus understood are not merely ideas, but *ideas expressed in words*. That usage is consistent with the chapter's beginning definition, and limits concepts to human experience (except for a very few exceptions all relating to the possibility of language in animals). Such a definition emphasizes how important language is to the formation of any but the simplest concepts. It is by learning language that Mary really starts building concept upon concept, as we will see in a moment.

Nonetheless, the non-linguistic roots of conceptualizing are also significant. That such roots exist is evidenced by the fact that pre-language children conceptualize very much as animals do, and afterwards they conceptualize differently. The similarities are not surprising, since our growing brains pass through similar stages as do those of higher animals, and since both we and they learn by establishing or modifying memory pathways. Neither should it surprise us that differences soon emerge, since human imaginative powers are greater than any animal's, and since language adds a powerful new dynamic. A brief example will look deeper into this relationship between animal and childhood thinking, and further demonstrate how learning and language converge to produce concepts. The example can be presented in three short paragraphs:

A. One sort of learning which is much like concept-formation, is called associative learning. It is based on concrete characteristics which we learn to associate with an object. Mary associates blue uniforms with police, and thinks anybody in a blue uniform is a police officer.[6]

B. A simple kind of associative learning is used in animal experiments. For example, suppose a dog is taught that it can find food by going through the right-most door of three doors; it associates that door with food. Then a change is introduced, to see if the dog can adjust. Surprisingly, the dog adjusts to dramatic shifts easier than to simpler ones. Thus, if food can now be found behind the left-hand door instead of the right-hand one, the dog has trouble figuring it out. But if the rules completely change, so that food can be found behind any door with a blinking light (and only behind that door), the dog figures it out pretty easily.

C. Interestingly, pre-language children respond to these tests like animals do, but post-language children respond differently, at least after kindergarten age. Apparently the reason such shifts are easier for older children (and adults) is that they look for general rules, rather than focusing on details of the problem like young children and animals do.[7] And the reason adults and older children make general rules is probably that language helps them define and formulate concepts out of which the general rules arise.

It is not universally agreed whether associative learning is something that precedes concepts, or whether associative learning and concepts are actually the same thing; some writers believe concepts are merely complex associations.[8] They say words and symbols are specific facts like the blue in a policeman's uniform, which are learned and subsequently built upon in the course of intellectual growth. For our purposes either argument leads to the same conclusion, namely that a continuum of complexity exists, with simple associative learning on one end of the continuum, and complex concepts on the other end. Each particular association or concept falls somewhere between these two extremes.

Seen this way, any difference between associations and concepts is approximately the same as between the concrete and the abstract. So when young Mary and a lawyer call the same person an officer, the child is making a concrete association (blue + uniform = officer), whereas the lawyer has learned abstract legal relations which fit a police officer and a judge into the same conceptual category "officer of the law," whether or not in uniform. Yet Mary would have trouble seeing a plainclothes policeman or a judge as a law officer, because she cannot yet abstract to that extent.

In fact there are many cases where the difference between concrete and abstract (and between simple associations and concepts) is hard to grasp. That is partly because we are often dealing with a continuum rather than with completely different things, and a concept serves to "freeze" a point on the continuum, which is quite an abstract thing in itself.[9] Thus concept-learning is not always easy; although Mary grows and matures in her conceptual abilities, some abstractions may be difficult for her even when she is grown.

Modifying Concepts

The policeman example serves as a good summary of the idea of concept-formation this chapter is pursuing. *It is the process by which we take observations from our general experience that seem to be related, construct verbal categories to identify and summarize each set of related experiences, and then use the resulting words and phrases to mentally structure the world.* Of course this implies that our concepts can be skewed in one direction or another depending on our experiences and the words we learn. It also implies that concepts have to be modified over time, as broader slices of the world are encountered.

These implications turn out to be correct. As Mary learns language, it serves as a vehicle for further learning, and what she learns later depends on what she learned earlier. For example, she identifies the tail-wagging barking animal with the word "dog" after a few tries, and doesn't learn "bird dog" or "Championship bird dog" until later. And of course she learns many other concepts as she grows up. But always, simple concepts come first, and become foundations for more complex concepts.

Inevitably, however, it becomes necessary to modify and restructure concepts, and not just build upon them. Even as a small child Mary may modify her concept "dog" to include some few similar animals without tails or without a bark. And she may leave out animals she thought were included in "big dog," because they are horses. Another example is the concept "citizen," which Mary will first learn without much understanding, and probably not modify the concept until much later, when she has a fuller understanding about the group that grants her rights and requires responsibilities of her.

Although language is a great help in the process of learning concepts, it is even more necessary to modifying them. In fact, language and concepts interact in at least three different ways:

(1) We learn and modify concepts by learning new words. It is easier for Mary to exclude horses from her "dog" category once she knows the word "horse." Later, language permits the growth and modification of concepts by providing additional words, even coining new words when old ones seem inadequate. Thus even for adults, learning new concepts is often a matter of grasping new words or symbols. For example, it may be easier for us to remove certain ideas from supernatural categories and place them in scientific ones, once the concept "randomness" is learned.

(2) We modify our concepts by moving experiences from one word-category to another, as when the United States incarcerated its own Japanese citizens (during World War II) and later had to make amends because it confusedly removed them from the conceptual category "citizen" and erroneously placed them into the category "enemy."

(3) We modify our concepts by leaving our experience in the same word-category, but with a more complete understanding of what the word means, as when an adolescent conceptualizes "love" and later modifies what that concept means.

After she learns basic concepts, Mary continues to add concept upon concept throughout her life, modifying where she must. Yet once her basic concepts are established, she departs from them only with difficulty. Apparently that is because most concepts are built on simpler concepts, and disturbing any one of them may require disturbing more fundamental ones that have been part of her world-view for a long time. She (and we) must sometimes reroute mental pathways to overcome the psychophysical "rut" mentioned earlier, and that is not easy.

Imagination and Concepts

Eventually Mary can formulate quite complex concepts, modifying them not only because of the dynamics of language, but also because of learning and culture. Through learning she embeds each concept into a finite psychophysical form, as a memory trace or pathway, and culture contributes by supplying a beginning supply of concepts. Imagination adds to the mix by presenting new ways of looking at things.

One of imagination's impacts on concept-formation is especially interesting. You may notice that my earlier examples are really instances of concept-identification, rather than concept-formation. When Mary learned "dog," she was not forming a new concept, so much as identifying one already formed by her culture.[10] The mental processes in concept-formation and concept-identification are similar, so perhaps no harm is done by treating them as one and the same. However, completely new concepts are sometimes imagined, and new combinations of old ones frequently occur. These new concepts arise in every area of human life, and compete with previously existing concepts.

Later chapters will talk more about the tension between old concepts and new ones, but the imagination of new concepts brings up an immediate question. Since imagination seems to work somewhat independently of the "real world," to what degree must our concepts accord with external reality? Normally, evolutionary explanations imply that, since the entire brain evolved in response to the external environment, anything produced by it must be tested against that external environment. Thus if I imagine a "friendly nature," my imagination does not significantly change the fact that a tiger might eat me. So if early humans had evolved an imagination that did not interact with rationality to test their concepts, they would have perished by doing foolish things like petting tigers.

Yet we do develop concepts at odds with reality. Sometimes that causes harm. But sometimes imagination turns out to be more consistent with reality than its critics suppose. How can this dilemma be resolved?

Must imagination, and the concepts it produces, be consistent with reality, or not?

The answer seems to lie in the ways non-rational and rational thoughts interact. As already noted, there are extensive ties between each kind of thinking, so that there probably is no such thing as "pure emotion" or "pure logic."[11] Thus there seems to be a premium on using imagination to help construct concepts, even though they may not be consistent with reality, and another premium on using rationality to test and modify those concepts. As I said earlier, imagination can cause harm, but its dangers were outweighed by its adaptive advantages over evolutionary time, and therefore it remained in our mental makeup. So imagination-produced concepts do not necessarily have to be consistent with reality. We can only determine whether they are consistent with reality by testing them against other processes, especially the processes of culture and rationality.

* * *

The next chapter begins Part Two of this book, and explores the interaction of concepts and imagination further, adding a theological dimension and examining the special kinds of concepts produced by "the theological imagination." Later chapters will explore specific theological concepts, trying to determine which ones are consistent with reality, and how to modify those that aren't. But before leaving Part One, a summary of these first four chapters is in order.

A Summary and Model of Part One

In one way or another, most components of the evolved thinking/language/culture mix which these four chapters has discussed is important to biotheology. Ancient emotional impulses impact on its ethical aspect; concept-formation is important to its idea of God; and so on. While we cannot exhaustively examine every part of the mix, a structure is needed to fit the basic components into a unified schema. From that structure, particularly relevant components can then be elaborated.

The model presented below is just such a structure. Like all models, it omits many details,[12] but it aims to include enough specifics to let you assess the validity of conclusions that flow from it. Coming as it does from an evolutionary perspective, the model is simpler than most of those offered by classical philosophy, and it may be presented in seven short paragraphs:[13]

a. We are born, not with a blank slate as John Locke claimed,[14] but with many innate mental pathways produced by eons of trial and error evolution.[15] Those pathways, and the tendencies they produce, are common to the species. Yet as in evolution always, there is significant variety within the species.

b. Some of our innate tendencies are probably rigid at the individual level. This is certainly true in such physical characteristics as eye color, but it is also true in some psychological phenomena. For example, bonding may affect humans (though less profoundly than it affects other animals[16]), and sex roles seem to form early and not change easily.[17] Thus just as one's eye color is the same every day because of stable underlying physical characteristics of the iris, stable nerve construction may produce rigidity in a few behavior patterns.

c. To a remarkable degree, however, innate mental tendencies are not rigid; they can be modified through experience, imagination, and learning.[18] We are born partly programmed, but mostly programmable. Most brain processes are plastic throughout life, and many of the rigid ones do not solidify until long after birth.[19] This flexibility is a trade-off, evolution-wise, because there is great advantage to rigidity (witness the success of the robot-like insects); yet for our niche there have been even greater advantages to plasticity and flexibility. So these malleable tendencies reside in our brains, ready to be developed and formulated and released by personal experience and learning.

d. In general, learning and experience begin by gathering information through the senses, and storing that information in memory along with the innate and modified-innate information already in the brain. Language and culture significantly influence the way information is perceived and stored by contributing to the formation of concepts, which in turn influence our further learning and experience.[20]

e. Once in the brain, incoming information is processed. Concept-formation is a major part of that processing, but other processes are working too. Dreaming, for example, may be an "off-line" processing of the day's incoming data,[21] or a junking of irrelevancies. In general, only a small part of brain processing rises to consciousness, the vast majority of it happening subconsciously.

f. The acquisition, processing, storage, and use of information are seldom perfect. On its way to or from storage, traffic jams or organic problems may disturb the synaptic flow.[22] Genetic disabilities of the brain afflict a large number of us in minor or major ways. Our differing experience, some of it pathological, may cause us to store or retrieve the wrong things or different things,[23] or to form dangerous concepts.

g. In daily living each situation triggers responses based on information which has been stored and processed, and on the concepts formulated from it. A lot can go wrong, but the brain does a reasonably good job of yielding up responses at appropriate times. At least it has worked well enough for our ancestors to survive and prosper over long periods of time, in a variety of circumstances and environments.

Although this model may seem to describe a disorderly group of processes, with information going in randomly and coming out unpredictably, the processes are actually quite efficient. Overall they provide an adequate adaptation to the real world, where information comes to us at odd times and is needed on short notice. By structuring these dynamic processes into the simple model above, we can now identify specific concepts for further discussion and analysis, without denying either their rational or non-rational origins.

My thesis is that imagination, mixed with other thinking processes and language and culture, produces **concepts**, on which more complex concepts are built. These first four chapters themselves provide an example of that building process, because biotheology's foundational concepts were presented in a specific order. I have tried to place the concept of "concepts" on evolutionary and psychophysical piers, so that further concepts can now be built on that understanding of concept-formation.

From this point we could pursue many different disciplines, because many different sets of concepts could be placed on the foundation of these chapters. Psychological concepts, scientific concepts, anthropological concepts, theological concepts, mathematical concepts, and hundreds of others could be identified and examined. The choice of which area to concentrate on determines which parts of the model we will develop further. Because I have chosen to examine theological concepts, some products of the thinking/language/culture mix will be pursued and others left behind. Yet many of those concepts impact on biotheology and will be discussed, worked into the rooms of our understanding as wall-stones if not foundation stones. Some such concepts might have been squeezed into the discussion above, but I reserved them for later chapters to avoid the confusion of trying to work in too much at once.

Part Two will depart slightly from our biological perspective and take a historical approach, but each chapter will contain at least one biological section. Thus the chapter on theological imagination will ask whether our tendency to make explanatory systems is innate; the chapter on writing will discuss the relation of scripture to biology; the chapter on faith will examine the ways belief enters brain pathways; and the chapter on God-concepts will ask whether such concepts have any roots in language and culture. In Part Three the biological approach will be explicitly resumed, with special emphasis on individual psychological dynamics.

Notes

[1] For example see *Concept Formation*, by Neil Balton (Oxford and New York Pergamon Press, 1977).

[2] Actually she starts to learn *before* birth. See Gilbert Gottlieb's "Conceptions of prenatal Development: Behavioral Embryology," Chapter 17 in *Learning Development, and Culture*, ed. H.C. Plotkin.

[3] See Vol. 1 (*Attachment*) of John Bowlby's *Attachment and Loss* (London Institute of Psycho-Analysis, 1982); or "The smiling response: A contribution to the ontogenesis of social relations," by Rene A. Spitz and Katherine M. Wolf, in *Genetic Psychological Monographs* 34:57-125 (1946), more accessibly excerpted in Spitz's *Dialogues From Infancy* (New York: International Universities Press, 1983), ed. Robert N. Emde.

[4] Hunt, Earl B. *Concept Learning*. New York: John Wiley and Sons, 1966

[5] *Concept Development and the Development of Word Meaning*, ed. Th. B. Seiler and W. Wannenmacher. Berlin and New York: Springer-Verlag, 1983.

[6] The literature makes two distinctions between associative learning and the related discriminative learning. From an evolutionary perspective one distinction is superficial, in that it calls animal learning associative and human learning discriminative. The other distinction is unimportant for our purposes, though it properly confines associative learning to *similarities*, and discriminative learning to *differences*. I treat them together, because the distinction seems more confusing than helpful. For more discussion see Chapters 8 and 9 of *Conditioning and Associative Learning*, by N.J. Mackintosh (New York: Oxford University Press, 1983).

[7] This isn't absolutely true—some animals in some kind of problems seem to generalize, just as older children do, though never in as complex a way.

[8] B.F. Skinner points out the learning aspect of operant conditioning in *Verbal Behavior*. New York: Appleton-Century-Crofts, 1957. The other side of the argument is well represented by H.C. Plotkin and F.J. Odling-Smee in "Learning in the context of a hierarchy of knowledge gaining processes," chapter 20 of *Learning, Development, and Culture*, cited earlier.

[9] Although experiences fall over a continuum, Richard Gregg argues that our symbol-defining nature requires us to think in discrete categories. Words and concepts draw boundaries between segments of experience, and profoundly influence our perceptions. See his *Symbolic Inducement and Knowing: A Study in the Foundations of Rhetoric* (Columbia: South Carolina: University of South Carolina Press, 1984).

[10] Jean Piaget expands this theme in Chapter VIII of *Play, Dreams and Imitation in Childhood* (London: William Heinemann Ltd., 1951). His ideas are readably presented in Chapter 3 of Henry Perkinson's *Learning from our Mistakes* (Westport, Connecticut: Greenwood Press, 1984); the idea presented in the text is found on p. 64. For a partly but not entirely contrary view, see Konrad Lorenz's description of a child, who called all mammals (and only mammals!) bow-wows;

at p. 307 of *Studies in Animal and Human Behavior*, vol. II, "Gestalt Perception as a Source of Scientific Knowledge."

[11] Although there has been much publicity about the right brain/left brain dichotomy, they actually communicate with one another every few seconds through the corpus callosum. In healthy people, therefore, neither one operates in a "pure" fashion without influence from the other.

[12] An excellent and interesting book on the nature and limits of modeling is *The Search for Solutions*, by Horace Freeland Judson (Baltimore: The Johns Hopkins University Press, 1987).

[13] I consider this model entirely consistent with the "general theory of learning" proposed by Plotkin and Odling-Smee, (Chapter 20 of *Learning, Development,, and Culture,* ed. H.C. Plotkin), though my derivation is from Lorenz.

[14] In his *Essay Concerning Human Understanding.*

[15] See Franz Wuketits' lead article in *Concepts and Approaches in Evolutionary Epistemology.*

[16] See p. 323 of *Human Evolution*, by Bernard Campbell. For some promising research in finding the mechanisms involved in bonding see "A Role for central vasopressin in pair bonding in monogamous prairie voles," by James T. Winslow, Nick Hastings, C. Sue Carter, Carroll R. Harbaugh and Thomas R. Insel in the October 7, 1993 issue of *Nature* (Vol 365, no. 6446), pp. 545-548.

[17] The most readable discussion I have found was written in 1977, namely *Sex Roles*, by Shirley Weitz (New York: Oxford University Press).

[18] Yet even this ability to learn takes place through evolved innate processes such as the gestalt process. See p. 41 of Konrad Lorenz's "Innate bases of learning," in the 1969 book *On the biology of learning*, ed. K. Pribram (New York: Harcourt, Brace and World). Incidentally, I hope my representation of the Lorenzian scheme helps to show why we do not need to go so far as Timothy D. Johnston wants to go, by abandoning the conceptual distinction between learning and instinct. While he offers fine insights arising out of many examples of something between the two, they can be effectively expressed as emendments of the two existing categories, or as a third category intermediate between instinct and learning. See his "Learning and the evolution of developmental systems," Chapter 19 of *Learning, Development, and Culture*, ed. H.C. Plotkin.

[19] For example, puberty completes certain brain processes which permit ranges of thought not theretofore possible. Newcombe, Nora S. and Baenninger, Maryann: "Biological Change and Cognitive Ability in Adolescence," in *Biology of Adolescent Behavior and Development*, ed. Gerald R. Adams, Raymond Montemayor, and Thomas P. Gullota (Newbury Park, California: Sage Publications, 1989). See also *Developmental Plasticity: Behavioral and Biological Aspects of Variation in Development*, ed. Eugene S. Gollin (New York: Academic Press, a subsidiary of Harcourt Brace Jovanivich; 1981).

[20] In fact, the more enriched our experience is, the more massive our cortex will apparently be. We will actually have more "brain-power." See Marian Diamond's

"The Aging Brain: Some Enlightening and Optimistic Results" in *American Scientist* for January/February 1978 (Vol. 66 no. 1, pp. 66-71).

[21] Palombo, Stanley R. *Dreaming and Memory*, pp. 53-36 (New York: Basic Books, 1977). See also *Sleep, Dreams and Memory*, ed. William Fishbein (New York: Spectrum Publications, 1981). For a review of other research on dreaming, see James Ashbrook's "Making Sense of Soul and Sabbath: Brain Processes and the Making of Meaning;" *Zygon* Vol. 27 No. 1 (March 1992), pp. 39-45.

[22] For a technical discussion see Chapter 1 of *Chemical Pharmacology of the Synapse*, by D.J. and C.R. Triggle (New York: Academic Press, 1976).

[23] Doubtless there is an evolutionary background for this. Storage or retrieval appropriate to one environment might be inappropriate to other environments, for example. Evolution favors preparation for alternatives, but that can make us respond as if we are in one situation when we are actually in another.

Part Two: Theology

Introduction to Part Two

The next four chapters could be presented in a different order than the one shown above. Building on Part One, it would make sense to put Chapter Seven next, because "faith" arises from our tendency to group concepts into world-views. After that we could develop the cultural theme by inserting the chapter on scripture, and then we could look at the history of theological concepts. Following that organization, Part Two would finish in grand style by discussing the concepts of God arising from human thought processes and ways of viewing the world.

One reason I have not chosen that order is that Chapter Seven (on faith) is very important, and the evidence for its argument is found not only in Chapters One through Four, but also in Chapters Five and Six. I also wanted it in a strategic place to remind the reader of the central role of concepts, this time in a theological context. Another reason for the chosen order concerns idolatry; since it is a pathology which easily infects faith, I wanted to introduce it before discussing faith, and the logical place to do that is in connection with scripture.

Finally, this order shuttles us off the analytical path we have been walking, to a more restful narrative. We will resume the analytic approach later, but the narrative will allow us to briefly and efficiently trace theological concepts from the very beginnings of history, all the way down to our own century.

;hapter Five

[he Theological Imagination: A Brief Iistory of Concepts in the Judeo-;hristian Heritage

'rom the beginning of our existence as a species, we humans have tried to xplain the mysteries around us. Earlier species were undoubtedly urious, but until thought and language evolved to permit complex onceptualization, only the simplest mysteries could be resolved. So even 1ough apes and monkeys and pre-humans were quite intelligent, they ould not begin to process complicated mysteries without an adequate 1ought/language mix. Even if they dimly wondered at such questions as 1e cause of sickness or what follows death, they could not hope to find olutions until they evolved the ability to formulate questions and concep-1alize answers in words.

Early in human experience, however, such questions could be tackled. 'he mental architecture was there, wonder was there, imagination and anguage were there, and conceptualization had begun. Concocting xplanations requires time, however, and our forebears had to spend most f their time carving out a basic security from an insecure environment; 1us we shouldn't expect to find many examples of competent explanation ight at the outset. Yet the stage was set, and these ancestors grappled with difficult questions when they could. Over time they probably imagined hundreds of possibilities for each mystery. By 20,000 years ago (when we can first piece together enough history to find out which solutions tended to satisfy them) their imaginations were actively pursuing almost every category of explanation we pursue today, with only a few exceptions.

This ability to apply imagination to mystery and construct concepts to explain it, has been aptly called "the theological imagination."[1] We have already explored the biological origins of imagination, but "theological" is also appropriate in discussing early explanations, because they were part of theology until science became a separate discipline, and that was only 300 years ago. Thus primitive theology contained not only the concepts of God and religion, but other systems of explanation too. Indeed, almost everything produced by the human imagination was initially explained in theological terms.

This chapter traces the development of explanatory systems in terms of imagination's natural operation as influenced by this early theological spin. The development will be presented in rough chronological order from earliest times forward, with pauses to point out the emergence of significant new concepts. Then it will begin to ask what role "theological imagination" had in fashioning our current explanatory concepts.

From the Beginning of Humanity until History Began

If we could trace the evolution of the brain structures that permitted theological speculation, it would be no different from tracing the evolution of anything else. First, enough brain cells accumulated to allow dim theological pre-conceptualization; chimpanzees murmuring over their reflections in mirrors probably have enough neurons for that.[2] Then more cells gathered, perhaps those that led to language, enabling the translation of murmurs into words. Deep feelings of longing and sympathy and wonder evolved, and language translated those feelings into meaningful symbols too. Just as physical growth proceeded at 1/2,000,000th of an inch per year, our ability to construct theological concepts grew at correspondingly miniscule rates, occasionally punctuated by larger spurts of growth.

We do not know exactly how explanations first began, but the story of Imo[3] shows that insightful individuals arise in other primates, and surely must have arisen among early humans. But early explanations probably were not systematic; that is, they did not try to explain all the various mysteries in terms of any overarching system. That is partly because the number of interesting questions was almost infinite, including questions about thousands of things we take for granted. Rocks, rivers, tastes, and colors were just as mysterious as death or menstruation, thunder or comets. Thus the focus was probably on particular mysteries, temporarily solved by some competent deduction or sudden gestalt inkling. Doubtless many local reputations for explanation-construction were made, and returned to dust with the insightful individual.

From archeological records and from knowledge about today's primitive peoples, we can guess what happened next, and it paralleled the emergence of culture. Particular explanations became accepted by primitive groups, and were passed down to succeeding generations. Someone began elaborating explanations into systems, so that (for example) an explanation about the sun, perhaps perceived as a personality, became merged with an explanation for death or fertility or one of the tribe's other concerns.[4] We can easily imagine the advantages a tribe must have gained from having explanations, especially *systems* of explanation. Not only would such explanations tend to satisfy personal curiosity, but they would also increase the group's sense of unity against nature and other tribes.[5]

The impact of culture on these pre-historic developments cannot be overemphasized. Such dynamics as group solidarity, social hierarchy, and training the young all interacted with imagination and other mental processes, as we saw in Chapter Three. One specific result was that culture determined which new ideas got incorporated into the prevailing explanatory system. And since culture continued its growth from prehistoric times forward, we should expect to find its influence growing by leaps and bounds as we enter the historical era.[6]

From the Beginning of History Until Monotheism Emerged

At this point we can begin to talk on a different basis. Until now we have guessed at how theological imagination manifested itself, based only on what few artifacts we have, on our understanding of concept-formation and culture established in previous chapters, and on projections from primitive practices still current today. But now we can begin to draw conclusions from more concrete evidence. Since that evidence remains sparse until about 10,000 years ago, we must still be careful, but the story quickly loses much of its speculative nature and becomes history.

By the beginning of recorded history, explanatory systems typically featured supernatural components.[7] Each tribe had its own gods, but as in other aspects of cultural dispersion (tools or dialect, for example), theological systems of adjacent tribes were usually similar, whereas the systems of distant tribes were likely to be different. The range of gods was extraordinary, which should be expected, given the range of imagination and the depth of environmental mysteries humans faced. This implies a direct relationship between each locale's gods and its immediate environment, and that implication is validated by actual cases.[8]

For example, a storm god loomed large in areas of violent weather. If weather was more predictable, but floods were not, a river god took prominence. Nomadic tribes often worshipped a god of the herds, while hunting tribes revered a god of prey. Fertility was always an issue, and with the advent of farming, fertility gods grew even more important. Most fascinating of all were the sun, moon, and stars, which often served as gods or the homes of gods.

From the perspective of these early people it was nearly impossible to conceptualize a single unified god relevant to all of life's problems. With seemingly little relationship between various aspects of life, each aspect seemed subject to the whims of particular gods. It was easy to imagine these gods disagreeing with one another, which explained why things sometimes went wrong even after sincere worship of one of them.

Indeed, the idea of collecting all the systems into a single Godhead at first shocked the theological imagination. When it was proposed in the midst of ancient history, it seemed downright obnoxious and intolerant. The proposal was first made in Egypt,[9] which is by tradition a tolerant area, but after a few years of flirtation with the idea, the Egyptians rejected it.

Yet think how powerful such an idea would have been. Not only did it simplify life, which must have attracted the busy Egyptians; it also resolved confusion over all the gods that had infiltrated Egyptian society from the north and south and west, and especially from the east. Worshipping a single god who was either superior to, or incorporated, all the others, gave an internal satisfaction and an empowering that we still feel today.

Here is how it seems to have happened. Egypt had long been a melting-pot society, and the resulting pressures were often expressed in theological terms, with different gods serving as rallying points for various political factions. Finally, one statesman/theologian proposed a great compromise, re-defining three important gods as a trinity manifesting separate aspects of an overarching spirit. This early trinity was composed of the visible sun-god Re, the sensed god of earth Ptah, and the hidden ultimate god Amun.[10] But the idea seemed too contrived; it would not become dominant for another 2000 years, and then in a different form.

A simpler idea was held by the religious-minded young king Amunhotep IV. He held that there was only one god, Aten. His strong monotheistic belief caused him to change his name from one honoring the old chief god Amun to Akhenaten, meaning "He who serves Aten." And since he had not only ideas, but also power, he didn't need compromise or tolerance; he *decreed* that Aten was the only god, and he suppressed other gods and references to gods, even striking references to them from monuments.[11]

Nonetheless, his ideas didn't stick either. It didn't take long for Amun's priests and their allied traditionalists to organize a response to this revolting development. Akhenaten's successor was forced to change his name from Tutankhaten back to the old style honoring Amun, not as the only god but as first in a group of gods that also included Re, Ptah, and Seth. The court of Tutankhamun (King Tut in popular stories, whose tomb was found intact in 1922 and toured the world in 1977) returned to a polytheistic belief and practice which greatly influenced the later pantheons of Greece and Rome.

* * *

In a moment we will pursue the Egyptian experiment with monotheism further, but let us pause to examine the psychological aspects of the imaginative pluralism which could produce (and can still produce) so many potential explanations for a given phenomenon. It will be a small interruption, but one which ties earlier chapters to this one.

The Relation of Explanatory Systems to Culture and Psychology: Schemas and Paradigms

Once scholars became aware of the historical reality of explanatory systems as introduced above,[12] a question arose: Is this tendency to produce explanatory systems inherent in humans, or is it a superficial tendency only rarely encountered, dependent on the particular history of particular explanations?

The answer seems to be that producing such systems is inherent in us.[13] A built-in program forces us to construct order out of sensory input which would otherwise seem chaotic. We playfully find order in clouds by seeing faces or scenes in them; similarly but more seriously, we construct consistent world-views from the random observations of daily life.

Psychologists see this tendency as part of concept-formation. Instead of maintaining hundreds of isolated concepts, our brain groups them into explanatory units linked by mental pathways. Grouping concepts into *sets* apparently gives us each a sense of integrity or psychological cohesiveness. One writer distinguishes between a culture's more generalized world-view and the specific world-view held by each individual, by calling the first a "paradigm" and the second a "schema."[14] For now we are concentrating on cultural systems of explanation, but individual systems will become important when we discuss faith.

Another word for paradigm is "myth," which anthropologists use to describe the underlying concepts and beliefs of cultural systems, including our own. That does not mean all myths are equal or that we may as well worship rocks, but it adds a needed dose of humility to our own conceptualizations. By realizing that our explanatory system constitutes a paradigm, or a way of looking at the world, we might avoid the tendency to think we have discovered the last word in explanation.

Although it is true that every known culture has imagined a system of explanation, I promised to point out exceptions to what I say, and there is a partial exception to what I am saying here. Eskimos in the western part of North America, perhaps because of their monochromatic environment, developed a very unimaginative set of explanations. Though they had a loose concept of spirits inhabiting animals and weather, they didn't have much of an over-arching system or god-concept, and no idea of creation at all.[15]

However, all other primitive groups known have had a more complicated mythology, and most of us are only a few generations removed from such mythologies within our respective heritages, be they European or African or Asian or Indian. That bolsters the conclusion that explanation-constructing is inherent in the theological imagination, which evolved naturally in our species.

My own background is Irish, and thus a mythology based on leprechauns has a certain claim to my allegiance. Nowadays "little people" are taken lightly, but they were once taken seriously all over Europe,[16] producing local variations such as elves, gnomes, and fairies. After feeling the satisfaction that came from explaining mysteries in terms of little people, and after reports of rare evidentiary glimpses of them, it is no wonder it became heresy to deny their existence or power. In the rest of Europe other explanatory systems finally pushed them into storybooks, but Ireland's island isolation kept leprechauns alive as a serious part of the national identity, even into the 20th century.[17] Thus even though we know leprechauns as charming myth, they were an honest attempt by a relatively modern people to explain things otherwise mysterious.

Understanding the theological imagination as an invariable part of our psychological makeup helps us accept leprechauns and other such symbols as genuine creations arising out of our experience of wonder, which we develop into reasonable hypotheses for explaining that wonder. Yet we will see that even good Irishmen must modify our language and symbols, if we wish to grow from new knowledge as it unfolds. And we must sadly withstand the poignant impulses that draw us backward.

Although human imagination invariably produces some sort of explanation for the mysteries it encounters, particular histories are important too, because they determine the details of given belief systems. In the judeo-christian case the history is rich and accessible to those really interested in it. Because that history influences our beliefs and concepts in profound ways that impact on our daily decisions, it is important that we know about it, so let us now return to the story of that heritage.

The History of Yahweh

Until now I have avoided any direct suggestion that the theological imagination might apply to our own understanding of God, but we would be fooling ourselves to think we are immune to such a basic process. Many people wish to avoid that thought, because they fear it would undermine their concept of God. We will not avoid it here, however. Instead we will be brave enough to look at the history of our own gods, confident with Emerson that "when half-gods go, the Gods arrive."

The Old Testament was put into its present form only 125 years before Christ. Even conservative scholars now acknowledge that each rewriting edited earlier sources to accord with theological concepts current at the time of editing.[18] Moreover, researchers have uncovered convincing evidence that the Hebrew concept of God was not always the same.[19] Instead, it slowly developed over the centuries, and changed significantly from time to time.[20] During the Old Testament era, we can identify at least four different stages of the unfolding conceptualization of God (called "Yahweh" by later Old Testament writers), which are as follows:

1) Before Abraham

We usually think of the Hebrew experience as primitive and original, but the truth seems to be that it developed as part of a city-based intellectual experience, and was neither primitive nor original.[21] It was influenced primarily by Babylonian concepts of nature and God, as well as by other local theological systems, and it was in its earliest stages polytheistic.

The evidence for this conclusion comes from archeological and textual research, but it is hinted at in the surviving Old Testament, even after the editing mentioned above. For example the first biblical word for God is the plural "Elohim."[22] Despite the tradition of translating this word as singular, and later efforts to call it a reference to the trinity, the greater probability is that early Hebrews thought in terms of many gods.[23]

Additional evidence of that probability may be found in Biblical references to various gods, even after the Hebrews embraced monotheism. Before Moses, the Bible uses several words for God, such as El Shadday,[24] Elyon,[25] Olam, Bethel, and Ro'i,[26] but never Yahweh.[27] It is possible for present theological imaginations to believe the Hebrews always had a strong concept of a single God, and therefore to see these other god-words as merely identifying characteristics of that one God, but it seems more likely that early Hebrews were polytheistic.[28] And as to the time period between Adam and Abraham, it is impossible to make a persuasive case in favor of an unbroken chain of monotheistic belief.[29]

2) From Abraham to Moses

Abraham's contribution to theological history was to choose a particular god of the many available in his region, and to give it priority. His father (tradition says) was an artisan who specialized in idol-making for the many cults of the fertile crescent,[30] and we can imagine young Abram struggling over the diversity passing through his father's shop. Finally he resolved any confusion with a clear declaration in favor of a particular god, by making a covenant with that god. Since covenant-making was characteristic of religious experience in the region,[31] Abraham's covenant directly reflected his culture.

Neither Abraham[32] nor any subsequent descendant before Moses claimed that other gods weren't real. Their belief was rather that such gods were valid for the people who chose them.[33] The only question was which god was more powerful, more protective of his people, and more capable of keeping his promises and threats.

I do not mean to imply that the idea of God was stagnant after Abraham. Instead a consensus slowly emerged about specific theological concepts, clarifying both God's nature and related issues like ethics and cosmology. For example a lively controversy arose as to whether the gods encouraged human sacrifice. This discussion (depicted in Abraham's struggle over sacrificing his son Isaac[34]) led most Semitic tribes to reject human sacrifice as part of what they could imagine in God's character.[35] Although a later passage shows respect for rival gods who still required that ultimate sacrifice,[36] animal sacrifice took the place of human sacrifice in Israelite understanding, and this change satisfied most theological imaginations until the time of Christ.

I have identified particular points in the history of our heritage because seminaries have not emphasized them enough, and because pastors have not discussed them in public sermons. A sound theology requires an understanding of imagination's contribution to our beliefs, even if that is uncomfortable to think about. Let us now add to that understanding, by discussing how a strong monotheistic concept made its way into our heritage.

3) Moses

The time of Moses was a great clarifying time for Hebrew conceptualizing.[37] This may have been because of Moses's powerful personality and intellect, but it must also have been because of the context in which the Hebrews found themselves.

When Moses was born the Hebrews had been in Egypt about 400 years.[38] King Akhenaten, introduced above as the first known adherent of monotheism, came to power not long before Moses was born.[39] It is highly probable that the Hebrews were thus exposed to Akhenaten's ideas, and that unambiguous monotheism entered our heritage then and there.[40] In fact, Psalms 104, a song to monotheism recorded in Hebrew literature only much later, is very similar to one of the hymns composed to Aten for Akhenaten.[41]

Since the idea that monotheism entered Judaism through Egyptian influence was initially controversial, I want to pause and give a further explanation of the controversy. Freud's book *Moses and Monotheism* argued that Moses was an Egyptian who got his ideas from Akhenaten, and may have been Tutmose, a deposed successor who took his followers into exile (the name Moses can easily be derived from Tutmose).[42] This stimulated some scholars to dispute Freud's claim,[43] and others suggested alternate sources for Mosaic monotheism, such as the Kenite theory (which says he got it from his father-in-law). After all the controversy, the most balanced view is one written by the Christian scholar William Foxwell Albright, which comes at the end of a thorough discussion of the controversy and the evidence. "In the light of the now available data," he says, "it is perfectly clear that the period between 1350 and 1250 B.C. was ideally suited to give birth to monotheism."[44] He concludes by acknowledging Moses's role in bringing monotheism into our particular tradition.

Thus when Moses led his people out of Egypt and created such institutions as the priesthood and a system of justice,[45] his genius also distilled a clearer concept of God than the Hebrews had ever known. Hereafter Yahweh, not El Shadday or the others, was the only god of Israel.[46] Moses clarified and solidified this development in concrete ways like making God portable at a time when most gods were tied to a specific geography; in abstract ways like rejecting direct symbolic representations of Yahweh; in social ways like the elaborate ritual dedicated to Yahweh; and in philosophical ways like making each person responsible for his own conduct.[47]

4) After Moses

Even though the concept of Yahweh developed greatly under Moses, it was not finished. It was only later, for example, that the Hebrews began to say that other gods were no gods at all, but mere wood or stone.[48] After securing Canaan the Hebrews reverted to an earlier belief, that each god is supreme over specific territory. This meant that Moses's idea of a portable god was redefined as a temporary expedient, housing Yahweh only until a temple could be built for him on his conquered land.[49] Later, when the Israelites left Canaan to fight the Moabites on Moabitic territory, Moab's god Chemosh was respected and even feared. (Chemosh had many characteristics in common with Yahweh, and reading surviving records of him like those in the Bible or on the Moabitic Stone[50] reminds us that adjacent gods are likely to be similar.[51]) Many other Biblical passages (as well as secular accounts of the time) demonstrate the belief that each God, including Yahweh, was supreme over specific real estate.

Other developments occurred during the second Babylonian captivity, after the Hebrews migrated back eastward, and after Egypt lost power and Babylon regained it. Only then did certain stories and concepts enter our heritage, among them the flood story, the garden of Eden, heaven and hell, the devil, and the attachment of creation to God.[52] Each addition rendered the explanatory system broader and more comprehensive, and each was thus included in new editings of the Old Testament. Some new ideas were still controversial during New Testament times, and became a basis of disagreement between Pharisees and Saduccees.

Indeed, the conceptualization of Yahweh *never* solidified completely, either in the Jewish system or in the early Christian system, but it greatly stabilized once reduced to writing. By the time the Old Testament reached its present form God was seen as one, polytheism was rejected not only for Hebrews but as an acceptable belief for anybody, and various characteristics of Yahweh had been worked out.[53] For example there had been some confusion over whether he was a sun-god or a fertility-god,[54] but he was now conceived as having broad power to command both nature and human activity. He was merciful when mercy was called for, vengeful where vengeance was called for, and wise enough to know the difference. He had earlier been a jealous and even vindictive God, but this came to be understood as part of his magnificent concern and compassion for his children. Above all, he looked after his children, punishing them only for their own good like an exceedingly competent parent.

In coming chapters we will see how these concepts were further modified by Greek and Christian thought, the latter already incubating when the last Old Testament book reached its present form. Though we won't take a strictly chronological approach to later God-concepts, Chapter 8 will build on the facts presented here to look more deeply into such concepts.[55]

Chapter 4 introduced concept-modification as an important principle of Biotheology. Yet this section was not meant to persuade you of the need for ongoing modification just yet—that task will be delayed until Chapter 6, which is next. It should already be clear, however (and despite traditional pronouncements to the contrary), that the concept of God has gone through many modifications in our own heritage. Even if we later conclude that God himself is "the same yesterday, today, and forever," it is certain that our understanding of Him is not.

Theological Imagination in the Modern World

Giambattista Vico, early in the 1700's, was the first philosopher to describe the role of creative imagination in the construction of explanatory systems.[56] Since then a long and unbroken tradition has seen such systems as the result of imagination interacting with other mental and social processes.[57]

The study of theology in such terms is not limited to ancient or primitive or foreign theological systems. Its main importance is to help us understand our own theology. That approach is well represented today by the Harvard theologian Gordon Kaufman, whose earlier cited book furnished part of this chapter's title. Kaufman develops his ideas in a different way than Biotheology does, but his observations about imagination's role in constructing religious concepts seem accurate and worth quoting. In a well-worded paragraph that is consistent with the entire post-Vico tradition, he says (at pp. 11-12):

> . . . theology is (and always has been) essentially a constructive work of the human imagination, an expression of the imagination's activity helping to provide orientation for human life through developing a symbolical "picture" of the world roundabout and of the human place within that world. In the course of history the fertile human imagination has generated, in the great religious and cultural traditions of humankind, a number of very diverse views of the world and of the human. Among these are those monotheistic perspectives—largely descendant from ancient Israel—in which the symbol "God" provides the ultimate point of reference and orientation for human life, indeed for understanding all of reality. This symbol has provided meaningful and effective orientation for human existence through a very long past, partly because, from the eighth-century prophets to the present, it has been continuously reëxamined and reconstructed, as new historical situations placed new demands upon it. Thus, a diverse and sophisticated theological tradition has developed over the centuries, analyzing and exploring the meanings, uses, and functions of the symbol "God"; and along with it has gradually emerged what I call "the theological imagination." The theological imagination devotes itself to the continuing critical reconstruction of the symbol "God," so that it can with greater effectiveness orient contemporary and future human life.

Biotheology traces theological imagination from much earlier than Kaufman does, seeing it as "gradually emerging" from the dim reaches of time, and not merely from the 8th Century before Christ. Biotheology also places less of a burden on the word "God" than Kaufman does, by considering evolution "the ultimate point of reference and orientation for human life." Otherwise, biotheology's view is much the same as Kaufman's. Even when we branch in different directions from Kaufman three chapters hence, we will honor his idea that we must "continuously reëxamine and reconstruct" the content of the language symbol "God."

Notes

[1] See Gordon Kaufman's book by that name (Philadelphia: The Westminster Press, 1981). See also *Religious Imagination*, ed. James P. Mackey (Edinburgh: Edinburgh University Press, 1986) and *Nature and Religious Imagination from Edwards to Bushnell*, by Conrad Cherry (Philadelphia: Fortress Press, 1980).

[2] See pp. 58-61 of *Chimpanzee Material Culture*, by W.C. McGrew (Cambridge: Cambridge University Press, 1992).

[3] To review the story of Imo the genius monkey see paragraph E. in the section entitled "Imagination and Learning" in Chapter Two.

[4] See Charles H. Long's *Alpha: The Myths of Creation* (Toronto: Ambassador Books, 1963).

[5] For anthropological support of this suggestion see I.C. Jarvie's "Understanding and Explanation in Sociology and Social Anthropology," in *Explanation in the Behavioral Sciences*, ed. Robert Borger (Cambridge at the University Press, 1970).

[6] For other examples of the interaction between the theological imagination and culture see pp. 14-29 of Sallie McFague's *Metaphorical Theology* (Philadelphia: Fortress Press, 1982). Also see *Naming God*, ed. Robert P. Scherlemann (New York: Paragon House, 1985).

[7] See *Before Philosophy: The Intellectual Adventure of Ancient Man*, by H.A. Frankfort et. al.

[8] See Joseph Campbell's *The Masks of God: Occidental Mythology* (New York: The Viking Press, 1964).

[9] See pp. 61-66 of Jaroslav Cerny's *Ancient Egyptian Religion* (Westport, Connecticut: Greenwood Press, 1952). Also see James Henry Breasted's *Development of Religion and Theology in Ancient Egypt* (New York: Charles Scribner's Sons, 1912); Chapters 18-19 of Breasted's *History of Egypt* (New York: Charles Scribner's Sons, 1910); and Chapter 6 of *The World of the Past*, ed. Jacquetta Hawkes (New York: Alfred A. Knopf, 1963).

[10] For a fuller discussion of several trinities in Egypt see pp. 396-402 of James Bonwick's *Egyptian Belief and Modern Thought* (Indian Hills, Colorado: The Falcon's Wing Press, 1956).

[11] See Vol. 2, p. 125 of Joseph Campbell's *The Masks of God.* See also pp. 244-50 of *Conceptions of God in Ancient Egypt*, by Erik Hornung (Ithaca: Cornell University Press, 1982).

[12] Braithwaite was probably the first scholar to analyze explanatory systems from a psychological perspective, in 1953. See Honans, George C. "The Relevance of Psychology to the Explanation of Social Phenomena," in *Explanation in the Behavioral Sciences*, ed. Robert Borger. See also Dick Ruimschotel's *Explanation, Causation, and Psychological Theories* (Amsterdam: Swets and Zeitinger, 1989).

[13] We owe to Freud the first modern formulation of that case (*Civilization and its Discontents*, London: Hogarth Press, 1939), though his arguments seem outdated. For a modern view of the interactions between explanation and culture see *A Theory of Religion*, by Rodney Stark and W.S. Bainbridge (New York: Peter Lang Publishers, 1987).

[14] Jones, James W. "Personality and Epistemology: Cognitive Social Learning Theory as a Philosophy of Science," in *Zygon: Journal of Religion & Science* Vol 24 (March, 1989), pp. 23-38.

[15] See p. 181 of Kaj Birket-Smith's *Eskimos* (New York: Crown Publishers, Inc., 1971).

[16] Briggs, Katharine M. *An Encyclopedia of Fairies, Hobgoblins, Brownies, Bogies, and Other Supernatural Creatures* (New York: Pantheon Books 1978).

[17] See pp. 100-108 of *Priests and People in Pre-Famine Ireland*, by S.J. Connolly (New York: Gill and Macmillan, St. Martin's Press; 1982).

[18] See p. 198 of Sir Leonard Wooley's *Abraham* (London: Faber and Faber Limited, 1936).

[19] See Erich Fromm's *You Shall Be as Gods* (New York: Holt, Rinehart and Winston, 1966).

[20] The editing and eventual canonization of the Old Testament is authoritatively set forth in Chapter 6 of Fred Gladstone Bratton's *A History of the Bible* (Boston: Beacon Press, 1959). For a description of several shifts in Hebrew history see Chapter 4 of Loyal Rue's *Amythia* (Tuscaloosa: The University of Alabama Press, 1989).

[21] See Chapter 1 of *A History of the Literature of Ancient Israel*, by Henry T. Fowler (New York: The Macmillan Company, 1922).

[22] For the relation of El (singular of Elohim) to other Gods in the area, see pp. 81-2 of Fred Gladstone Bratton's *A History of the Bible.*

[23] See p. 1 of Jeffrey H. Tigay's *You Shall Have No Other Gods* (Atlanta: Scholars Press, 1986).

[24] For a note on El Shadday's history as a Canaanite God with a consort (probably Asherah) and an analysis of his role in Genesis 49, see David Noel Freedman's "Who is Like Thee Among the Gods? The Religion of Early Israel," Chapter 16 (pp. 315-335) in *Ancient Israelite Religion*, ed. Patrick D. Miller Jr., Paul D. Hanson, and S. Dean McBride (Philadelphia: Fortress Press, 1987,

especially pp. 322-327). See also Raphael Patai's *The Hebrew Goddess*, 3d edition (Detroit: Wayne State University Press, 1990).

[25] Elyon was ultimately merged with Yahweh, but not until early in the reign of David. See pp. 48-50 of *In the Shelter of Elyon*, ed. W. Boyd Barrick and John R. Spencer (Sheffield, England: JSOT Press, 1984).

[26] See Albrecht Alt's *Essays on Old Testament History and Religion* (Sheffield, England: Sheffield Academic Press, 1989). Some of the scriptures where these gods are mentioned are: El Olam of Beersheba in Genesis 21:33; El Bethel of Bethel in Genesis 21:13; El Roi in Genesis 16:13; and El Berit of Shechem in Judges 9:4 and 46. See also II Kings 16:10 and 21:1 and Joshua. 24:2. For evidence that some Hebrews worshipped Molek see *The Cult of Molek: A Reassessment*, by George C. Heider (Sheffield, England: JSOT Press, 1985).

[27] See Exodus 6:2-3. See also p. 137 of *Adam to Daniel*, ed. Gaalyahu Cornfield (New York: The MacMillan Company, 1961).

[28] It is somewhat arbitrary to identify a "beginning" of Hebrew concepts, since semitic theological history already stretched back for centuries at the point usually thought of as the beginning. And like the Egyptian and Babylon histories on each side of them, semitic theology had contained various combinations of polytheistic, trinitarian, dualistic, semi-monotheistic, and other concepts of God. Nonetheless, because the Hebrews have had a large influence on history, searching for the concepts they entertained at the beginning of *recorded* history is appropriate and useful. For some further background see *The Death of Gods in Ancient Egypt*, by Jane B. Sellers (New York: Penguin Books, 1992).

[29] For a recent and fair-minded effort to show how the earlier El was incorporated into Yahweh, see *The Rise of Monotheism*, by Johannes C. DeMoor (Leuven-Louvaine, Belgium: Leuven University Press, 1990).

[30] See p. 199 of Sir Leonard Wooley's *Abraham*, cited above.

[31] See pp. 71-2 of Georg Fohrer's *Introduction to the Old Testament* (Abington Press, 1968). That Abraham's covenant was a late addition to stories about Hebrew origins is persuasively argued by Ernest Nicholson in *God and His People: Covenant and Theology in the Old Testament* (Oxford: Clarendon Press, 1986).

[32] Robert R. Wilson presents the probability that the patriarchs before Moses were not real people, but only symbolic characters who represent the general and partly mythical history of a people. See pp. 54-66 (Chapter 4) of his *Sociological Approaches to the Old Testament* (Philadelphia: Fortress Press, 1984).

[33] Some scholars have argued that this idea persisted until the time of the prophets. See p. 226 of *Hebrew Origins*, by T.J. Meek (University of Toronto Press, 1950).

[34] Genesis 22.

[35] See Jo Ann Hackett's discussion of child-sacrifice in the Transjordan area on pp. 125-136 of *Ancient Israelite Religion*, ed. Patrick D. Miller et. al.

[36] II Kings 3:27; Judges 11:24-30.

[37] See p. 288 of Ninian Smart's *The Religious Experience of Mankind* (New York: Charles Scribners Sons, 1984, 3d Edition).

[38] They had begun their sojourn by some combination of migration and captivity, which was a common pattern in the area. Distances were short enough, conflicts were frequent enough, Egypt was powerful enough, and motivations were straightforward enough to help us see how earlier Hebrew ideas came under the influence of the Egyptian theological imagination. See Chapter 6 of *Adam to Daniel*, ed. Gaalyahu Cornfield.

[39] The weight of evidence is that the Exodus was in the 13th Century B.C., which means Moses probably grew to maturity not long after the death of Akhenaten in about 1347. See p. 156 of Nahum M. Sarna's *Exploring Exodus* (New York: Schocken Books, 1986).

[40] See pp. 312-370 of James Henry Breasted's *Development of Religion and Thought in Ancient Egypt*. Actually many writers claim that no Hebrew document is monotheistic until the 6th Century b.c.: see pp. 111-2 in *A History of the Bible*, by Fred Gladstone Bratton.

[41] See pp. 369-371 of J.A. Wilson's *Ancient Near Eastern Texts Relating to the Old Testament*, ed. James B. Pritchard, 3rd edition (Princeton, New Jersey: Princeton University Press, 1969).

[42] Vintage Books, New York: 1955.

[43] See Cyril Aldreds' *Egypt: The Armana Period and the End of the Eighteenth Dynasty* (volumes I & II of the *Cambridge Ancient History*, Cambridge University Press 1971), which claims that Akhenaten's monotheism may itself be a rehash of earlier Egyptian themes.

[44] *From the Stone Age to Christianity: Monotheism and the Historical Process*, p.12 (New York: Doubleday 1957).

[45] See pp. 259-9 of *From the Stone Age to Christianity* (cited just above).

[46] P. 6 of Albrecht Alt's *Essays on Old Testament History and Religion.*

[47] I say "his" because the responsibility for women's conduct was formally laid on their husbands or fathers, though men were always willing to shift the blame back.

[48] See p. 226 of T.J. Meek's *Hebrew Origins.*

[49] P. 56 of *Biblical Faith*, by Gerd Theissen (London: SCM Press, 1984), citing pp. 341-8 of "Temporary Henotheism" by A. van Selms, in *Symbolae Biblicae et Mesopotamicae* (Leiden: FS F.M.T de Liagre Böhl, 1973).

[50] For some Biblical references to Chemosh see Numbers 21:29; Judges 11:24; I Kings 11:1-8 and 33 (where Solomon in his diplomatic fashion made a "high place" for Chemosh); II Kings 23:13 (where Josiah broke it down); and Jeremiah 48:7-46. For a secular discussion of the Moabitic Stone (also known as the Mesha Stone) see Mattingly, Gerald L., "Moabitic Religion," in *Studies in the Mesha Inscription and Moab*, ed. Andrew Dearman (Atlanta: Scholars Press, 1989), especially pp. 218-219.

[51] Indeed, the Moabs were so closely related to the Hebrews that it would be surprising if their gods were very different. Genesis 19:37 says the Moabs descended from Lot's daughter, and Lot was Abraham's nephew.

[52] See pp. 120-121 of David Damrosch's *Narrative Covenant: Transformations of Genre in the Growth of Biblical Literature* (New York: Harper & Row, 1987).

[53] A very interesting and readable book that traces the editing process of the Bible is *Palestinian Parties and Politics That Shaped the Old Testament*, by Morton Smith (London: SCM Press Ltd., 1987).

[54] See p. 50 of *In the Shelter of Elyon*, edited by W. Boyd Barrick and John R. Spencer.

[55] In fact Chapter 8 is actually where I formally deal with god-concepts, and this discussion was meant primarily as an example of theological imagination at work.

[56] *Vico's Science of Imagination*, by Donald Phillip Verene (Ithaca: Cornell University Press, 1981).

[57] After Vico, Fontenelle theorized a universal human predisposition toward mythology; thereafter the tradition was kept alive by Schelling, Freud, and Eliade. On Fontenelle see Chapter 5 of Frank E. Manuel's *The Eighteenth Century Confronts the Gods* (Cambridge, Massachusetts: Harvard University Press, 1959).

Chapter Six

Scripture and the Modification of Concepts

<u>An Outline of the Chapter</u>

A. A Brief History of Writing and Scripture
B. Scripture as a Mechanism of Culture
C. Scripture's Relation to Biology
D. Scripture's Relation to Language and Concepts
   1. Words and Myth
   2. Scripture's Pathological Possibilities: Idolatry
   3. Avoiding Idolatry by Modifying our Concepts

Concepts originating in theological imagination have been reduced to writing by many cultures, and have achieved special status within those cultures. Well-known scriptures like the Bible and the Koran were formalized into canons, but some eastern cultures have not clearly declared which writings constitute scripture. This chapter will explore scripture in four sections, using examples from both eastern and western heritages. The last section contains the main substance of the chapter, which concerns concept-modification.

A Brief History of Writing and Scripture

The Oral Tradition

Before writing was invented, we had already made many cultural advances by dispersing tools and other inventions.[1] Exciting new ideas were spread along with this crude commerce, and also by more formal means. Through feast-days that brought tribes together, through storytelling, and through other institutions of pre-writing cultures, many concepts were transmitted, including those produced by the theological imagination.

We must not overlook this oral tradition. Not only did it determine the initial content of written forms, but it also permitted a long period of development for each concept. Wisdom could be distilled over many generations, without undue influence from any one person or period. Storytellers were probably admonished to tell stories exactly, just as they are in primitive societies today; but oral transmission also permitted occasional changes, as major events like earthquakes and floods worked their way into the stories, along with imagined explanations for them. Noteworthy murders or adulteries or wars or great acts of kindness were also included, as early story-tellers learned to enthrall and entertain, powerfully communicating the culture's ideas and explanatory concepts. A modern parallel, though less profound, is stand-up comedy, where good practitioners modify basic routines very slightly with each retelling, until they are guaranteed to strike responsive chords in any audience.

Pictures in the Sand

Several cultures developed picture-writing, and children spontaneously draw pictures in the sand today. These facts lead to the reasonable guess that hieroglyphics were a natural extension of language, and probably developed sooner than our earliest records show.

One disadvantage of hieroglyphics is the difficulty of constructing an easily-taught system of writing with them. Even when pictures become stylized and simplified into "characters" (as in modern Chinese), it taxes the memory to learn all the characters necessary for effective written communication. Yet that difficulty can certainly be overcome by a body of scholars, as the great Chinese civilizations demonstrate. Thus writing's influence would have been felt even without the alphabet.

The alphabet's advantage, however, is that it uses a small number of symbols, each corresponding to a simple verbal sound. These "letters" can be combined in almost infinite variety to represent words. Because of the small number of letters, alphabets are much easier to learn than hieroglyphics are. So while the alphabet was not an obvious invention, and arose only once that we know of,[2] it probably enhanced cultural expansion more than hieroglyphics alone would have.

About five thousand years ago, writing began to have broadly similar effects in both alphabetic and hieroglyphic traditions, allowing communication and organization on a scale that would have been impossible without it.[3] The effects were pervasive, sparking a cultural expansion unmatched in prior history except by horticulture and fire; therefore writing is considered one of the great human inventions. Just as spoken symbols had expanded our ability to communicate, writing now expanded our ability to preserve ideas and to send them across great distances.

A Definition of Scripture

The tendency to fashion explanatory systems received a great shove forward by writing. At first writing merely transferred old stories into new forms, but it also permitted a closer look at those stories. A written tradition gradually developed which could not be contained in simple oral styles; it began to explore and correlate mysteries and explanations, adding subtlety and complexity to the simpler concepts of pre-literate cultures.

The fact of writing was itself a mystery that had to be explained, and as usual this was accomplished in theological terms. In almost every culture, writing was seen as a gift of the gods. Those who could write were believed to have special powers, a belief that persisted far into recorded history. Indeed, it seemed reasonable to say "all script is given by inspiration of the gods,"[4] and several scriptures contain that idea. Special training for selected leaders allowed favored ones to learn reading and writing, which further encouraged the belief in writing as mysterious or even magical.[5]

These observations, combined with insights from earlier chapters, help us construct a working definition of scripture. Since canonization is a political process, with each side claiming God's support, we cannot easily accept definitions of scripture depending on canonization. Therefore we will use a broader definition, defining scripture as *any set of writings embodying the main concepts of a culture's theological imagination, distilled over time.* That definition provides a basis for examining how scripture influences culture, how it influenced citizens of particular cultures (especially in the judeo-christian culture), and how it still impacts on us today.

Scripture as a Mechanism of Culture

Several theologians, and especially George Lindbeck, have persuasively shown how thoroughly our theology is embedded in a cultural framework.[6] In the case of the judeo-christian culture, this can be clearly seen by discussing the role of priests, prophets, scribes, and other leaders of society.

Priests and Prophets

Even though it was often seen as magical, writing served culture's normal functions, emphasizing both its conservative tendency and also its ability to institutionalize new ideas. We can see this normal dynamic at work in the Old Testament conflict between priests and prophets.

Then as now, priests tended to be conservative. And because writing stabilized concepts more than oral forms did, scripture helped them maintain tradition. Once written down, for example, the ten commandments helped cement society. This conserving function would operate even if scripture only codified human wisdom; seeing it as recording the words of God made it even more immune to change. The problem was to determine how to decide which writings came from God, and priests usually opted for writings that supported tradition.[7]

On the other hand, writing had an effect that aided change. In the normal interaction between conservatism and innovation, prophets arose to challenge the status quo, and they often cited scripture in support of their challenges. Kings were humbled, for example, when prophets used scripture to show how they had abandoned the national "constitution." Ironically, successful challenges once recorded, absorbed prophets into the next edition of scripture, where priests protected them from new challenges.

Yet scripture was more than simply a resolution of tensions between priest and prophet. It also reflected the tensions and concerns of such leaders as Saul or Samuel and the groups supporting them. Eventually, out of a large catalogue of concerns, these leaders and their allies forged one of history's great cultural traditions and preserved it in scripture: Concepts of ethical principle, social interaction, sexuality, medicine, diet, and international politics all found their ways onto its pages. Even when time strained some of these ideas, the cultural strength was supple enough to permit a "New Testament" that kept many traditions, abandoning those that were causing trouble.[8]

Scriptural changes were not always in a forward direction, however. Priests were conservative, but prophets often harked back to an imagined "golden age" of God's blessing, thus acting as super-conservatives rather than innovators. For example they were adamant that jewish practices not be adulterated. But even with this additional conservatism from an unexpected source, the body of scripture continued to develop, and yet remained the focus of the culture's identity. It served as a living textbook by which every citizen learned an ethic and a world-view, establishing heroes and standards for young and old alike. Scripture thus served as the primary mechanism for transmitting culture, routinely used by families, schools, and armies for teaching and motivation.

Scribes

Priests and prophets and political leaders were not the only occupations influencing scripture. The scribes had a powerful impact too. Sometimes they performed merely secretarial functions, but as specialized possessors of the writing magic they tended to pore over their books, developing personal interpretations and private philosophies.[9] At crucial points this privileged position allowed them to exercise their own theological imaginations; to originate scripture instead of merely transcribing it. Indeed, scripture's role as a record of their theological imaginations sometimes eclipsed its role as a record of military history, medical knowledge, and the like. Chapter Five presented several instances of re-editing by the scribes, and it would be a fascinating detour to explore the philosophical disputes behind specific changes. The editing usually took place quietly behind the scenes, but sometimes (I am thinking especially of Ezra), it was direct and unambiguous and pervasive.

Over time these scribes proved how mighty a pen can be. Their editing influenced the culture's self-definition as much as did priestly caution or prophetic indignation. Even today, editors of new versions of the Bible exert considerable influence on our concept-formation with seemingly minor word-choices.[10] None of this is to deny that other influences on scripture can also be powerful; public sentiment or slow inexorable change also have profound effects, for example.

As a conservative agent of culture, scripture can provide justifications to thwart innovation, but its brilliant prophetic passages can also inflame the imagination to forcefully overcome culture's lethargy,[11] as we saw earlier. But it has more subtle effects; for example it can become a counterpoint for new ideas, inadvertently propelling them forward by its resistance.[12] Even when it seems to lose influence altogether, gathering dust in the attic, scripture usually comes back to life.

These great cultural capacities of scripture urge us to explore it further. The first section below will look at scripture's nature in terms of its biological power and range of accuracy, from possible absolutes to clear error. The second will explore a dangerous aspect of scripture, even while praising its wisdom and practical knowledge.

Scripture's Relation to Biology

Modern anthropology has shown the relativity of each culture's scriptures, and we must acknowledge the truth of anthropology's discoveries in specific cases. However, we must also acknowledge some culture-free components within each set of scriptures, resulting in passages that transcend culture. Such passages represent distillations from the experience of a people, so similar to distillations by other peoples, as to constitute what one writer calls "cultural universals." That writer goes on to show how such universals evidence biologically based truth arising from our "deep structures," rather than just culturally relative truth sliding off our "surface structures".[13]

To the extent that scripture makes such distillations, we must abandon the anthropologist's tendency to call it culturally relative. On the contrary, we must affirm such distillations as marvelously well-developed articulations of our basic nature. Pending more scientific understanding of that nature, and more systematic descriptions of it, scripture remains an excellent repository of truth about our nature, precisely because it was written over such a long time period, and is so much in agreement with other explanatory systems.

But what portion of scripture is comprised of such reliable universals, and how should we view the rest of it? We can answer that question by dividing scripture into four natural categories, unless we are committed to such an unreasonably narrow view of scripture that we "strain at gnats and swallow camels." Those four categories of scripture are: a) Innate behaviors codified by scripture into cultural universals, which arise out of our biological commonality; b) Scriptural concepts that are biologically based but variable within the human population; c) Culturally relative scripture; and d) Errors in scripture.

Innate Commonality. About matters for which there is essential agreement, scripture serves as an honored repository of principles distilled over time. "Thou shalt not kill" is such a principle, and prohibitions against incest may be another. Furthermore, the next category below may contain dozens of inbuilt principles valid for everyone, if those principles are stated with the sort of qualifications that accompany most biological truths.

Even within areas of agreement, however, humility requires us to let imagination and culture work in their normal dynamic, as one tries to make inroads and the other cautions against too rapid change. In my own morally conservative mind, "Thou shalt not commit adultery" seems biologically based, which would place it in this first category. Yet I acknowledge the possible relativity of such a rule, and wonder at the neural dynamics that underlie its frequent violation. Thus I see scripture as presenting an intellectually supportable view of our sexual nature, but I remain alert to other ways of understanding that nature. In the meantime, scripture's long history would seem to establish a rebuttable presumption, requiring the tentative placement of many moral principles in this first category, or at least in the second category, and prohibiting an easy placement of them into the other two categories.

Biological Variables. One category of scripture lies between the biologically innate and the culturally relative, and includes especially those passages which arise from genetic diversity.

Of course all scripture arises within a genetically diverse population, because all populations are always genetically diverse. Canonization is often a victory of one part of that population over the other. There is nothing inherently wrong with scripture arising like that, but understanding that element of its origin should add a measure of tolerance and humility to our efforts at understanding specific scriptures.

Thus alcoholism and other "immoral" tendencies must be seen in light of their genetic components. If further research supports initial indications that they have such components, that research must surely influence our view of particular passages. While that may not (for example) change our commitment to sobriety (indeed it may deepen it), it should show us that scripture must be applied in different ways to different genetic endowments.

This category of scripture will probably turn out to be a very large one, once we understand more details of our various genetic heritages. In the meantime, we can again stay open to the possibilities suggested by such a category, and still pursue the implications of the other three categories, without too easily rationalizing our conduct in terms of this one.

Cultural Variables. On many matters there was once agreement about scripture which now seems to be eroding. But even in those cases we can honor scripture as a record of past understanding, and stand ready to either 1) reaffirm it, 2) modify our understanding of what it says, or 3) modify our understanding notwithstanding what it says. In deciding which choice to make we should listen carefully to priestly voices supporting tradition, and also to prophetic voices trying to make a case for cultural relativity.

Several past issues turned out to be culture-bound, and thus fit clearly in this category. Rules permitting ill-treatment of women come to mind. That is not to say we must always transcend culture, and indeed there is wisdom in accepting its answers where they do not impinge on our moral sense. But acknowledging that not all scripture is appropriate for all situations for all time can free us in important ways, even though we might continue to accept most scripture as valid.

Errors in scripture. Where scripture seems clearly in disagreement with truth, and especially where we can understand how its writers came to misapprehend truth, we must reject specific scriptures. While this may seldom happen in matters of personal morality, it often happens in matters of science. That is because human nature hasn't changed much since scripture was written, whereas our ability to measure and observe has changed. So even if the Bible says the world is square or flat, or that humans were created by fiat rather than by evolving naturally, we may reject the Bible's understanding without guilt, and without accepting fallacious either/or accusations that we are thereby rejecting the entire Bible. Neither does our rejection of errors in scripture remove the honor we give it as a historical record of the theological imagination, which served as a foundation for our current understanding, as a "schoolmaster to salvation."

Although it may only happen infrequently, there are also scriptural pronouncements in the moral area which turn out to be wrong. Nowadays neither racism nor slavery can be accepted as charming cultural anomalies, and perhaps we will come to see war as immoral regardless of scriptures that support it. Modern theology often tiptoes around such issues by arguing that scripture meant to say the right thing after all, but it seems more honest to say that scripture was just narrow and wrong, and that it took us awhile to learn more broadly valid concepts.

The examples above were merely meant to explain the four categories of scripture, and different people may place different examples in each category, pending further evidence and analysis. Thus someone might put abortion in the first category, while someone else puts it in the third, and so on. Eventually, a biologically-based psychology may identify which scriptures articulate innate truth common to all people and which arise out of biological variables, and anthropology can clarify the cultural components of scripture. In the meantime we must each make our own judgments about the categories and their contents, cultivating and nourishing those judgments by studying the scripture, and by learning about new scientific discoveries and new approaches to moral concepts.

In making such judgments, it is important for us to recognize scripture's role as a record of its culture, and to calmly acknowledge the range of its content, from solid biological universals all the way to downright errors. If we are unable to do that, I fear we are confusing language's function of constructing symbolic representations of reality with reality itself, and that is an important matter which we will now explore in some detail.

Scripture's Relation to Language and Concepts

Words and Myth

Scripture is frequently described as the "word of God." Since we have already talked about words and language, we can think about the "word" of God from that perspective, if we first detour to explain what anthropologists mean by "myth."

In his excellent popular books,[14] Joseph Campbell notes the trouble people have with the concept "myth," because they think it refers to untrue fables or lies. To anthropologists, a myth is a set of stories and rituals and explanations, which serve as symbolic structures to represent deeper realities. Since language is more limited that the underlying realities are, these "ways of saying something" are inevitably incomplete or even erroneous. Yet the fact that stories have limitations does not keep them from containing profound truths, provided we keep them vibrant and useful by continually redefining our myths. That means the "word of God" is made up of the words of humans, and our understanding of it must be modified as we increase our knowledge of ourselves and our language and our concepts.

In biological terms we could say the same thing a different way, and avoid the pitfalls of trying to keep a new definition of myth in our heads. We would first recognize the need to express our deepest internal experiences, and acknowledge that God-words are linguistic approximations derived from such experiences. These approximations are expressed in concepts, and shaped by the processes of theological imagination in a cultural context. What we thereby learn of "God," and how we express it, becomes incorporated in a set of neural pathways, which we access whenever we think about God or talk about God or perform acts related to our god-concepts.

Chapter Eight will explore concepts of God further, but scripture contributes to the shaping of our neural pathways by serving as a ready reservoir of traditional concepts. Yet since scripture can record error and perpetuate cultural bias, we need to understand how it influences our brain pathways, and how we might avoid its dangers. The remaining two sections will address that need, by showing how religious concepts can become idols, and how we can avoid idolatry by modifying our concepts.

Scripture's Pathological Possibilities: Idolatry

Sometimes I wonder how people who claim to worship the "spirit of truth" could be so adamant in clinging to old errors. Protestants who desperately cling to errors are especially hard to understand, because the Protestant Reformation was supposed to establish the principle that we worship a dynamic God who continuously reveals Himself. While the puzzle presented by such clinging is partially resolved by thinking of scripture as an agent of culture's conservative function, I think there is a more subtle mental process that underlies the puzzle. This mental process is a common one among humans, which I call "the confusion between symbol and reality." It has a shorter name which we will get to in a moment.

There is perhaps no clearer example of this confusion than the golden calf story. You remember that while Moses was on the mountain receiving the ten commandments, Aaron succumbed to the people's desire for a symbol, and manufactured a golden calf, which they began to worship. The children of Israel were thus guilty of focusing too finitely on symbol, confusing it with reality and granting it powers it could not have. Indeed, Aaron's reason for giving them a physical symbol was his belief that they were concrete in their thinking, and needed a "hard" symbol to propel their thoughts godward. Moses' anger was not based on the fact of symbol, because he also used symbols to express his understanding of reality. But he knew that finite symbols can too easily distract us; his own symbols were more abstract, and ultimately more powerful. Thinking back to how brain pathways work, one may legitimately suggest that abstract Mosaic symbols might be harder to embed into the memory pathways, but remain very stable once they get there. Worshiping golden calves might work for awhile, but abstracting human moral behavior into ten short commandments is even more powerful.

The golden calf story introduces a new word, which helps us consolidate this concept of confusing symbol and reality, and that word is idolatry. As you may recall, idolatry was a concept that developed along with God's oneness, in the declaration that other gods were no gods at all. But as the notion of idolatry developed, it began to refer to any departure from Mosaic concepts of God, including moral and intellectual departures. This expansion of the concept meant that conflicts within Jewish life, and later within Christian life, often centered on what idolatry means.

Centuries of that conflict played out in thousands of minor skirmishes and a few major ones. I can condense one kind of skirmish by recounting the schoolboy arguments I had with a Catholic friend (I was Protestant) as to whether Catholics worship idols. I had been taught that statues of Mary and the apostles, and even of Jesus, were essentially idols, being made of wood or stone. This was consistent with the usual Protestant antipathy, in my subculture at least, to excessive art. My friend patiently explained that such artifacts were merely symbols to aid in worship, which nobody confused with the underlying reality. "Oh," I said. Our subsequent conversations helped us agree that he was more correct than I, but that it is nonetheless possible to confuse a symbol with its underlying reality. We agreed that such confusion is the essence of idolatry, made grosser by pious reinforcement of the confusion.

This muddling of symbols and reality is most clearly seen in the confusion between *artistic* symbols and reality, and that is why I began with them. But the same confusion can exist between reality and *word* symbols, and between reality and those complex word-sets called concepts.

Before passing to more subtle kinds of confusion, we might pause to notice how pervasive such confusion is in everyday usage. We regularly confuse words with their underlying reality: We accuse politicians of doing it purposely by saying things they don't mean, hoping we will mistake their words for reality. Adolescents think *saying* "I love you" is the same as loving you. And each of us makes the same mistake in many unrecognized ways, trying to avoid the opposite mistake of cynicism, which denies that words have any connection to reality whatever.

The confusion of word and reality is also illustrated by Abraham Lincoln's favorite riddle. "If you call a dog's tail a leg," he asked, "how many legs does it have?" The answer is four, because calling the tail a leg doesn't make it one.

The confusion between words and reality is easily expandable to a confusion between concepts and reality. In daily life, consider the concept "clean." If I try to win a domestic argument with my wife by calling our house clean when it is merely not unsanitary, and if by "clean" she means spotless, we may unwittingly open the door to this kind of confusion—we are each using the same word, but we mean to describe different states of reality.

Writing, despite all its positive contributions, has an inherent tendency toward this kind of confusion. It helps us nail down complex concepts, but it can nail them down too tightly, and yet leave the categories fuzzy. It is thus that writing can give confused ideas such finite form that they become impossible to change. It is thus that writing takes on a life of its own, so that a false reality is created. A word becomes not only a symbol, but a construction of the imagination that seems to have actual existence when it does not. Let me further explain how this happens with some non-scriptural examples, and then focus again on scripture.

Economics is the study of resources and their allocation among the population. It is a difficult study, and its concepts are not entirely satisfactory. For example there is no clear connection between microeconomics (the study of individual households) and macroeconomics (the study of decisions and trends at government and international levels). It is widely believed that economic concepts need reworking, making some of them more specific and clarifying the connections between them (see Chapter 13). But we seem so "locked in" to current economic concepts that we can't see the overall picture because it is obscured by too many details—a classic case of not being able to see the forest for the trees. Said another way, our concepts are preventing us from seeing reality clearly, and the resulting confusion is stifling our further understanding.

This confusion happens not only in art and in the words and concepts of daily life or in practical economic analysis; it also affects our philosophical and theological understanding, which then circle back to influence our approach to practical problems. Two examples from philosophy will further illustrate that point.

First, consider the concept of duality. This is the idea that two kinds of reality exist in the world, usually listed as matter and spirit. We know this is an ancient belief, and must have occurred to many early thinkers independently of each other. Once it became written, however, and bolstered by respected writings that made it seem more reasonable, it became an unchallenged part of cultural understanding, and remained so for centuries.[15] It now wears a semblance of reality that obscures the basic question of whether such duality actually exists.

Similarly, a few philosophers today contend that scientific concepts of causation have received undue weight, because they lend themselves to writing, and especially to that pernicious form of writing called scientific notation. They say that writing and mathematics lead us to believe in ranges of reality that may not exist, and that we connect things in cause-effect relationships that may not actually have such relationships.

In each of these two examples—duality and causation—truth may or may not be contained in the writing; the only point I am making here is that writing can "freeze" an idea into a solid concept. This may often give traditional concepts more weight than they deserve. The other side of that coin is that writing can expose an idea to greater scrutiny, so I do not mean to say that the potential for freezing is the whole story, though it is the part we are discussing here.[16]

Whether we commit idolatry with wood or with words and concepts, mental processes are involved. In terms of brain circuitry, we can imagine how the same pathways which represent a given word might get fused with the pathways which represent the experience independent of the word. Or perhaps the confusion occurs some other way in the circuits—maybe some small but important reminder circuit is supposed to "kick in," furnishing feedback as to the difference between reality and symbol.[17]

However the confusion occurs, we can be sure it has always occurred, and still does, whenever we use symbols, and especially the complex symbols and concepts of abstract language. Since scripture is a record of our attempts to symbolize deep and powerful experiences in complex language, it would be very surprising if we humans avoided idolatry in our use of scripture.

And indeed we do not. By freezing early explanations into written form, and insisting on those explanations as having priority over the spirit of truth as it unfolds, we confuse symbolic representations of reality with reality itself, and thereby commit idolatry. One way we do that is by squeezing all four categories of scripture examined above—biological absolutes, biological variables, cultural norms, and clear errors—into the first category. We will examine the reasons for that tendency in Chapter 7, but for now I hope you can accept the idea that we do it, and begin to wonder how we can avoid it.

Some scriptural examples of such idolatry seem obvious—insisting on seven-day creation or a 6000-year history of the earth, for example. Yet ere are countless instances where we do the same thing more subtly, but ist as dangerously. If I take a concept like "salvation" and give it a finite eaning which develops in a way contrary to psychological reality, and I then insist on the classical explanation even when it conflicts with a ontrary developing truth, I am confusing symbol and reality. My further isistence is a form of idolatry. In New Testament terms, I have abandoned e "spirit of truth"[18] in favor of the "spirit of tradition" or the "spirit of ulture."

Through the centuries, the conflict over what idolatry means involved inumerable discussions of what it meant to be a "good Jew" or a "good hristian." Later, such conflicts centered on differences between Catholic nd Protestant doctrines, as my schoolboy pal and I discovered. Within rotestantism, denominationalism became a form of accusing one another f various idolatries. Modern discussions center on what words like "God" ean, and that leads to the most subtle forms of idolatry in modern eology. We will review those modern discussions in Chapter 8, but this iscussion of scripture should demonstrate that words and scriptures can iverge from truth but still demand our allegiance, and thus become lolatrous.

Interestingly, we seem able to see the scriptures of other cultures as roducts of theological imagination, and we see why those cultures should y aside old errors or else be guilty of idolatry. Yet we hold to the idea at our own scripture is revealed truth, without any mixture of error. That not as strange as it may seem, given culture's conservative function and iven scripture's role as an agent of culture. Yet acknowledging our own ultural bigotry does not mean we should become cynical and reject ripture. On the contrary, we have seen that it contains profound truth. he question is whether we can find ways to use scripture appropriately, tting it express our beliefs in language true and beautiful, but without lolatry. I think we can accomplish that task by learning to modify our oncepts, so let us now resume our earlier discussion of that topic.

voiding Idolatry by Modifying our Concepts

Previous chapters explained the origins of human thought and language patterns, which constitute deep structures underlying our evolved ommonality. Making concepts is a very relevant part of that commonlity, and so is the tendency to group concepts into systems of explanation.

We have also discovered already that concepts change over time, icluding those concepts produced by the theological imagination. What e found was that our theological (and other) concepts are always being

modified, sometimes in minor ways, and sometimes in major ways. Th
intellectual creativity has teamed up with curiosity, with rational unde
standing, and with writing, to slowly unravel nature's mysteries. B
unless we were able to modify concepts, we would stay hopeless
committed to old concepts, and could not fashion new ones to act as vesse
to hold new understanding. In fact, we often do remain deeply committe
to old concepts as a part of our innate conservatism encouraged by cultur
but it isn't a "hopeless" commitment because we also have innate drive
toward new concepts, and especially toward modifications of our old one

Perhaps you can begin to see the importance of biology to this concep
modifying ability, and specifically the importance of evolved ment
pathways which are capable of being rerouted. Such pathways underli
little Mary's initial acceptance of concepts provided by her culture, an
also explain how she modifies them. When she modifies her simpl
concept "Dog" into the fuller concept "Championship Bird Dog" and the
into her personal conception of what an "even better" bird dog is, we ca
imagine new dendrites growing and reaching out, and new pathways bein
established. If she is born into a dog-breeding family, those pathway
likely become strong ones, ready to serve as templates or metaphors f
viewing important parts of her unfolding world.

Similar mental pathways are the basis for all individual learning an
culture, including our theological understanding and personal moralit
and they constitute the physical foundation for concepts as they are forme
and consolidated.[19] Later we will see that many pathways can be r
consolidated and re-routed when their present construction causes prob
lems, and we will see how that impacts on our ultimate success an
happiness. For now, however, I only mean to reiterate the point that thes
neural pathways are the physical basis of our concepts.

Scripture contains many concepts that have gone through the modif
cation process, and every such modification was accompanied in ind
vidual brains by new or modified neural pathways. These modificatio
occurred not because there was something wrong with scripture, b
because the nature of our unfolding discovery always requires modific
tion of concepts. That would happen if there were no scripture, or even
there were no writing, and it happens in each individual life as well a
within the life of a culture and across cultural lines. It is a very bas
process. For us to deny it happens in scripture is idolatry, a clinging t
childish understanding of the nature of scripture, and a refusal to see
biological truth over cultural tradition.

On the other hand, if we can modify our concepts, including thos
concepts embedded in scripture and traditional theological doctrine, w
have a vehicle for keeping the good and avoiding the bad. Scripture ca
serve as a reservoir of living waters, without becoming stagnant from ou

ailure to rework its inlets and outlets. By being conservative, but not gidly so, we can keep the old truth until we clearly see how to replace it ith modern understanding, allowing a slow but steady growth.[20]

Every once in a while, scripture gets so out of date that it must be estated, either smoothly or radically. It happened radically when the New estament was canonized. It happens, more or less awkwardly, in all ultures. The Japanese Buddhists, with characteristic efficiency, have ound a particularly smooth way to restate scripture, which I will explain ot so much as a model for our culture, but to illustrate another variety of e modification that occurs in every culture.

I was staying in a Tokyo hotel, and discovered that some oriental vangelicals had furnished each room with a copy of the Buddhist criptures. I read enough to become interested, and asked the demure ostess downstairs if I could purchase the book upon departure so I could nish reading it. She said she would ask the manager. The next morning, hen I was checking out, she approached me with a courteous bow. "Sir," he said, "I am very sorry to report that the management owns no extra opies of the scripture, and must leave a copy in each room. However, I ave purchased a copy of the precious book from my own funds, and I ould like you to have it as a gift from me." I recovered my composure uickly enough to avoid total gracelessness, and quietly thanked her for ie gift. I know when I have been witnessed to.

The Buddhist scripture was interesting, and like others who have read , I was struck by its similarity to ours. The preface intrigued me more than e scripture, however, because it explained how the book in my hand was ompiled.[21] Buddhists, you see, have an enormous scripture, including everal sub-traditions and many prolific writers. Their oriental culture has evered writing far more thoroughly and for much longer than ours has, roducing thousands of volumes that are considered holy. I began to onder how that literature could have been condensed into the normal-ized book I held. The preface answered my question.

The Japanese Buddhists keep a committee, learned and yet chosen for ensitivity and piety, who frequently revise popular versions of scripture ke the one I held. The vast library of scripture allows the committee to ind selections that address current issues, and yet retain a comprehensive pan of Buddhist thought within the covers of a single book. Stories similar o ours are chosen because of their universal message, and others are eleted when it appears that they lack universality. For example, the Buddhist prodigal son story reads as if it were taken from our Bible, though f course it wasn't—it is simply a great story full of wisdom, which found s way into both sets of scripture because it touched hearts in both cultures.

The Japanese Buddhist experience shows another example besides the Hebrew one of changing theological concepts. Both together demonstrate that modification happens, it happens for good and understandable rea sons, and it will continue to happen. Let us accept that process, and affirm it as part of the normal working of the theological imagination, as we tur our attention to specific theological doctrines, seeking to modify ou moded concepts into new ones that deserve our confidence.

Notes

[1] See Chapter 8 of Creighton Gabel's *Analysis of Prehistoric Econom Patterns* (New York: Holt, Rinehart and Winston, 1967) and other sources cite therein.

[2] See pp. 2-10 of *The Origin of Writing*, ed. Wayne M. Senner (Lincol University of Nebraska Press, 1989).

[3] Pp. 60 and 80 and 88 of *A History of Writing*, by Albertine Gaur (New Yor Charles Scribner's Sons, 1984). See also *Cadmean Letters*, by Martin Bern (Eisenbarauns: Winom Lake, 1990). On the tendency of young children to dra and even write spontaneously see Chapters 9 and 11 of *The Psychology of Writte Language*, ed. Margaret Martlew (New York: John Wiley and Sons, 1983 Chapter 9 (by Cecilia de Góes and Margaret Martlew, pp. 217-236) is entitle "Young Children's Approach to Literacy." Chapter 11 (by Lev S. Vygotsky, p 279-292) is entitled "The Prehistory of Written Language."

[4] Scripture originally meant "writing." See II Titus 3:16 and entries 1123-112 of James Strong's Greek Dictionary, in *Strong's Exhaustive Concordance of th Bible* (Nashville: Abington Press, 1974 edition).

[5] In the United States, slaves were prohibited from learning to read or write, s well acknowledged was writing's power, though by that time it had lost i supernatural mysteriousness. See pp. 74-79 of *The Struggle for Freedom: Pha. I*, by Margaret Y. Jackson (Chicago: Adams Press, 1976). Note that about 5% the slaves learned to read and write by 1860 notwithstanding the law; pp. 131-13 of Thomas L. Webber's *Deep Like the Rivers: Education in the Slave Quart Community 1831-1865* (New York: W.W. Norton & Company Inc., 1978).

[6] See his *The Nature of Doctrine* (Philadelphia: Westminster Press, 1984).

[7] See Aelred Cody's *A History of Old Testament Priesthood* (Rome: Pontific Biblical Institute, 1969). For more on priests and prophets, see Daniel Da Williams's "Priests, Prophets, and the Establishment," pp. 309-326 of *Zygo Journal of Science & Religion* for December 1967 (Vol. 2 No. 4); and also th comment on Williams's article by Sanborn C. Brown immediately following (p 327-330).

[8] The symbolic tradition of circumcision, interestingly enough, may have bee the "straw that broke the camel's back," forcing Christianity to become a separa religion instead of a jewish sect. More substantive issues survived the transitic unchanged. See Peder Borgen's *Paul Preaches Circumcision and Pleases Me and Other Essays on Christian Origins* (Trondeim, Norway: Tapir Press, 1983

For a skillful argument for a smooth transition from circumcision to the newer symbol of baptism, see Meredith G. Kline's *By Oath Consigned* (Grand Rapids, Michigan: William B. Eerdmans Publishing Company, 1968).

[9] This is how the Talmud grew to almost equal importance with the Old Testament. See *The Talmud* (especially Chapter 4), by Isaac Unterman (New York: Record Press, 1952). See also *Talmudic Thinking: Language, Logic, Law*, by Jacob Neusner (Columbia: University of South Carolina Press, 1992). For the role of scribes and the relative freedom of their renderings in antiquity, see authorities cited at pp. 353-359 of Christopher D. Stanley's *Paul and the Language of Scripture* (Cambridge, England: Cambridge University Press, 1992).

[10] For example, the preface of the popular *Living Bible* frankly acknowledges an editorial slant, and after reading dozens of important passages in it, one can see that the editing influences the meaning in significant ways. (*The Living Bible* is published by Tyndale House Publishers: Wheaton, Illinois; 1971.)

[11] This is happening today in Third World Liberation Theology. See Part II of *Religion and Politics in Latin America: Liberation Theology and Christian Democracy*, by Edward A. Lynch (New York: Praegor, 1991). Also see *Third World Liberation Theology*, by Deane William Ferm (Maryknoll, New York: Orbis Books, 1986).

[12] My perspective on the Creationism controversy, for example, is that it ultimately backfired on the creationists, causing people (biology teachers especially) to think seriously about the issues and begin to teach evolution in more persuasive and interesting ways.

[13] See *Human Nature & Biocultural Evolution*, by Joseph Lopreato (Boston: Allen & Unwin, 1984), especially pp. 38-42 and Section 2.3 (pp. 56-65).

[14] Shortly before he died Campbell wrote a very readable book with Bill Moyers in conjunction with their television presentation on myth. It is *The Power of Myth*, ed. Betty Sue Flowers. New York: Doubleday, 1988. For a comprehensive critique of Campbell's theories of myth see *Joseph Campbell: An Introduction*, by Robert A. Segal (New York: Penguin Books, 1987).

[15] Although Descartes is usually given credit for the clearest explication of duality, the concept itself predates Plato, and has a prominent role in his dialogue "Phaedo." In *The Dialogues of Plato*, vol 1, translated and analyzed by B. Jowett (New York: Charles Scribner's Sons, 1907).

[16] For further discussion on "freezing" concepts, see p. 189 of Robert A. Segal's "Paralleling Religion and Science: The Project of Robin Horton," pp. 177-198 in *Annals of Scholarship* Vol. 10 No. 2 (Spring 1993).

[17] I do not mean to imply that either of those possibilities are very likely to be the actual neural mechanisms of idolatry or other kinds of confusion. Yet few biologists doubt that every mental process, including idolatry, is supported by psychophysical pathways that are capable of being understood, once enough research is done. So even though this paragraph's specific speculations are unlikely to be true, it illustrates a more general and well-established truth.

[18] See St. John 14:16-17. For more on the specific example of salvation, se Chapter 11.

[19] Saint Thomas Aquinas would have been happy to learn about moder neurology and the rerouting of mental pathways, because it is so consistent wit his emphasis upon rationality as a primary source of theological understanding See Chapter 7 of Book One of his *Summa Contra Gentiles*.

[20] For another argument that religious concepts must be modified over time, se pp. 251-279 of James Gustafson's *Ethics from a Theocentric Perspective*, Volum 1 (Chicago: University of Chicago Press, 1981). See also pp. 79-89 of Dougla F. Ottati's *Meaning and Method in Richard Niebuhr's Theology* (Lanham Maryland: University Press of America, 1982). Of course Augustine's approac is also very relevant; he resolves conflicts between science and scripture by givin a presumption to scripture, but if scientists "demonstrate" the truth of their claim one obviously has to reinterpret scripture, in order to affirm the truth of both. Fo a fuller discussion of Augustine on this point (and Galileo's interpretation of him see pp. 307-309 of Ernan McMullin's "Evolution and Special Creation" in Vol. 2 No. 3 (September 1993) of *Zygon: Journal of Religion & Science*.

[21] *The Teaching of Buddha*, published by Bukkyo Dendo Kyokai (Buddhis Promoting Foundation); Tokyo 1981.

Chapter Seven

Faith, Belief, and World-Views

Of the many human mental processes, those related to belief are among the most intriguing. Why do people believe what they believe, and why is there such a great diversity of belief? More importantly, can we have confidence in our beliefs, and on what basis?

Theology has usually approached these questions in terms of the concept "faith." Although faith originally meant something like loyalty (as in the word "faithfulness"), it has taken on many different meanings over the centuries, and nowadays any group of friends is likely to discover that they mean different things when they use the word. Some use it as a form of belief that disregards evidence (as in blind faith) while some use it to describe beliefs consistent with evidence (as in the usage "faith in science"); some limit its use to religious questions or even to the basic question of whether there is a God, while others use it to refer to their overall framework of belief, which includes non-religious beliefs as well as religious ones. Some use it to refer to simple trust (as in childlike faith), while some use it to refer to a conscious and knowing commitment. Some use it to refer to traditional belief (the "faith of our fathers") while some use it to refer to the most current and cohesive body of concepts (as in "the humanistic faith"). Some mean it as the mental foundation for actual behavior, and some mean beliefs which may or may not be consistent with behavior. The real problem is that we all use the word in all of these ways (and more), without understanding the nuances of our own meaning. As a writer, I found I could not use the word consistently without very artificial constructions, because of the many overlapping and conflicting meanings which the word has come to carry.

To overcome the confusion resulting from so many usages, this chapter will do two things. First it will search the language for other words besides "faith" to describe the mental processes involved in belief. Towards the end of the chapter faith will be discussed more particularly, but we can get some conceptual clarification by also using parallel language to describe the mental operations we are talking about.

The second way the chapter will seek clarity is to divide the discussion into four sub-chapters. The first one examines the social origins of belief. The second one discusses the emergence of personal belief systems from those we inherit socially. The third explores the possibility of evaluating beliefs objectively, and the fourth sub-chapter points out several important implications of the chapter for theology.

Sub-Chapter 1: The Origins of Belief in Biology and Sociality

The previous four chapters—on culture, concepts, theological imagination, and scripture—imply that beliefs come primarily from one's group. To a large extent that is true, and it has been true ever since our species first developed enough brain complexity to formulate beliefs. Born helpless, we are taught by our parents, their substitutes, and the rest of the people in our culture, whatever they themselves have learned. And we are taught it in the particular stories and structures of belief that were handed down to them, along with whatever modifications or new concepts they have added. Beliefs thus enter our individual brain pathways through early acculturation.

Early acculturation fulfills many needs in the young child's life, which carry over into later life. One such need is the need for security, which young Mary fulfills by identifying with the family group. Of course she makes no conscious decision to become more secure by identifying with the family, but instead feels an internal need to do that as a part of her biological pre-programming.[1] Her need for security creates strong pressures toward group beliefs and practices, and she therefore begins her life identifying with the family faith.

These early pressures toward group beliefs are part of normal healthy living. Although they soon begin to interact with other pressures which may challenge belief, we cannot easily cope with life's later problems without a foundation of personal psychological security, and that comes largely from early identity processes. Fortunately there are ways to reconstruct this foundation if we miss it in childhood,[2] as by adopting particularly strong beliefs that provide security from aimless drifting. Cults often rely on this dynamic to attract followers,[3] but even in more ordinary experience, the group faith makes a powerful contribution toward meeting the need for security, supporting social interactions at every turn.

This emphasis on the social and need-meeting origins of belief is not the whole story, of course, and the rest of the chapter will discuss other dynamics of belief, including the development of personal systems of belief. Yet there is clarity in seeing belief as beginning with inherited cultural traditions which we modify personally. Life's later experiences either reinforce earlier beliefs, change them, or sit idly in storage ready for those functions. Actually a more subtle and dynamic thing is probably happening, namely the consolidation of some beliefs and concepts, even while new ones are being formed.

One mystery that is resolved by seeing belief as beginning in social life, concerns the relation of truth to belief. It often happens that people believe things that aren't true, and it can be disconcerting to catch oneself or another person holding such beliefs. Yet our need for truth (which sets us free, the scripture says) may not be as basic as the need for identity, which binds us securely to the group. So if my group believes in leprechauns, it is adaptive for me to believe in them too, at least in the early stages of my growth—at that stage it is better for me to be bound than set free.[4] Thus, because of the need for identity and early acculturation, belief is at its earliest stages simply an acceptance of what we are taught. It is not logical, and it is part of our necessary trust and reliance on others.

Biologically, a specific process is operating as a person first accepts, and then modifies, the cultural or family beliefs, and that is the development and modification of mental pathways. Since those structures are basically stable, we wake up each morning believing approximately the same things we believed when we went to sleep the night before, except insofar as nighttime processing has changed or rerouted them.

This basic biological observation places clear boundaries around our understanding of faith, and applies both to inherited faith and personalized faith, so let me rephrase and repeat it: we believe what we believe because of the particular structure of our memory tracks. By realizing that at any point in time, a person operates from a set of pathways--some of which came with birth, some of which were rigidly imprinted soon after birth, some of which developed through family and cultural influences, and some of which originated from personal decisions--we more clearly see our own struggles as natural and understandable.[5]

Sub-Chapter 2: The Development of Personal Belief-Systems

Family beliefs usually remain with us in one form or another our whole lives, but because of individual experiences and individual orientations[6] toward experience, we eventually develop a personalized set of beliefs. Some people's experience dovetails perfectly with their inherited beliefs, so they find meaning and satisfaction with minimal modification to their beliefs. In our stimulating and diverse culture, however, inherited beliefs often undergo significant modification.

But whether a belief system is adopted or personalized, it grows out of more than our social context. I mentioned our genetic endowment earlier, and it will be mentioned again later. But there is another powerful influence on our beliefs which is better understood—personal experience. It influences belief in many obvious common-sense ways, but some ways are anything but obvious, and this sub-chapter will emphasize those. It will do that by stressing the great variety of experience, by exploring the ways we process experience, and by outlining the framework of belief that results from the interplay of experience with society and genetics.

The Variety of Experience

It would be trite to point out the variety of individual experience, if it were not so profound. It is almost impossible to over-emphasize it. Not only do people in different cultures and sub-cultures have different experiences, but people in near-identical situations have very different experiences. Siblings often take different approaches to life, partly because "minor" differences in experience can have major results.[7]

Freudian psychologists insist that some experiences are particularly formative, and although their views have been criticized as being too sweeping, they contain many helpful insights. The trauma or ease of our particular births, our toilet-training, our sexual initiation, and other highly specific experiences can have a significant effect on our world-view and belief-system.

Yet our more mundane daily experiences are also formative, and assuming that no dramatic event affects us too strongly, such experiences are the largest contributors to the formation of our beliefs. The things we observe, the things we read, and the ritual we participate in,[8] all influence our world-view.

Finally, we are influenced by our experience with cultural trends. These may be minor or major, including the large "paradigm shifts" already discussed in Chapter 5. As western cultures shifted from a monarchical system of government to a democratic one, for example, it influenced the beliefs of everyone in the culture. Similarly, the United States has shifted from its 1700's conception of equality between propertied white males to a conception of equality among all humans. That particular shift is still subjected to frequent challenges, whereas the democratic paradigm isn't—indeed it has recently renewed its relentless march, this time in communist societies.

Many other kinds of experience also influence our beliefs (the influence of sickness or malnutrition comes to mind), and different belief systems will doubtless dovetail with different kinds of experience. Despite this variety, however, a profound commonality underlies our ***processing*** of experience. This commonality results from the similar ways in which individual brains interpret experience and establish neural pathways. So having acknowledged the great variety of experience, let us now look at some of the commonality which underlies our processing of it.

Processing Experience

Whether we affirm, slightly modify, or significantly alter the cultural faith we inherit, particular mental processes are always involved, as I have repeatedly said. The three processes that will be discussed here are the gestalt, trial-and-error, and hypothesis-construction. There are other important mental processes that should be noted in passing, however, because they influence our beliefs in important ways too. Ancient emotional processes were acknowledged in Chapter Two, and I have also mentioned identity processes, which lead us to see ourselves as separate individuals, though still in the group context.[9] So when this section emphasizes three particular processes, it does not deny the importance of other deeply embedded ones. It only means to discuss those which are most directly involved in the development of belief-systems. Two of them are conscious processes, but even conscious processes originate in subconsciousness, so the first task of this section will be to describe the gestalt.

1. The Gestalt. In the late nineteenth century, German psychologists identified a mental process which answered perplexing questions about concept-formation and other aspects of thought, and they called that process the gestalt. Gestalt theory has so completely merged into mainstream psychology that one doesn't hear much about it nowadays,[10] but it has two important implications for this chapter. One implication concerns the initial stages of a belief, and the other concerns the overall framework of belief.

The first part of a gestalt operates like this: With eyes and ears and other senses we gather information and store it in the brain, where it is sorted into various trial groupings and tentative conclusions. These sub-conscious[11] conclusions eventually bubble up into consciousness, usually without our awareness, but sometimes accompanied by a dramatic "Aha!" sensation. In practical matters the process is seen when one slowly becomes aware that a friend is having financial problems, even before he says so. In more philosophical matters one may suddenly see a new way of looking at prayer.

The concept of the gestalt helped resolve a question that had baffled philosophers and scientists for a long time, namely the question of where ideas come from. Scientists especially prided themselves on logical analysis, but the mystical nature of an idea's initial formulation forced them to admit that ideas often begin in hunches, dreams, and other non-respectable places. Two giants in psychological history helped overcome the bafflement by explaining how the gestalt is related to imagination, and how it contributes to faith and other parts of concept-formation. Because of their work, it is quite unnecessary to attribute gestalt realizations to "psychic" intuition or other kinds of magic. The two giants I refer to are Sigmund Freud and Konrad Lorenz.

Most people identify Freud with theories of hidden sexual motivation, but his contribution to the broader question of subconscious processes is probably more important.[12] His writings introduced subconsciousness into popular conversation, and his ideas provoked more study of the relation between conscious and subconscious processes.[13] Nobel laureate Konrad Lorenz then built on Freud's insights to explain that relation further.[14]

As Lorenz explains it, one function of the subconscious is to tug at consciousness, pointing out possible correlations between pieces of experience. Sometimes the correlation is between small bits of experience, and sometimes it puts whole chunks of experience into a framework of new understanding, but the earliest inklings of the correlation may be felt as a hunch (which Lorenz calls an "inspiration"). This early stage of concept-formation is explored further by gestalt psychologists, but Lorenz emphasized the biological interaction of conscious and sub-conscious processes.

He helped hunches gain respectability as fundamental operations of human mental life, by showing their relation to animal thought-processes which are much like hunches, resulting from the evolution of neural pathways which underlie both human gestalt processes and similar processes in animals.

I will present a broader picture of the gestalt in a moment when we look at the "framework" of belief. There we will see how hunches and inspirations get worked into a comprehensive world-view. My purpose here was to show that each of us processes experience at subconscious levels in similar ways. Now let us look at another common mental process, which is perhaps the most basic process of all.

2. Trial and Error. When gestalt inklings bring family or cultural beliefs into question, we often begin to tinker with them through trial and error. Trial and error is such a basic process that some philosophers place all other mental processes as sub-sets of it. Some give it even more importance, declaring it to be fundamental to all physical life, all subconscious process, and all conscious process.[15] That would not be inconsistent with the approach here, but in order to show how belief develops, I treat trial-and-error as a separate concept rather than a master concept. Even as a separate concept, trial and error is so important that we must return to the beginning, and review its place in the evolutionary scheme outlined in earlier chapters, before looking at its role in the development of beliefs.

The first chapter listed several important principles of evolution, all of which function through trial and error. Most significantly for our purposes, trial-and-error produced the human brain. But trial and error continues to operate in the daily lives of virtually all species. For example plants send out shoots here and there, some of which find light and some of which wither; most animals use trial and error more or less continuously, sniffing in this direction or that to test for food or danger. We humans frequently try out ideas or concepts on one another, knowing some of them will wither and some flourish.

We should not assume, however, that trial and error is always successful—"error" is half the process. Even when evolutionary trials succeed, the fit between resulting structures and the physical environment is often quite loose, as the section on jury-rigging explained. A horse's hoof evolved to fit the ground upon which horses customarily trod, over eons of trial and error evolution. But because each stage of evolution is built on what was there before, the fit is never exact, and a horse may go lame when its hoof encounters certain kinds of challenges. Similarly, our brains evolved to permit a fit between our concepts and the reality of our experience, but the fit is rough, and it would be naïve to think our brains can produce perfect fits between the two, by trial-and-error or otherwise.

Indeed, since language can only reflect experience in representational symbols and concepts, the fit between a horse's hoof and the ground is likely to be better than the fit between external reality and the systems of concepts which constitute our world-views. That is because there are several steps between reality and belief (one from experience to memory pathways, another step or two from memory into concepts expressed in language, one from concepts into a set of beliefs,[16] and so on), whereas the fit between the horse's hoof and the ground is direct and unambiguous. Thus the section on idolatry is re-ëmphasized: we should be humble about our concepts, and acknowledge that language may not always express complex experience in precise and unchanging symbols.

Nonetheless, there is every reason for a humble confidence in our beliefs. That is because language and conceptualization have a characteristic not found in specialized structures like hooves. Hooves do not easily adapt to new experiences (say a steep rocky slope) without long evolutionary time, whereas our ability to modify concepts through learning and rationality gives us the opportunity to make closer and closer fits between external reality and our depiction of it in language, even during a single lifetime. Through the community of culture especially, we have the opportunity to test and modify concepts, both through the theological imagination as described in earlier chapters, and through specific techniques discussed later in this chapter. Thus trial-and-error can propel belief forward, helping it to develop rather quickly, and giving a basis for confidence in beliefs that pass its tests.

Trial and error is applied at every level of belief, and whether or not the belief is tested in more formal ways. For example when creationists confront the evidence of fossils, they cast about for explanations that fit with the rest of their beliefs. They experiment with the idea that fossils were created in place to test our faith, and they experiment with the idea that fossils aren't really very old. That kind of trial and error is just what we all do in the process of constructing a cohesive set of beliefs, though the creationist example demonstrates the common tendency to "strain at a gnat while swallowing a camel." Such straining and swallowing is what usually happens when we stretch the evidence and arguments to make old belief systems work, and refuse to modify the belief system itself. There are ways to avoid this tendency, however, and to articulate our beliefs so as to discover errors sooner. Foremost among these is the construction of formal hypotheses.

3. Hypothesis-Construction. Hypothesis-construction is an elegant form of trial and error which humans have used from the dim reaches of time, though we have only begun to formalize it and obtain its maximum benefit in recent centuries.[17]

In earliest childhood we begin to manipulate objects, babbling forth our delight (or frustration) with the outside world, as we try to see how objects and ideas fit together. Even these early efforts are accompanied by mental processes which consider alternative possibilities for the fit or lack of it; and with language we start making rudimentary explanations for our manipulations. It is not long before we start to verbalize hypotheses, combining the glories of language with trial-and-error. (For example "I bet this peg will fit here.") By age seven or eight we even begin to hypothesize about abstract ideas,[18] and there is a middle ground where younger children sound as if they are hypothesizing about abstract things even though they are actually thinking concretely (They may ask, for example "Does God live on the moon?"). Adults hypothesize constantly, both from our own experience and vicariously. For example we readily construct hypotheses based on the mistakes, successes, and arguments of others, adopting or adapting their hypotheses consciously or sub-consciously.

When hypothesis-construction is augmented by rationality, intelligence, and language, a particularly powerful characteristic of hypotheses appears—they can be tested for flaws even before being put into practice. This testing can take place through formal or informal methods such as discussion, literature, brainstorming, gossip, committee meetings, and cocktail-party arguments, as well as through safety-conscious physical experimentation or other scientific methods. Popper says this allows erroneous hypotheses to die in our stead; we ourselves survive to test other hypotheses, if we are wise enough to avoid hasty implementation. Yet testing new hypotheses is not always safe. Icarus perished by testing his father's flight hypothesis, and Walter Reed gave his life for us by courageously testing his hypothesis that mosquitoes carry yellow fever.

Hypothesis-construction applies not only to physical inventions or medical theories, but also to social systems and structures, and it has powerful implications for theological belief. Imagination and gestalt processes constantly push toward new ideas within cultures and individuals, by suggesting new hypotheses for discussion and testing. Hypothesis-construction thus influences belief as it influences other parts of life, by suggesting new concepts which stimulate a response.

But just as it can be risky to try out some physical hypotheses, exploring new theological concepts can be risky too. Any list of martyrs reminds us of the many people who have died for unpopular beliefs, not because of risks from sunshine or mosquitoes, but because of their own species. When philosophy and theology allow safe consideration of new hypotheses, they encourage individual development even while maintaining the body of culturally approved ideas. But when philosophy and theology penalize hypothesis-construction, they usually undermine belief even while appearing to strengthen it, primarily because they interfere with the development of a cohesive framework of belief.

The Framework of Belief

Because we all process experience in similar ways, using the gestalt, trial-and-error, and hypothesis-construction, we end with a similar product, though its details vary from person to person and from culture to culture. That product is a framework of belief, and it results not only from processing experience, but from the ways the brain organizes concepts.

The complexity of the brain is so great that no simple understanding of that organization is possible. Just as homing pigeons are thought to use several different means of navigation, it is likely that our brain employs several different kinds of organization. Pending further neurological research, however, it seems justified to say that we do organize our concepts, and the challenge is to find effective ways of explaining the resulting framework of concepts and beliefs.

We can meet that challenge by using an analogy to expand our grasp of the gestalt. So far I have only talked about its first stage, in which the subconscious mind gathers seemingly unrelated observations and pulls them together into new insights and beliefs. But a fuller understanding complicates the picture, because beliefs are held within a structure, and not as isolated or free-floating concepts. We might think of the structure as a giant jigsaw puzzle holding concepts together,[19] but more complex than that because the concepts come in groups, which fit into the puzzle like large modules or super-pieces.

These modules don't fit perfectly together. Instead they overlap or have gaps between them, and a given concept may act as a component of several different modules. It is as if the jigsaw puzzle is three dimensional, with pieces that don't perfectly fit together, and yet belong in a certain spot anyway.

The nature of the gestalt is to strive for a better fit. The "Aha" sensation is part of that, not so much because of our delight in finding a piece that fits, but because we momentarily see how the piece fits into the whole picture. In fact, what gestalt really refers to (though it translates awkwardly from German), is the interrelation of the whole and its parts in such a way that the sense of the whole momentarily obscures the parts.

It isn't often that we have such a cohesive picture of our own world-view. While it is probable that we each operate from an identifiable world-view, its boundaries are hazy, the fit of one module with another is either too loose or too tight, and there is a great deal of internal conflict between concepts.

Even so, the framework is usually stable. While the puzzle-piecing process permits many pathologies that support a large mental health industry, it is by no means necessary or possible to maintain a structure of belief with all its pieces fitting perfectly together. On the contrary, some inconsistency is necessary as we work new ideas into the old framework, both within concepts inside a module, and between the modules of belief themselves.

Incidentally, these modules of belief go by many names. I call them concept-sets, but traditional religious faith calls them doctrines. Having doctrines is a formal way of dividing belief into parts, so that we can experiment with and argue about and hypothesize about individual components, without toppling the whole framework of belief. It is a necessary accommodation to ongoing development in belief, allowing changes even within a highly stable framework. For example the "Doctrine of Man" is a set of concepts about humanity, which may slide neatly into a modular space within the larger world-view, but which can get disrupted when adjacent concept-sets change. Right now many of us are changing our concept of ecology, and that necessarily changes our view of humans in general, and of ourselves in particular.

A strong culture maintains the stability of its concepts or doctrines in various ways. One way is to convince its members that their inklings toward change are pathological. But actually the range of health is very broad, and the flexibility of the human mind is great. It is not pathological for belief-systems to change, even continuously, provided we keep them stable enough to allow adequate functioning. Changes can also be slow or fast, provided they are not so slow as to induce rigidity, nor so fast as to induce psychosis or instability. Nor is it harmful to have moderate degrees of certainty or wishy-washiness in our beliefs. Finally, it is not necessarily destructive to select a belief-system (or some module of one) off the shelf of pre-packaged beliefs which are persuasively presented in today's competitive supermarket of belief. If one is too busy to construct a personal belief-system, that might be a reasonable thing to do. In fact we *must* do it to some extent, though psychologists warn that such vicarious living can be harmful, if personal decisions are too completely avoided.

This "golden mean" approach to mental health will be pursued further in Part Three. I present it here to illustrate the forms a framework of belief can take, and the various schedules for changing beliefs. While we have an innate tendency to gather beliefs into doctrines or concept-sets which make up a world-view or faith,[20] and while this framework of belief is usually stable, it does not have to be rigid. In fact it is usually undergoing changes that are more or less dynamic, depending on the person and the culture.

Although we process experience in similar ways, and although we each develop a framework of belief, we can still have problems confronting the great variety of beliefs that exist in the world. Thus the question arises as to whether there are any reliable processes for evaluating beliefs. In a busy world where beliefs do come pre-packaged, is there a way to exercise good consumer judgment in picking among beliefs? That is the subject we turn to now.

Sub-Chapter 3:
The Role of Theory in Evaluating Beliefs

Think for a moment about the many different beliefs that pull on us in today's world. At subtle and not-so-subtle levels, advertising strives to lead us toward beliefs about ourselves, our culture, and what makes life worth living. Self-help books confront us with potential changes in our beliefs about child-rearing, relations with parents or spouses, our self-concept, and so on. Cults and religions urge us to accept belief-systems which can lead to dramatically different conclusions, depending on which of them we accept. Political parties solicit our votes.

Sometimes we find ourselves drawn to any belief-system which would easily resolve all these pressures, and allow us to make a thousand decisions at once by simply accepting a particular world-view. That may work for some people, at least temporarily. If one's framework of belief is to be consistent with one's experience, however, it is unlikely that any pre-existing belief-structure can be accepted without modification, because of the great variety of experience. Furthermore, we are always having new experiences and insights, and thus we must be prepared to modify our beliefs in the future, even if they seem satisfactory today.

The process of modification, which has been a theme of Part Two, requires some means of evaluating beliefs, both those we already hold and those we encounter. Of course we could rely on sub-conscious gestalt evaluation, and that is a powerful kind of evaluation that happens whether we want it to or not. But this sub-chapter will recommend a more objective and conscious kind of evaluation, which will supplement and direct subconscious evaluation, allowing us to modify our beliefs toward cohesiveness and consistency. If we can find a process of evaluation that does that, it should add great confidence to our beliefs, allowing our faith to grow in a smoothly dynamic way.

From Hypothesis to Science

Throughout history a few people have tested their beliefs against the evidence of experience and against the framework of beliefs already held. Some of them developed formal systematic approaches for testing and evaluating belief, and some of the resulting systems are still quite useful today. For example the "rules of evidence" became central to judging between competing legal claims, and they have been adapted to other kinds of evaluation.

About 400 years ago certain social changes paved the way for a new approach to testing and evaluating beliefs, and that approach has now become widespread. It was not a political change, but it could not emerge until the breakup of Europe's monolithic empire reduced the intellectual barriers between Europe and the Arabian world. Those developments released curiosity from a straightjacket, heightening interest in inventions like the microscope and telescope. In turn those inventions raised many new questions, which the culture was finally able to tolerate.[21]

In retrospect we lump those discoveries and questions into a single category and call it the birth of science. But it took a long time for those early thinkers to find common threads in their explorations. I will not trace the history of science here,[22] but I do want to reiterate that "science" is a kind of inquiry—a historical and methodological process—more than it is an isolated or specialized body of knowledge. Let me explain that further by outlining the recent history of science's most important concept, the theory.

I noted above that a well-stated hypothesis lends itself to testing. Science is founded on that idea, and constantly tries to restate beliefs in ways that invite testing and challenge. Once a hypothesis is stated in a way that does that, science digs in and tries to disprove the hypothesis, using such methods as experiment, prediction, observation, and comparison.

If a significant hypothesis passes various kinds of tests, it starts to be called a theory instead of a hypothesis. Until recently, a theory had a chance of graduating further and becoming a scientific "law," but in this century the philosophers of science have refined their idea of what a theory is and what it is not. One result is that the concept of "law" is almost dead. How did that happen, and what adjustments has it required?

By the middle of the nineteenth century, scientific achievements had become very impressive, leading some people to give science more weight than it deserved. Instead of seeing science as an effective way of evaluating beliefs in a relatively objective, simple, and straightforward way, these people considered science the path to absolute knowledge. Calling a scientific conclusion a "law" was a subtle expression of that mindset, and it was more directly expressed in the arrogance which scientists seemed to have about their way of thinking. By the early twentieth century, one group of scientist/philosophers even waxed bold enough to suggest that we ought not to believe anything unless it could be established absolutely. These "logical positivists" were soon subjected to serious criticisms from several different directions, including psychology, the humanities, and philosophers of science themselves.

Psychologists saw that this position denies the role of gestalt or other non-logical processes. That denial leads to a sterile belief-system, since limiting belief to absolutely substantiated concepts would leave so few beliefs that we could not function. In fact many—perhaps most—of our beliefs are in some respect incompletely supported by clear objective evidence. In other words, the evaluation of evidence is not entirely cognitive and logical, but is instead partly emotional, pre-logical, and subconscious. And even if science can lead to dramatic extensions of our knowledge and understanding, we cannot claim anything close to absolute knowledge.

Humanities scholars made another kind of criticism, arising from their skill in hermeneutics. In Greek mythology Hermes was the messenger of the gods, sent to deliver their secrets to humans. Therefore his name came to refer to scriptural interpretation—the discovery of God's secrets—and thus "hermeneutics" was born. More recently hermeneutics has expanded to include all interpretation, including the interpretation of secular literature. By the 20th century hermeneutics scholars had developed real expertise in interpretation, and it offended them to hear positivists say their methods were the only kind that counted. They correctly pointed out that we approach every question in the light of previously existing concepts, or "hermeneutical pre-knowledge." Our acceptance or rejection of a new theory is as much a function of these pre-existing beliefs, as it is a function of objective evaluation.

Finally, objections came from philosophers of science themselves,[23] who were beginning to define science in humbler terms. They pointed out that science is always tentative, serving to state reasonable conclusions until new facts or theories come along. A theory's goal is not to define absolute truth, so much as to organize and explain the evidence that has been discovered so far. Thus philosophers of science affirm the concept of hermeneutical pre-knowledge,[24] as part of the subconscious processing discussed above.

So even though logical positivists did have an influence, especially in the analysis of language, the argument between them and their critics was eventually won by the critics, and science became more humble.

Unfortunately, science's new humility exposed it to another attack which is largely unwarranted. This criticism says that if science cannot lead to absolute knowledge, then it must be just a kind of faith. Or said another way, every belief is "just a theory." This seems to imply that all beliefs are on the same footing, depending for their validity on social acceptance or mystical revelation, and perhaps having no general validity at all, but only personal validity.

That criticism has forced science to further refine what a theory is. Even though it is not absolute, and even though the "law" language is out of vogue, some theories are more strongly supported by evidence than others. Any theory could turn out to be wrong, but some are wronger than others.[25] A theory's status depends on the strength of its supporting evidence and the extent to which it organizes evidence in an objectively demonstrable way. With each new advance in its supporting evidence, beliefs based on a theory march toward greater reliability.

Of course scientists can be as petty as anyone else, and individual scientists may exercise blind faith in their favorite theories, refusing to see evidence that does not support them. But the centrality of testing and objectivity in science eventually overcomes such pettiness, significantly reducing the impact of blind allegiance on scientific theory. Scientists make their reputations by challenging old theories and building new ones on accepted foundations. This building and re-building has kept the vibrancy in scientific theories. Indeed, it is this lack of blindness that makes theories so valuable as a tool for evaluating beliefs. Scientific methods succeed where many disciplines fail, because they focus on objective methods for overcoming our tendency to fool ourselves.[26]

To say a theory is just a kind of faith not only shows a misunderstanding of theory, but also of the nature of faith. It requires one to ignore the role of evidence. This chapter, on the other hand, presents faith as a framework of belief based on culture but modified by evidence. The "just a kind of faith" argument is inadequate precisely because it doesn't distinguish between beliefs supported by different amounts of evidence.[27]

Even though we cannot support the idea that all statements are "just a theory" or "just a kind of faith," it is true that faith and theory are related. That relationship furnishes the grain of truth on which the popularity of the "theory is just a kind of faith" argument is based, so let us spend a paragraph or two looking at it.

In discussing the framework of belief, I pointed out that faith is built of modules, each comprised of one or more sets of interrelated concepts. Since a theory *is* a set of interrelated concepts, it may constitute a module of belief within the jigsaw puzzle. Putting it in place may require modifying adjacent modules to make a better fit, so there is indeed an intimate relationship between faith and theories.

Furthermore, in order to compensate for the loss of the "law" concept, scientists have refined the concept of theory to distinguish between minor and major theories, and between narrow and broad theories. That results in calling some theories "general" theories or "grand" theories. These are not individual theories so much as groups of theories which tie large chunks of evidence together. They are still modules of the puzzle, but those modules can be quite large, supplying the borders or organizational structure for large parts of the framework.

This section has pointed out some characteristics of theories, to show how they fit into the chapter's earlier language of "modifiable belief-structures." I have not gone into detail about the glories of scientific method,[28] except to underline its heavy dependence on evidence and on evidence-based theories. It is tempting to explain theory further, but other books do it well,[29] and I am confident that most people already acknowledge the great necessity of evaluating beliefs in light of evidence. And that, after all, is my main point.

As science developed its methods, other disciplines adopted and adapted them, so that most disciplines today depend upon some variation of those methods. Humanities uses the three c's to teach critical thinking;[30] in philosophy the study of logic has been profoundly influenced by science and mathematics; even rhetoric, an erstwhile champion of subjectivism, has recently begun to adopt objective and realistic criteria for evaluation.[31]

Over most of its recent history theology has resisted objective evaluation, but as science has slowly unraveled the biological bases of morality and uncovered facts relevant to traditional theological understanding, even theology has had to adopt more objective ways of thinking. We will explore that development further in a moment, but let us first summarize the chapter as it has unfolded so far.

Modification Revisited

This chapter has proposed a formal approach to belief. The approach begins in sociality, acknowledges the impact of gestalt and trial-and-error processes, and ends by showing how belief-structures can be modified based on an adequate appreciation of hypotheses, theories, and evidence. We can summarize and restate this approach in 7 short paragraphs:

1. We must begin with whatever belief-structure we already have. That may be the structure we learned as children; or one derived from our occupation, a charismatic leader, or some other source.
2. Because of our need for stability and security, we resist evaluations which call on us to modify our belief-structures. But modification is ongoing, whether we acknowledge it or not. The only question is whether it is based on conscious evaluation, on sub-conscious evaluation, or on a combination of them.
3. Good conscious evaluation must remain open to new evidence and to new interpretations of evidence.[32] I have said that some people are fortunate enough to live successfully without much change in their belief-structures; but if they are not open at all, they would have to be very fortunate indeed to avoid living a frustrating life of reactivity,[33] at least in our world.

   But continuous conscious evaluation is neither possible nor desirable, and it is not necessary to modify our belief-structure at the first hint of new evidence. We can rely on subconscious processes to initiate modifications, even without conscious involvement.[34] To return to the jigsaw analogy, it is as if we leave the puzzle on the table while we pursue other things. If we have programmed the subconscious with attitudes of openness and humility, we can be confident it will do its work quite well.

4. From time to time we must give conscious effort to the belief-structure—we must return to the table. Our subconscious will have put a few pieces in place while we were gone, or perhaps tried some things that didn't work; but with effort we might now be able to put more pieces into place, or at least work on one module. We sit down and try to work on new evidence or interpretations, trying to work them into our previous structure of belief. This may require shifting some beliefs around, discarding others, or combining pieces into new concepts.
5. Openness and humility require that we be sensitive to the belief-structures of other people. Interaction with an immediate circle of friends allows society its proper influence,[35] and interaction with the broader culture avoids the dangers of tunnel-vision. Tunnel-vision is dangerous because it can operate even when one is intently focusing on evidence. This danger has been explored by the philosophers of science[36], and if it can happen in science it can easily happen in other areas of belief. Especially must we compensate for our tendency to misinterpret evidence and even to see evidence where there is none.
6. Evaluation depends not only upon evidence, but also upon the way evidence is organized into concepts and structures. To the extent such structures are consistent with evidence (and vice-versa), the puzzle pieces fit better. As we continue to uncover the order that exists in the universe and within the brain,[37] we are coming closer to a good fit than ever before. It still eludes us sometimes, but that is often because we are unwilling to make the evaluations called for by evidence, and because we don't modify belief-structures to accommodate new evidence.
7. At any point in time, we live from our belief-structure as it then stands. Ideally, continuing openness allows change to happen so slowly that the belief-structure is never capsized. Thus we can live joyfully and enthusiastically out of our respective belief-systems, relaxed but relishing the idea of change. That is not to deny that some changes challenge one's faith, but realizing the inevitability of change should help us embrace it, and minimize the anxiety inherent in change.

The evaluation process outlined above does not produce a perfect belief-structure. Every such structure has inaccuracies, awkwardness, ill-fitting pieces, or downright errors built into it. To expect otherwise is inconsistent with the evidence of every historical belief-system and with the nature of human conceptualization, and also encourages a neurotic perfectionism that can lead to serious psychological problems. On the other hand, admitting that a belief-system is imperfect does not imply that it is untrustworthy. With each passing year of growth and refinement, a belief-system can become more cohesive and useful.

This sub-chapter has focused on rational evaluation of belief, not because other kinds of evaluation are unimportant, but because theology stands poised to leap forward, if it can incorporate new evidence and concepts. Indeed, new evidence and concepts have already begun to modify theology, not just through the slow influence of time or the aggressiveness of science, but through the active initiation of theology itself. This is a very exciting development, which deserves the attention it has received. Our discussion of faith will end by addressing it.

Sub-Chapter 4: Faith and the New Paradigm

In most fields outside theology, it is now well accepted that beliefs are most reliable when they accord with evidence, and most productive when they seek new evidence and new conceptual structures to organize that evidence. At various times in the past theology has also been interested in evidence, but modern theology has resisted overt efforts to modify its concepts toward a greater consistency with evidence. (It could not of course resist the long-term effect, as we have seen.) Yet theology has recently begun again to affirm faith as a belief-structure which honors evidence, rather than one which fights against it.[38] How did that happen, and what are the implications?

Theology Embraces Science

The ancient myth of the phoenix paints a graphic picture of culture adopting new ideas over time. It tells the story of a wise and sacred bird (called the phoenix) that dominated the landscape during its 500-year life. At the end of its life, the phoenix built a funeral bonfire, climbed on it, set it afire, and perished in the flames. But out of the fire flew a new phoenix, who lived and ruled another 500 years. Each new phoenix died the same way, and the race of phoenixes thus ruled the world for thousands of years. Phoenix, Arizona was named after the myth, because it was built on the ruins of a native American city. In early Christianity, the myth was used to explain resurrection to the pagans, to show them that God had returned in a new form.

The phoenix myth also helps explain paradigm shifts. Several examples of the shift have already been mentioned, such as the paradigm of kings which gave way to democracy, the paradigm of slavery which gave way to human equality, and the paradigm of human sacrifice which gave way to animal sacrifice.

Because faith always has its roots in the past, it tends to resist paradigm shifts. Actually that is a part of its role, as we saw when discussing the conservative function of culture. Nonetheless, faith is not immune to paradigm shifts, and is instead intimately tied up with them. Sometimes it is dragged along by a culture-wide paradigm shift, and sometimes it shows the wisdom of the old phoenix, by actively initiating the change.

Nowadays science has become such an integral part of our world-view that it influences our thinking at every turn. We should not be surprised to see faith changing in its direction. Indeed, there are already several theological journals which explore the interplay between science and theology. Even outside such ivory towers, one frequently hears sermons which deal favorably with science. Another shift has begun.

No clear conception of the new paradigm has yet emerged, but it will probably take one of these shapes: 1) The broad paradigm of scientific method might merely influence theological thought like it influences the rest of culture; 2) The paradigm might turn out to be a partnership between science and theology; 3) Theology might serve as the spiritual foundation for a paradigm growing out of the ecological movement. I will refer to all of these possibilities in the discussion below, leaving open-ended the ultimate direction in which the new paradigm might go.

Perhaps the most honest and prestigious workplace for fashioning the new paradigm is the quarterly journal *Zygon: Journal of Religion & Science*. It reports on dozens of efforts to correlate evidence, theory, and theology. I will describe some of its work in the next chapter, as we try to fashion a new understanding of God. But a simpler example can begin to show how theology is already embracing science.

Decades ago, a few thinkers began to understand how badly we treat our environment, and what we need to do about it. But the rest of us, including most theologians, didn't worry about it much. Yet in the recent crescendo of environmental awareness, every segment of society has had to change its way of looking at the world and its ecology, and that includes theologians.

Actually this shift has not been as difficult for theology as have other shifts, such as the shift toward racial equality or the shift toward international peace. That is because the stewardship commandment of Genesis is in harmony with ecological sanity. Genesis still contributes to the problem because its "be fruitful and multiply" command underlies a western philosophy of growth which contributes to the worldwide destruction of our environment, but theologians are saying stewardship is the more general command. We cannot be good stewards without addressing human reproductive urges.

Some people argue that "stewardship" language is itself a dangerous part of the old paradigm. By thinking of ourselves as "in charge," instead of as partners with other life-forms, we may create mindsets that abuse the environment. The analogy would be to a paternalistic government that takes responsibility for "our Indians;" this attitude guarantees abuse of native Americans, whereas a concept of fellow-citizenry offers some hope.

Nonetheless, the shift toward a new model of stewardship is a step in the right direction, and might easily give way to further modification toward a partnership paradigm. Even in its current form, the refurbished stewardship paradigm illustrates my point, which is that theological understanding expands and changes to accommodate and participate in paradigm shifts. Whatever the details, we are clearly struggling toward a new and broader concept, which affirms whichever of our previous beliefs it can, while shifting toward scientific ways of thinking. The question is not so much whether the new paradigm is developing, for that seems well established; the question is rather how far the shift will take us, and what the implications are for theological belief. In the words of theologian Loyal Rue, we are looking for a new myth.

Four Implications of Accepting the New Paradigm

Even if we accept the need for a belief-structure which incorporates evidence and scientific theory into theology, we know that people have to function, whether or not they strive for a comprehensive world-view and whether or not they actively modify it toward greater consistency with evidence. This has been true from historic times, when farmers planted by the moon or relied on other folklore because there was no such thing as scientific agriculture, and witch-doctors did the best they could at formulating concepts out of their theological imaginations. But it is also true today, both in science and in daily life, notably in medicine. Indeed, because there is no complete knowledge, there must always be some component of unknowledgeable faith in our living.

Nonetheless, the first implication of the new paradigm is that ***faith gives way to knowledge***. When we can only "see through a glass darkly" it is reasonable to operate on what we can see, but when our concepts become clearer and our evidence more reliable, we use it. This is true in both secular life and religion, as illustrated by the recent tests on the Shroud of Turin, which caused most religionists to stop calling it a miracle once that usage was contradicted by compelling physical evidence. In the 17th century it may have been reasonable to think that each lightning bolt came from the finger of God, but once we understood the causes of lightning, our theological concepts had to be modified, and they were.

A second implication of the emerging paradigm is that ***faith and knowledge are not opposites***. While new evidence and traditional belief may conflict, evidence eventually modifies all but the most rigid faith. This observation has led modern theology to reject an older tradition which saw faith and knowledge as opposites,[39] instead seeing faith and evidence as interrelated parts of a complete mental process. One of Jesus's metaphors tells what will happen if we insist on seeing them as opposites; the new wine of modern experience and evidence will burst the old wineskin of traditional concepts.[40] In other words the old belief-structure will be modified radically if we don't find ways to modify it gradually.

A third implication of accepting science into theology is a negative one: ***it is <u>not</u> easy***. There will be discomfort and confusion as we try to figure out how new evidence fits into or modifies the traditional framework of understanding. The discomfort and confusion are well illustrated by Huckleberry Finn's agony over whether to surrender the runaway slave Jim. The formal faith said he should do so, but a contrary gestalt was rising to consciousness which he could not quite articulate. Fortunately, Huck wasn't unduly concerned about a completely consistent faith, and his final conclusion thus shows how conceptual frameworks can be used even before they are completely understood. He decided not to surrender Jim, because "Every now and then you got to forget about heaven and hell and do the right thing." While he saw himself as abandoning faith, Huck was actually responding to the urgings of a new and deeper faith than the one he thought he was supposed to embrace.

This leads to a fourth implication. Whether the new paradigm is incorporated smoothly or radically, and whether we can fully articulate it or not, ***the change is likely to have widespread consequences for belief***. As theology moves toward science we can imagine all kinds of impacts on traditional belief: The biological understanding of territoriality gives new meaning to commandments like "Thou shalt not steal"; the possibility of genetic determinants of fidelity gives new insights into "Thou shalt not commit adultery;" sociobiology has dozens of implications for community, validating our grouping of ourselves into cooperative units and exploring the biological bases for morality; studying the impact of belief-structures on behavior can help us understand sin and redemption; subconscious processes obviously have something to do with prayer. Examples are myriad, and Parts Three and Four of this book will develop many of them further.

This chapter has presented an approach to faith that begins in cultural acceptance, develops through trial and error and other basic processes, and finds its fullest expression in systematic belief-structures which grow out of the accumulation and evaluation of evidence. The chapter has not analyzed particular tenets of faith, nor has it talked about the relationship between belief and action; that will be the task of Part Three. But before turning to that practical discussion, let us end Part Two by discussing our concepts of God.

Notes

[1] See pp. 330-346 of *Imprinting: Early Experience and the Developmental Psychology of Attachment*, by E.H. Hess (New York: D. Van Nostrand Reinhold Company, 1973).

[2] The disruption of identity processes is discussed in *Transcending Turmoil: Survivors of Dysfunctional Families*, by Donna F. LaMar (New York: Plenum Press, 1992).

[3] Halperin, David A. "Group Processes in Cult Affiliation and Recruitment," Chapter 13 in *Psychodynamic Perspectives on Religion Sect and Cult*, ed. David A. Halperin (Littleton, Massachusetts: John Wright PSG, Inc., 1983).

[4] For further discussion of the fascinating idea that it may be adaptive for us to believe things that are not true, see "Illusion and Well-Being," by Christine Bates and Darryl Stevens, in *Journal of Counseling and Development* Vol. 67, no. 10 (June 1989), pp. 590-591. See also *The Biology of Moral Systems*, by Richard D. Alexander (New York: Aldine DeGruyter, 1987).

[5] See Chapter 9 for more on the nature of personal decisions.

[6] Apparently we approach the world in a variety of ways which are not yet well understood. Many writers even argue the existence of a specific number of "ways of knowing," such as five or seven or four or six. See for example "What Kind of Thinker are You?" by Robert and Susan Bramson, in Vol CII No. 10 (October 1985) of *Ladies' Home Journal*; László Mérő's *Ways of Thinking* (London: World Scientific 1990); *Six Thinking Hats* by Edward DeBono (Boston: Little, Brown and Company; 1985); and especially *Multiple Intelligences*, by Howard Gardner (New York: Basic Books, 1993). Some personality theorists claim there are four broad types in the human family; see for example *Manual: A guide to the development of the Myers-Briggs Type Indicator*, by J. B. Myers and M. H. McCaulley (Palo Alto, California: Consulting Psychologists Press, 1985). The Myers-Briggs approach is based on Jung, according to Robert J. Drummond and Ann H. Stoddard in "Learning Style and Personality Type," pp. 99-104 in Vol. 75 No. 1 (August 1992) of *Perceptual and Motor Skills*. Therefore see also *Personality Types: Jung's Model of Typology*, by Daryl Sharp (Toronto: Inner City Books, 1987). The concept of learning styles has gained a great deal of attention in the educational literature; for example the entire October 1990 issue (Vol. 48 No.

2) of *Educational Leadership* is dedicated to it. Finally, Lawrence Kohlberg has built on the insights of developmental psychology to propose several "Stages of Morality" through which we pass as we attain greater degrees of maturity; see his *Moral Stages: A Current Formulation and a Response to Critics*, with Charles Levine and Alexandra Hewer (New York: Karger, 1983.)

[7] For a persuasive argument that birth order by itself cannot determine personality, see *Birth Order: Its Influence on Personality* by Cécil Ernst and Jules Angst (New York: Springer-Verlag, 1983).

[8] Nobody has treated ritual with more competence and subtlety than Victor Turner. See his *The Anthropology of Performance* (New York: PAJ Publications, 1987) and *The Anthropology of Experience*, edited by him and Edward M. Bruner (Chicago: University of Illinois Press, 1986).

[9] Erik Erikson has done more than anybody to explain identity processes. See his monograph entitled "Identity and the Life Crisis," No. 1 in *Psychological Issues* (Vol. 1, No. 1: New York: International University Press, Inc., 1959). Chapter 2 of the monograph ("Growth and Crisis of the Healthy Personality," pp. 50-100) is especially readable, and the monograph also includes an excellent historical survey of psychoanalytic ego psychology by David Rapaport (p. 5-17).

[10] For a fuller discussion see "The Legacy of Gestalt Psychology," by Irvin Rock and Stephen Palmer, in *Scientific American* v. 263 no. 6 (December 1990): 84.

[11] As with other divisions in this book (such as the one between rational and non-rational thought processes), I want to reiterate that the distinction between conscious and subconscious processes is not a perfectly discrete one. They both exist as a continuum, and even the mental processes which take place during sleep, which would seem completely subconscious, have been shown to vary in the *depth* of subconsciousnesses.

[12] Freud did not invent the concept of the sub-conscious, though he re-conceptualized it in new and important ways. See the excellent and acclaimed *The Unconscious Before Freud*, by L.L. Whyte (New York: St. Martin's Press, 1978).

[13] The following quotation from p. 10 of the book just cited is especially noteworthy. Whyte says: *For today faith, if it bears any relation to the natural world, implies faith in the unconscious. If there is a God, he must speak there; if there is a healing power, it must operate there; if there is a principle of ordering in the organic realm, its most powerful manifestations must be found there.*

[14] See especially "Gestalt Perception as a Source of Scientific Knowledge," pp. 281-322 in Vol II of *Studies in Animal and Human Behavior*. For a fuller discussion see my article entitled "The Precursors of the Eureka Moment as a Common Ground Between Science and Religion" pp. 191-204 in Vol. 29 No. 2 (June 1994) of *Zygon*.

[15] Karl Popper, a famous philosopher of science, has said that all learning and all concept-formation, including both the basic processes already discussed, and the later development of a more complex faith, arise from trial and error. At the end of his lecture "Towards an Evolutionary Theory of Knowledge," which can

be found in his *A World of Propensities* (Bristol, England: Thoemmes, 1990) he even goes so far as to say (p. 51) "from the amoeba to Einstein there is only one step. Both work with the method of trial and error." But others have suggested that he is squeezing too many concepts into one corset. See for example pages 32 and 272 of *Not in Our Genes*, by R.C. Lewontin et. al. While there are good insights in such criticisms, it should be clear that I find Popper, Lorenz, Wuketits, Bartley and other evolutionary epistemologists fundamentally persuasive.

[16] Nancey Murphy has urged me (personal communication April 8, 1995) to distinguish between concepts and the beliefs expressed in them, so as not to collapse the *truth* of beliefs and the *adequacy* of concepts. I have not overtly adopted that distinction, instead emphasizing how the two intertwine. Nonetheless, I think the insight is fertile for analytical purposes, and I hope this paragraph and others (for example in Chapter 12 where I discuss the relation of values to beliefs, and the whole of Chapter 4) show my recognition of its value.

[17] For a good discussion of hypothesis-construction, see "Rationality in Science," by Henry Harris, Chapter 3 (pp. 36-52) in *Scientific Explanation* (Oxford: Clarendon Press, 1981), ed. A.F. Heath.

[18] The ability to think abstractly doesn't develop very completely until adolescence or even later. This seems to be the biological foundation for the protestant idea of "the age of accountability," the doctrine which says persons cannot make a personal commitment to God until they reach an age where they can understand what the commitment means. In my sub-culture at least, children are encouraged to make the commitment as early as seven or eight, which would seem to water the doctrine down considerably.

[19] For a fuller development of the jigsaw analogy see p. 40 of *The Coherence Theory of Truth*, by Nicholas Rescher (Washington, D.C.: University Press of America, 1982). This analogy is criticized and another analogy (the crossword puzzle) substituted by Nancey Murphy in her fine article "Truth, Relativism, and Crossword Puzzles," *Zygon,* Vol. 24, No. 3 (September 1989), pp. 299-314.

[20] I hope it is clear that "faith-system" or "world-view" may also go by other names, such as "philosophy of life" or "value-system." For more on the latter term, see Chapter 12.

[21] Actually the interaction of experience and political change was more dynamic than my brief summary indicates. During the European Dark Ages, the Arab world experienced a virtual enlightenment, and Europe built on its achievements even after Arabia slipped into its own Dark Ages. See Will Durant's *The Age of Faith*, pp. 239-249 (New York: Simon and Schuster, 1950).

[22] For three books that do, see *The Struggle to Understand: A History of Human Wonder and Discovery*, by Herbert C. Corben (Buffalo, New York: Prometheus Books, 1991); *Conceptual Revolutions*, by Paul Thagard (Princeton, New Jersey: Princeton University Press, 1992); and especially John A. Moore's *Science As a Way of Knowing: The Foundations of Modern Biology* (Cambridge, Massachusetts: Harvard University Press, 1993).

[23] See, for example, Hilary Putnam's "Philosophers and Human Understanding," Chapter 6 (especially pp. 103-104) of *Scientific Explanation*, ed. A.F. Heath (Oxford: Clarendon Press, 1981). The most famous critique is of course Thomas Kuhn's *Structure* (University of Chicago Press, 1962) and its progeny. However, in Kuhn's own opinion he has been widely misinterpreted as an apostle of radical subjectivism. Fifteen years after *The Structure of Scientific Revolutions* he set the record straight in *The Essential Tension* (Chicago: University of Chicago Press, 1977) by saying at p. 325 "My point is, then, that every individual choice between competing theories depends on a mixture of objective and subjective factors, or of shared and individual criteria. Since the latter have not ordinarily figured in the philosophy of science, my emphasis upon them has made my belief in the former hard for my critics to see."

[24] For an up-to-date view of hermeneutical pre-knowledge, see Gunther Stent's "Hermeneutics and the Analysis of Complex Biological Systems" in *Evolution at a Crossroads: The New Biology and the New Philosophy of Science*, ed. David J. Depew and Bruce H. Weber (Cambridge, Massachusetts: The MIT Press, 1985).

[25] This has been delightfully demonstrated by Isaac Asimov in *The Relativity of Wrong* (New York: Windsor, 1989).

[26] Incidentally, another invalid criticism of scientific method deserves a footnote because one still encounters it quite frequently. It is that any experiment (measuring for example) distorts the results, so that we really cannot understand reality in any competent way. For a further explanation of the argument and its resolution, see Chapter 12 (pp. 167-175) of Nobel laureate Murray Gell-Mann's *The Quark and the Jaguar* (New York: W.H.Freeman and Company, 1994).

[27] For more on theory from a well-known physicist and theologian, see Ian Barbour's "Commentary on Resources from the Physical Sciences," in *Zygon,* Vol. 1 No. 1 (March 1966), pp. 27-30.

[28]Nobody has done that more persuasively than John Dewey, as in his influential *Experience and Education* (New York: MacMillan, 1938).

[29] For an explanation that places theory in a broader philosophical context see Chapter 3 of *Religious Explanations: A Model from the Sciences*, by Edward L. Schoen (Durham, North Carolina: Duke University Press, 1985). See also *Explanation from Physics to Theology* by Philip Clayton (New Haven, Connecticut: Yale University Press, 1989).

[30] See Wayne Booth's *Critical Understanding: The Powers and Limits of Pluralism* (University of Chicago Press, 1979), especially pp. 80-92.

[31] See "Rhetoric, Philosophy, and Objectivism: An Attenuation of the Claims of the Rhetoric of Inquiry," by James W. Hikins and Kenneth S. Zagacki, in *Quarterly Journal of Speech* 74 (1988) 201-228.

[32] The philosophical position called pancritical rationalism explains why, and is best described in W.W. Bartley III's book *The Retreat to Commitment* (LaSalle, Illinois: Open Court Publishing Company, 1984).

[33] The dangers of living reactively in a domestic context are demonstrated in Chapters 8-10 of *The Evaluation and Treatment of Marital Conflict*, by Philip J.

Guerin Jr., Leo F. Fay, Susan L. Burden, and Judith Gilbert Kautto (New York Basic Books, Inc., 1987).

[34] For an interesting (and biologically sophisticated) theory of how dreaming contributes to this process, see James B. Ashbrook's "Making Sense of Soul and Sabbath: Brain Processes and the Making of Meaning," previously cited.

[35] The Biblical concepts of exhortation and edification contain great truths for the operation of this interaction, and it is with great discipline that I avoid chasing a rabbit to discuss that aspect of community here. But since Part 4 will be riddled with the idea of community, I will forego the opportunity.

[36] See Irving Langmuir's "Pathological Science," in *Physics Today* Vol. 42 No. 10 (October 1989), pp. 36-48.

[37] Despite the current popularity of the subjectivist fallacy, the evidence reveals order and not chaos. See *Solid Clues*, by Gerald Feinberg (New York: Simon & Schuster, Inc.. 1985). For an excellent criticism of Rorty's subjectivism, see Peter Munz's "Philosophy and the Mirror of Rorty" in *Evolutionary Epistemology, Rationality, and the Sociology of Knowledge*, pp. 345-398.

[38] A good overview of the development may be distilled from *Science Ponders Religion*, ed. Harlow Shapley (New York: Appleton-Century-Crofts, Inc.). More recent books include *Christian Theology and Scientific Culture*, by Thomas Torrance (New York: Oxford University Press, 1981); *Science, Reason & Religion*, by Derek Stanesby (Dover, N.H.: Croom Helm, 1985); *Theology in the Age of Scientific Reasoning*, by Nancey Murphy (Ithaca, New York: Cornell University Press, 1990); *Reason and Reality: The Relationship Between Science and Theology*, by J.C. Polkinghorne (London: SPCK, 1991); *God and the New Biology*, by Arthur Peacocke (London: Dent, 1986); and *The Biology of Religion*, by Vernon Reynolds (London and New York: Longman, 1983). See also Langdon Gilkey's article "Nature, Reality, and the Sacred: A Meditation in Science," in *Zygon: Journal of Religion & Science* Vol. 24 No. 3 (September 1989), pp. 283-298. Other recent works will be pointed out in Chapter 8.

[39] The scripture's condemnation of the "wisdom of the world" should be viewed (in my opinion) as condemning *absolute* knowledge, not *any* knowledge, else much of Proverbs is contradicted. Claims of certainty are a form of secular idolatry, and rightly deserve the scriptural condemnation. See I Corinthians 3:19 and almost any chapter in Proverbs (for example 1:7, 4:7, 8:5). It is not necessary to go so far as Rudolph Bultmann and his followers have gone, by declaring human existence to be in *opposition* to nature.

[40] Mark 2:21-27.

Chapter Eight

Concepts of God

hapter Six agonized over scripture's tendency to "freeze" theological oncepts into growth-hampering rigidity. Yet when we look at any specific oncept we find that nothing, not even scripture, can hold back concept-nodification.[1] Chapter Five saw it at work in the Hebrew culture, and hapter Seven showed how individual faith gets modified. This chapter ill focus on one particular concept, the concept of God, to see how it has een modified in the recent past, and how it is being modified today.

Over historical time the human theological imagination has produced undreds of God-concepts and thousands of different religions.[2] The roliferation of concepts has not ended; each generation produces its share, as new needs and new knowledge modify older concepts. In 1931 Shailer Mathews, Dean of the Divinity School of the University of Chicago, wrote a fascinating book called *The Growth of the Idea of God*, in which he traced the history of God-concepts in our heritage. Beginning with Old Testament developments, he quickly moves to the present era, showing in detail how Greek concepts and Jewish concepts merged to produce an alliance in early Christianity. Then he shows how each great social change (Roman legal systems, Middle Ages chaos, European monarchy, Democracy, and the Scientific Revolution) caused significant modifications in the concept of God.

Because most modifications happened in the distant past, we tend to be ignorant of their significance, thinking they were minor adjustments. But reading the book above (and others to the same effect[4]) erases that tendency, revealing how dramatically our concept has changed over time. The New Testament God arose out of millennial Judaism as modified by the Greek practice of deifying important humans; the God modeled on European kingship was a sovereign who could not be reasoned with but only pleaded with; democracy accelerated God's accessibility to common people; and so on. The nineteenth century began to modify God-concepts toward a greater compatibility with science, and we are still in the midst of that modification today.

Ancient concepts of God (such as the storm God and fertility God) no longer interest us,[5] but more recent concepts, especially those which were popular in the last two centuries, cannot be disturbed without straining our loyalties. Just as we identify more with our grandparents and great-grandparents than with more distant ancestors, it is difficult to separate ourselves from 18th and 19th century concepts which took the place of older concepts. After all, those were the concepts we were taught as children, and they are firmly embedded in our neural pathways.

Yet many people in our century—perhaps most—feel some tug toward modifying their concept of God. This tug is partly offset by a competing pull toward tradition, but historians will undoubtedly see our age as one which altered the concept of God in remarkable ways. Some of those ways arise out of criticisms of 18th and 19th Century concepts, and some arise out of efforts to imagine new concepts that are consistent with both classical ideas and new knowledge.

Criticisms of 18th and 19th Century Concepts

It is impossible to discuss all the multitude of human religions and their many concepts of God here, but four specific parts of the god-concept handed down to us require significant modification or abandonment, and this section will discuss those. They are: 1) The God of Design; 2) The First Cause God; 3) The God of Purpose; and 4) The Supernatural God.

1. *The God of Design.* William Paley, the great 18th Century theologian, proposed a "God of Design" whose existence is proved by the orderliness of biological diversity. If a primitive man found a watch in the desert, he would know it had to have a designer because of the relation between its intricate complexity and its operation; similarly, the profound biological order found on earth implies a designer/creator. Paley was a rigorous thinker, however, and thought it fair to say what kind of proof would invalidate his concept: he said the "God of design" argument would fall if it could be shown that intricate life-forms could have arisen naturally, "without any indication of contrivance" by a higher intelligence. Of course that is exactly what evolution shows, beginning with an explanation of how life itself arose, and how humans evolved from lower forms.

2. *The First Cause God.* Another concept became popular in the 19th Century, as a response to the scientific thinking which had begun to permeate the culture. This concept pointed out that order exists in other places besides life (in atoms and in planetary systems, for example), and that direct cause/effect relationships exist within and between each kind of order. This "God of First Cause" was based on the observation that, since everything around us has a cause, there must be a cause that caused all the causes, and that must be God, a "prime mover."[6]

Physics now seems to show, however, that this concept must also be discounted, because its premise is not warranted. While humans were caused by evolution from lower forms, and while the lowest forms were caused by the chemical combination of certain elements in a field of energy, there is no reason to believe that matter and energy were themselves caused.[7] Instead they always existed. This boggles our minds just as much as it does to try to conceptualize God as having always existed, but there is no compelling reason to trace back to a first-cause God if we can stop with matter and energy.[8]

In slightly different forms this concept continues to attract many believers. I will discuss one of them below in the section on modern concepts.

3. *The God of Purpose.* This concept argued that God must exist, because otherwise there would be no ultimate purpose in the universe. Like the other concepts of God, it focuses on *human* needs and perceptions. While the first two concepts arise from our perceptions of order and cause/effect relationships, this one is more directly psychological, arising from our need to feel that life is worthwhile.

Biologists believe there is a great deal of purpose in our individual and social lives, but they agree with most physicists that any "ultimate" purpose is lacking.[9] Physicists say this because of their understanding of what will eventually happen to the universe: If more than a certain amount of matter exists in the universe, gravity will eventually pull the expanding universe back together into a dense ball that will compress all of us (even space travellers) into a black hole, which may thereafter explode into another Big Bang. If there is less than that amount of matter in the universe it will continue expanding until there are miles and miles between each molecule, including the molecules that presently make up our bodies.

An old joke puts our anxiety over this fate into perspective. "I predict," droned the lecturer, "that the Sun will go out in five billion years, destroying the earth in its final explosion." This startled a back-row listener out of his stupor to shout "What did you say?" The lecturer repeated it. "Oh," breathed the relieved listener. "I thought you said five *million* years."

4. *The Supernatural God.* As Chapter 5 said, the idea of God as supernatural has been with us almost since our species began, and thus it is not so much an 18th or 19th century concept as it is a persistent concept that lasted into those centuries, and even into our own. It is therefore one of the hardest concepts for us to lay aside, and yet our century is the difficult one in which that task began to be completed.

The modification away from supernatural concepts has primarily risen from a rediscussion of the actual causes for things once considered supernatural, and from a new understanding of why humans tend to accept supernatural explanations.[10] For example, miracles have been subjected to greater scrutiny, and many theologians now consider miracles better explained on other bases than supernaturality. They accept Hume's observation that "it is always easier to believe that people are deceived or deceiving, than to believe the laws of nature have been suspended." Even the miracles of the Bible are now understood by most theologians as profound literature—like Aesop's talking animals—rather than as factual occurrences.[11]

Miracles are perhaps the hardest aspect of a supernatural god-concept for us to abandon. Therefore I would like to spend twelve paragraphs discussing categories of events which may seem miraculous, but which a moment's reflection will show to be better explained in some other way. I think these categories cover all claimed miracles.

A. One usage of "miracle" holds that *everything* is a miracle, such as the miracle of childbirth, the miracle of a sunset, the miracle of rain, even the miracle of breathing. Of course all these things are wonderful and thought-provoking, but because each of them operates according to known laws of nature, it would be a misuse of words to include them in the more usual definition of miracle. That definition only includes events which violate natural laws.

B. Another category includes events once thought to violate natural law, but which turned out not to. Lightning was thought to come from the finger of God, but as we began to understand the causes of lightning, it became less and less reasonable to call it a miracle. Today there are many events, especially in the area of healing, which it might once have been reasonable to think of as miracles, but which are now understood in terms of the body's natural responses.
C. One concept of God, called "the god of the gaps," assigns anything we don't understand to God. When something happens that we do not understand, and which there is no particular reason to assign to God, it would be more honest to say "I don't know" than to say a miracle has occurred. But we humans are constructed so as to assign reasons—even highly improbable reasons—before admitting ignorance.
D. Events which flow as a natural result of psychosomatic suggestion are not miracles. When the crowd tried to praise Jesus for his miracles, he often replied "Your faith has made you whole." Today it is clear that the power of suggestion is very strong in healing, though the so-called "placebo effect" is still poorly understood. Most doctors and healers are all too willing to take credit for "miracle cures," when something natural is undoubtably happening.
E. Similarly, self-limiting illnesses (estimated to include 90% of all illnesses) are not cured miraculously just because one finally goes to a doctor or a healer immediately before self-limiting mechanisms begin to operate.
F. The process of creating legends must be distinguished from miracle. Just as we assign amazing deeds to Billy the Kid that he could not possibly have done, so we assign events to famous religious characters that they did not do. I once participated in this process myself. A wonderful old lady in our church died, and immediately the story went out that she died in prayer beside her bed; I repeated the story twice before her honest and feisty sister set the record straight: "Bessie did not die in prayer. She fell dead on the way to bed coming back from the bathroom." Of course that is a mild example of a well-documented human tendency, one that easily produces cowboy legends, stories in scripture, and artifacts like the Shroud of Turin.
G. Stories that make clear on their face that they are allegory or parable, are not miracles. There should also be a presumption against miracle when the story was *probably* meant as parable, as in the stories of Job and Jonah and Balaam's talking ass.

H. Stories should not be considered miracles if they were written whe authors had a different view of their ethical duty than we do now Nowadays a writer who claimed to report a miracle would b considered unethical if he or she wrote that a miracle had happene knowing that it hadn't. In the first century, however, writers did no have that sense of their duty. Reported miracles were not factua claims so much as signals of the story's importance. Virgin births for example, were a well-known convention for identifying th central character, and dozens of them exist in ancient literature.[1] To hold those writers to modern standards would be like accusin a funeral speaker of lying who merely performed the duty of sayin something nice about the deceased. It might technically be a lie, bu not if one understands the context and makes appropriate allow ances.

   Incidentally, newspaper reporters even today have a peculia notion of their duty, which inclines a reader to believe miraculou events have happened when they haven't. Reporters say their dut is to report what someone *said* happened, not what did happen. S if two young farmers claim to see a spaceship or a tourist claims t see Jesus in a tortilla, the statements are duly reported, and a gullibl public thinks the newspaper stands behind the reports.

I. Misinterpretations or mistranslations of words may sometimes giv rise to a miracle report when none was intended, or when the write purposely chooses an equivocal word. If "almah" meant *youn woman* instead of *virgin* as many scholars believe, the majo miracle of the New Testament hangs on a mistranslation.[13] Like wise if a writer calls a vivid dream or insight a "vision," we shoul not automatically assume that a miracle has even been claimed much less happened. We will later see how concepts like "grace" were originally meant to refer to something natural, before the were romanticized into supernatural claims.

J. An obvious category concerns fraudulent and mistaken claims o miracle. It is easy for speakers or writers to rationalize the ethic of making miraculous claims loosely, especially if such claim support a message that seems critically important. Unfortunately we tend to believe such charlatans, especially if they are charming

K. A more subtle exception to the category of miracles concern reports that arose before we had any real concept of order. In th earliest days of recorded history there was no such concept, becaus *everything* seemed arbitrary and without natural cause. Thus to tal about violations of an order that was not perceived is to impose late conceptions on earlier observations. Similarly, a good understand ing of chance arose only recently, and thus many earlier reports o miracle are better attributed to chance or randomness. Even today this creates problems; for example every mother worries about her child having a wreck on prom night, but those whose worries prove groundless don't have any later occasion to claim pre-knowledge. Some teenagers do die in prom-night wrecks, however, and it is usually impossible to tell the unfortunate mother that her "premonition" was shared by all mothers on that terrible night.

L. Finally, it should be obvious that an event is not a miracle just because everyone thinks it is. Most of us want to believe in miracles, perhaps because of our insecurity with life as it is. Further, we tend to repeat miracle stories, with "harmless" embellishments that give the story more plausibility. Indeed, research into "urban legends" shows that this is a common process, with each embellishment attributed to the person we heard it from, even when we add the embellishment ourselves.

When any claimed miracle is considered in light of these 12 categories, it will almost surely fit into one of them or some combination of them. And because the claim of a miracle is itself an extraordinary claim, we should hold such claims to a standard of extraordinary proof. We began to do that systematically only in 1976, when the Committee for the Scientific Investigation of Claims of the Paranormal started subjecting such claims to scientific scrutiny. Since then nobody has claimed the $500,000+ in prize monies that awaits the successful demonstrator of any violation of natural law. On the other hand, CSICOP has shown over and over again, most recently in China, that such claims are embarrassingly easy to disprove.[14]

The discounting of miraculous or supernatural causes not only forces modification of magical God-concepts rooted in antiquity, but it directly challenges other concepts that survive into modern times. Even sophisticated ideas that tried to conceptualize God in non-material terms (such as vitalism or dualism) have had to fall, because science can now explain in purely material terms the phenomena that were once explained by vitalism or dualism. This modern and profound development has been accepted by most forward-thinking theologians,[15] leading them to a productive rethinking of their concepts of God. In fact several satisfying concepts have resulted, which envision God in terms entirely natural, but which preserve many of the insights developed by earlier cultures.

Modern Concepts of God

Although dozens of modern concepts have been constructed in order to overcome the criticisms above, I will list only a few of them here. While one of them is radical in that it proposes to lay aside the concept of God entirely, the others conceptualize God in ways that retain important parts of older concepts, especially biblical concepts. These new concepts show that concept-modification is alive and well, and each of them contain seeds for a new and vibrant understanding of the experience we have traditionally called God.

1. *God as word.* Some theologians have focused on human language, to point out that "God" is a human word.[16] Like every word, it summarizes and represents experience, and permits efficient thought. The question is, which of our many experiences are summarized in the word "God"? If science has preëmpted expressions of vastness and biological design, are there other experiences summarized by "God"? Since language evolved precisely to permit sharing of experiences, including the deep and abstract experience traditionally expressed in the specialized language of religion, what might "God" mean?

Pursuing this analysis, some theologians look to the New Testament to support their concept of God as Love or Brotherhood,[17] saying "God" is a summary of those deep human experiences and other related ones, like joy and peace and patience. Or we can think of God in terms of another kind of human experience, namely truth or "the Spirit of Truth." That spirit is profaned by pious reliance on traditional understanding when it conflicts with unfolding knowledge. While such usages may lead to the comment that God is "merely love" or "merely truth," some theologians would snort "Merely? how can you think of such glorious and profound experiences as mere? Surely a God who is bound up with love or truth is as worthy as one who is only bound up with sky and earth."

This analysis of God as word allows Christian theologians to understand Christ in more natural ways. For instance they might say that if God is Love, Jesus was a man who so perfectly embodied love (and other related virtues) that it is poetically accurate to say that he and love are one, or that he is the son of God, or even that he and God are one. Loyal Rue takes a slightly different approach, by saying that Christ partakes in the *activity* of God, by the sacrificial choices he made.[18]

Another meaning of "God" was suggested by Emile Durkheim, the great French anthropologist/sociologist. He observed that religion always calls people to reaffirm the group, and that our worship of "God" is an elaborately disguised way of doing that.[19] This is not necessarily bad, because the group's experience and interpretation of experience is more comprehensive and reliable than that of any individual, and we need a way to get individuals to accept group wisdom. Since individuals tend to rely on their own experience rather than on the group's larger experience, and since that often causes harm to individuals and to society, constant reaffirmation of the group is necessary. Indeed it is one of the basic challenges of social life, and requires the full support of our heaviest theological word, namely God.

2. *The God Beyond Words.* A completely different concept sees God as beyond the realm of words, even though words may act as conduits to the experience beyond words. Joseph Campbell's studies of myth emphasize this concept of God, claiming that every culture is aiming at the same experience, whether expressed in dance or story or sophisticated theology. Each particular manifestation is a "mask of God," and the only real danger is intolerance between those who wear different masks.

3. *God as the Sum of Personality-creating Forces.* Another modern concept neatly combines the forces of physics with human psychology. Shailer Mathews (in the book cited earlier) calls God the "sum of the personality-creating forces in the universe." That vision of God allows physics and biology to operate in their normal spheres, and allows us to accept their descriptions of basic creative forces. Those forces undeniably produced humans and contribute to each human personality. Our duty and privilege is to better understand such forces, molding our lives to them in a joyous merging of ourselves with ultimate universal reality, which is an entirely physical reality. Naturally this requires particular rituals and stories that will vary from culture to culture and from age to age and even from person to person.[20]

Since physics has not yet adequately explained what happened before the Big Bang, some theologians embrace this mystery to say that we can define what happened as God. It is entirely possible that He somehow embedded Himself in the universe at the time of the Big Bang, and that He embedded himself into life when it emerged. Thus God is not only the sum of the personality-creative forces in the universe, but he operates within that universe in definite (though poorly understood) ways.[21]

4. *God as Process.* One group of theologians sees God not as word or beyond words, nor as a force *per se*. Instead they see God as process. This is a subtle conception,[22] but it is very interesting, and it is consistent with the early Old Testament, where God defined himself to Moses as "I am."[23]

Process theology (as this discipline is called, building on the "process philosophy" of Alfred North Whitehead[24]), is best articulated by Charles Hartshorne, Ian Barbour, and John B. Cobb Jr.[25] They and others see God as a process of becoming, as persuasion not commandment,[26] as aesthetic rather than formal.[27] They solve the supernatural problem by seeing God as so intermeshed with matter and energy that they are virtually the same thing. This leads them to the unusual claim that matter itself has feeling.

We see a rock as motionless only because eyes evolved to see larger motions than the atomic motions going on within the rock, but those motions are the precursors and the companions of much higher levels of feeling and existence. For process theologians there is no need for a beginning, because everything grows out of what was there before; each new event is the transformation of a previous situation. It is not so much that our concept of God changes as that God himself changes, as part of his own becoming.[28]

Despite its subtlety, process theology has certain advantages in the effort to create a workable concept of God. It reminds us of several important agreements between theology and science, namely: 1) there was a time in history when our universe did not exist, and a later time when it did; 2) there was a later time when the universe existed but the earth did not, after which the earth came into being; 3) there was a yet later time when life did not exist, after which it came into being; and 4) there was still a later time when humans did not exist, after which they did. Christian thinkers go further, to say 5) there was a time before which a higher Christ-consciousness did not exist, after which it did, the transformational power of which will not be exhausted until we heed its message.

One intriguing process theologian (though not always interpreted as such) goes so far as to predict a dramatic development in the far future. Pierre Teilhard de Chardin says evolution itself has an innate direction toward spirit, so that we humans will eventually evolve away from our material construction and toward the pure process that God is. Thus even the ultimate separation or compaction of molecules in our bodies cannot prevent our eventual merger with God.[29]

5. ***God as a historical concept.*** Finally, one group of theologians points out that some concepts have simply ceased to exist over time. Even though there are still a few kings and queens in the world, for example, it is fair to say that monarchy is a concept of the past. This leads them to claim it is time for us to accept a theology without a God-concept.[30]

Pursuant to this line of thinking, "God" was a profound product of the theological imagination, abstracting several ideas into one Godhead. The ideas summarized in religion and included in "God" varied from age to age, depending on which things could not be explained in more adequate ways. In earliest times that included almost everything.

More recently, the number of things that could not be explained in more adequate ways became fewer. As science grew, concepts that linked God with weather were modified, and so were those that linked Him with other natural events. In our age physics is on the verge of explaining all natural forces in a "unified field theory," and there will be no need for a God-concept to explain things better explained by science.

Since theology means "the study of God," it seems strange to think of a theology without God. But suppose the study of government had begun when there was no government besides kings, and the study had been named "Monarchology." It might have happened that, even after monarchs disappeared, the study of government would still be called monarchology. Similarly, we could call a study of deep human experience theology, even without a concept of God. It would be a sub-discipline of psychology.[31]

A Summary of Part Two: What God-Concept Does Biotheology Accept?

Biotheology's effort is to understand. It is clear that humans are symbol-users, and as such we need language and concepts with which to express our deepest experience. But traditional concepts for expressing deep experience have become profoundly unsatisfactory, and thus we must applaud modern efforts to find new symbols.

One can surely understand claims that the God-word is a *synonym* for other words and phrases aiming at the same experience (such as love or community or the spirit of truth), but it does not appear likely that we can adequately express the full range of theology's traditional concerns without retaining a concept of God. Granted, some parts of theology will probably merge into psychology within the next century, but it would be an unwarranted extension of that probability to try operating without any concept of God at all.

There are two reasons for this conclusion, one cultural and one developmental. Culturally, it seems impossible for masses of people to give the time and energy necessary to consider radical modification of concepts, and thus it is natural and right for our age to make those slow changes that seem called for by the intellectual currents of our time, without trying to foresee what further changes may be required in the distant future.

Second and perhaps more important, children in our species (as in other vertebrates) learn from role-models. As a matter of developmental psychology, their role-model concepts are quite concrete.[32] This may be the biological source of hero-worship, especially the kind that symbolizes God in human form. Depending on the particular products of theological imagination, such role-models may be entirely human (Confucius or Moses), entirely divine (Zeus), or mixed (Christ, Buddha). Recent research shows that this concrete phase of biopsychological development may last well into young adulthood, which helps explain why so many concepts of God are so concrete. It also helps explain why successive concepts of God may be necessary over the course of an individual's life, in addition to the larger succession of concepts that occur over the life of a culture.

These reasons for keeping a God-concept are not sufficient to reject any changes in it, however, and we have already seen why some modifications are necessary. That leads biotheology to consider other conceptions of God, and to affirm the first four modern concepts listed above. They are all consistent with our nature as symbol-users, and with our culture's recent growth toward scientific ways of thinking.

Of the four, the one that underlies the following chapters is the first one, which sees God as a word summarizing our deep psychological experience, including not only love and truth but also patience, joy, peace, and justice. Traditional biblical doctrines (like sin, salvation, mercy, and grace) have particular relevance to that God-concept, and a biologically-based psychology relates human experience to those doctrines in quite productive ways, which the coming chapters will explore.

This new concept of God permits us to recall Part One and summarize Part Two by presenting an outline of biotheology. In a sentence, biotheology synthesizes biology and theology by deriving the study of God from the study of life, but a longer summary will require several paragraphs, as follows:

1. We are biological beings. We evolved from simple beginnings to become the possessors of bodies much like those of other animals. As a direct result of that history, the human brain is fundamentally similar to other ape brains, although several significant differences evolved between our brain and those brains.
2. These evolutionary differences permitted new mental operations and modified old ones, especially when combined with changes in the mouth and throat and hands. Logic and rationality became possible, language allowed the sharing and preservation of insights produced by imagination, and social life coalesced into culture.
3. Culture is a powerful form of the sociality which preceded humanity. It is produced and developed by the acute human imagination, and preserved by institutions that persist over hundreds of generations. Among the most remarkable products of culture was religion, which organized and unified society through such means as ritual and myth and the codification of morality.
4. Religious concepts originally performed many functions: they explained mysteries, they established the human place within the world, they regulated human relationships. Since human imagination is so powerful and prolific, these functions were accomplished in many different ways in many different cultures.

5. Nonetheless, one concept proved particularly attractive for explaining mysteries and for regulating human actions in almost all cultures, and that was the concept of the supernatural. Life was too complex to explain naturally, so a variety of supernatural concepts were necessary.
6. At any given time and place religious concepts (including supernatural concepts) may appear solid and unyielding, but in fact they have been constantly modified as new ideas emerged and as new knowledge was extracted from nature. Sometimes those modifications were small and seemingly insignificant, and sometimes they were large. In the judeo-christian heritage the shift from polytheism to monotheism was profound, as were the shifts from an Old Testament to a New Testament, and from a monolithic Catholic church to a pluralistic multitude of churches and philosophies.
7. More recently, another modification has been taking place, because it has finally become possible to imagine natural explanations for almost everything. Neither science nor the humanities has achieved that goal in any absolute sense, but they have established knowledge and methods so reliable as to eliminate the requirement for supernatural explanation. Medicine and psychology and sociology are especially productive in exposing the natural dynamics that were previously explained supernaturally.[33] Thus biotheology rejects supernatural explanations.
8. It would be a mistake to think religion is dead, just because its supernatural concepts are being dismantled. On the contrary, biotheology presents an exciting and constructive *natural* worldview out of which grows a compelling and positive morality, all based on the modern concepts of God discussed above. A transition will be required, but theology has gone through such shifts before, and we are in the midst of a significant shift today. Chapter 7 introduced the *theological* part of that transition, and Chapter 16 will show how it impacts average people in our churches.[34]

This idea of ongoing modification of theological concepts is by far the most important point of these last four chapters. There are of course other modern concepts of God besides those listed above,[35] but I hope the discussion adequately emphasized how necessary and even desirable ongoing modification is. Like any other concept of God, Biotheology's concept will itself have to be modified over time and indeed by each individual, because experience is never static and because dynamic interactions within a culture always change the meaning of any given concept.

That brings up the question of whether our faith is utterly dependent on the culture we are born into and the individual experiences we have. Are we robots that live out our biological lives in a predictable and predetermined way, or do we have some say in the matter? The next chapter, which begins Part Three, will address that question.

Notes

[1] Larry Greenfield has pointed out (personal communication, May 10, 1995) that it is not so much that scripture tends to freeze concepts, as that *we* freeze concepts in our traditions about scripture. I hope Chapter 6 demonstrates my agreement with him on this point.

[2] For a listing of gods and religions see the 2 volumes of *Historia Religioniam*, ed. C. Jouco Bleeker and George Widengren (Leiden, the Netherlands: E.J. Brill Publishers, 1969 and 1971). The actual number of them is unknown, and in any case it would depend on how one distinguishes between closely-related gods. Probably the most balanced view is by A. Eustance Haydon, who says simply the "Gods are numberless . . . almost as numerous as the families and types of men." See pp. vii-viii of his *Biography of the Gods* (New York: The Macmillan Company, 1941). See also *Gods of the Earth*, by Michael Jordan (New York: Bantam Press, 1992).

[3] Chicago: The University of Chicago Press.

[4] For example John Fiske's influential *The Idea of God as Affected by Modern Knowledge* (New York: Houghton, Mifflin and Company, 1890). Incidentally, this wonderful volume contains a charming dedication which belies the caricature of philosophers as boring. Fiske writes "TO MY WIFE, in remembrance of the sweet Sunday morning under the apple tree on the hillside, when we two sat looking down into fairy woodland paths, and talked of the things since written in this little book, *I now dedicate it.*" For a more recent discussion which concentrates on the judeo-christian tradition see Chapter 4 (pp. 108-159, entitled "The Limits of Distinctiveness") of Loyal Rue's *Amythia* (Tuscaloosa, Alabama: The University of Alabama Press, 1989).

[5] Of course this isn't strictly true, because they interest anthropologists, historians of religion, and others. Moreover, some of these gods are currently being resurrected by certain environmentalist theologies.

[6] Although it is true that the "First Cause" God gained new popularity in the 19th Century, the concept itself originated with Aristotle, and can be found in several of the medieval theologians.

[7] This is well expressed by Paul Davies, professor of mathematical physics at the University of Adelaide, in his book *The Mind of God: The Scientific Basis for a Rational World* (Simon & Schuster, New York, 1992). In reporting on the work of James Hartle and Stephen W. Hawking (and on Chris Isham's theological comment on their work), he says (p.68) "Is it therefore correct to say that the

universe has *created itself*? The way I would rather express it is that the universe of space-time and matter is internally consistent and self-contained. Its existence does not require anything outside of it; specifically, no prime mover is needed."

[8] Some physicists talk as if matter and energy began with the big bang, and this has led to definitions of God as the cause of the big bang. See how Frank J. Tipler expands upon Pannenberg to create a sort of judgment day, in "The Omega Point as *Eschaton*: Answers to Pannenberg's Questions for Scientists" in Volume 24 Number 2 (June 1989) of *Zygon: Journal of Religion & Science* pp. 217-253, followed by Pannenberg's favorable response. More generally, however, physics means that *matter* as we know it began in the big bang, and the big bang "singularity" was preceded by other physical processes yet to be adequately articulated. Indeed, Tipler does not depart significantly from that understanding, and neither he nor Pannenberg explains where God was before the big bang, or where he/she (Tipler's designation) will be after the final collapse.

[9] This does not lead to nihilism, however. That my actions do not "ultimately" matter does not mean they do not matter at all. They matter to me, to my family, to my society and to the human race, as well as to other life-forms. For a biological treatment of these issues see Konrad Lorenz's "Inductive and Teleological Psychology," which also corrects a tendency in biology itself to over-emphasize purpose. Unfortunately even a first-rate biotheologian like Loyal Rue (who is familiar with Lorenz) misses the importance of the distinction between ultimate and ordinary purpose, as evidenced by the gentle nihilism embraced in his 1994 book *By the Grace of Guile: The Role of Deception in Natural History and Human Affairs* (New York: Oxford University Press).

[10] See the excellent summary in John Moore's essay "Understanding Nature—Form and Function" in *Science as a Way of Knowing, Volume V (Form and Function)*. Published by the American Society of Zoologists, University of California, Riverside, 1988.

[11] See *The Miracle Stories*, by Herman Hendrickx (San Francisco: Harper & Row, 1987); *The Miracle Stories of the Early Christian Tradition*, by Gerd Theissen (Edinburgh: T & T Clark, 1983); and *Miracles and the Critical Mind*, by Colin Brown (Grand Rapids, Michigan: William B. Eerdmans Publishing Company, 1984).

[12] For a balanced discussion of the arguments for and against the Virgin birth by a Christian scholar see *The Historical Evidence for the Virgin Birth*, by Vincent Taylor (Oxford at the Clarendon Press, 1920).

[13] This would certainly be consistent with the apparent fact that the virgin birth doctrine did not exist in apostolic preaching and was not a matter of public knowledge during the time period covered by the Pauline epistles. See p. 21 of *The Historical Evidence for the Virgin Birth*, cited above, p. 21.

[14] "Testing Psi Claims in China," by Paul Kurtz, James Alcock, Kendrick Frazier, Barry Karr, Philip J. Klass, and James Randi in *The Skeptical Inquirer* Vol. XII, No. 4 (Summer 1988), pp. 364-375.

[15] Nowadays, even treatments which claim to preserve a *spiritual* dimension often use the word to refer to entirely physical phenomena. For such a treatment, and for a recitation of authorities that concur on the death of vitalism and dualism, see Roger W. Sperry's "Search for Beliefs to Live by Consistent with Science" in *Zygon: Journal of Religion & Science*, Vol. 26, No. 2 (June, 1991), pp. 237-258.

[16] See *Patterns of Grace: Human Experience as Word of God*, by Tom F. Driver (San Francisco: Harper & Row, 1977). For the related idea of God as metaphor see Chapter 7 (pp. 153-184, discussing John Dunne, James McClendon, and Sallie McFague) of Lonnie D. Kliever's *The Shattered Spectrum* (Atlanta: John Knox Press, 1981). See also *Linguistics and Theology: The Significance of Noam Chomsky for Theological Construction*, by Irene Lawrence (London: The American Theological Library Association, 1980) and *Metaphorical Theology: Models of God in Religious Language*, by Sallie McFague (Philadelphia: Fortress Press, 1982). An earlier book by McFague, published under the name Sallie McFague TeSelle, is also helpful, namely *Speaking in Parables: A Study of Metaphor in Theology* (Philadelphia: Fortress Press, 1975).

[17] For a particularly good example, see pp. 60-68 of John Studdert Kennedy's *I Believe: Sermons on the Apostles' Creed* (New York: George H. Doran Company, 1921). Incidentally, this delightful book and its companion *Lies!* (London: Hodder and Stoughton Ltd., 1919) written by a British chaplain and theologian, deserve to be re-published, being even more relevant today than when originally published.

[18] For more discussion on the role of Christ in God-concepts see *The Myth of God Incarnate*, ed. John Hick (Philadelphia: The Westminster Press, 1977); *The Debate about Christ*, by Don Cupitt (London: SCM Press, Ltd., 1979); and *Incarnation and Myth: The Debate Continued*, ed. Michael Goulder (William B. Eerdman's Publishing Company; Grand Rapids, Michigan: 1979).

[19] See p. 387 of *The Elementary Forms of the Religious Life*, trans. by Joseph Ward Swain [Glencoe, Ill.: The Free Press, 1968 (original copyright 1915 by George Allen & Unwin Ltd.)]. Incidentally, Loyal Rue's discussion of the early jewish covenant (at pages 108-134 of *Amythia*) resonates with Durkheim's concept, and the covenant is clearly a mechanism for consolidating the group, if not a symbol of the group itself.

[20] Perhaps the most current formulation of this idea is that of Lindon Eaves and Lora Gross, who define "spirit" in terms consistent with biological genetics. See their article "Exploring the Concept of Spirit as Model for the God-World Relationship in the Age of Genetics" in *Zygon: Journal of Religion & Science* Volume 27, Number 3 (September 1992), pp. 261-285.

[21] For a presentation of the anthropic principle (which holds that initial conditions of the universe are fine-tuned for intelligent life) see Richard Swinburne's "The argument from the fine-tuning of the universe" in *Physical Cosmology and Philosophy* ed. John Leslie (New York: Macmillan, 1990). For a criticism of the argument see Quentin Smith's "The Anthropic Coincidences, Evil and the Disconfirmation of Theism" in *Religious Studies* Vol. 28 No. 3 (September 1992),

pp. 347-50. Another concept of God, which might be considered a compromise between process theology and these physics-based theologies, is that of the mathematician John C. Polkinghorne, in which God is to be found in the epistemological uncertain but (perhaps) ontologically determined nature of reality as seen from a top-down perspective, as opposed to the usual bottom-up analysis of other science-based theologies. See his *Science and Providence* (London: SPCK, 1989) and the technical but lucid presentation in his "The Nature of Physical Reality" pp. 221-236 in *Zygon: Journal of Religion & Science*, Vol. 26 No. 2 (June 1991).

[22] Loyal Rue implies an opinion that process theology is *too* subtle, when he says it abstracts the personality of God beyond recognition (at p. 95 of *Amythia*). For a good criticism of process theology in general and Ian Barbour in particular see Holmes Rolston III's "Religion in an Age of Science; Metaphysics in an Era of History" in Vol. 27 no. 1 (March 1992) of *Zygon: Journal of Religion & Science*, pp. 65-87.

[23] Exodus 3:14. Note that this scripture also serves as an early forerunner of existentialism.

[24] *Process and Reality*, ed. David Ray Griffin and Donald W. Sherburne (New York: The Free Press, 1978).

[25] See *Hartshorne, Process Philosophy, and Theology*, ed. by Robert Kane and Stephen H. Phillips (Albany: State University of New York, 1989); Barbour's *Religion in an Age of Science: The Gifford Lectures, 1989-91* Vol. 1 (San Francisco: Harper and Row, 1990); and Cobb's *A Christian Natural Theology* (Philadelphia: Westminster Press, 1965). See also *The Re-enchantment of Science: Postmodern Proposals*, ed. David Ray Griffin (Albany: State University of New York Press, 1980).

[26] See p. 19 of David Ray Griffin's "Charles Hartshorne's Postmodern Philosophy" included in *Hartshorne, Process Philosophy, and Theology*, ed. Robert Kane and Stephen H. Phillips.

[27] See Charles Hartshorne's *Creative Synthesis and Philosophic Method* (Lanham, Maryland: University Press of America, 1983), especially pp. 303-321.

[28] I do not mean to imply that all process theologians see these issues in the same way. They don't, and neither does their thought accord exactly with other similar theologians discussed in this chapter. For an effort at harmonizing one of the basic issues see Daniel A. Dombrowski's "Must a Perfect Being Be Immutable?" at pp. 91-111 in *Hartshorne, Process Theology, and Theology*, ed. Robert Kane and Stephen H. Phillips.

[29] Teilhard was a Catholic priest well-known as a paleontologist. The church denied his requests to publish his theological ideas, so they were not subjected to criticism during his life. His ideas seem quaintly mystical today, and no disciples have carried his thought much further, in my opinion. But see *Interpreting Evolution: Darwin & Teilhard de Chardin*, by James Bix (Buffalo, New York: Prometheus Books, 1991); and the entire March 1995 issue (Vol. 30 no. 1) of *Zygon*, especially the article by Karl Schmitz-Moormann entitled "The Future of

Teilhardian Theology." See also Teilhard's *The Phenomenon of Man* (New York Harper & Brothers Publishers, 1959).

[30] Raschke, Carl A. *The Alchemy of the Word: Language and the End o Theology* (Missoula, Montana: Scholars' Press, 1979). See also *Radical Theolog and The Death of God*, by Thomas J.J. Altizer and William Hamilton (New York The Bobbs-Merrill Company, Inc., 1966).

[31] A related idea emphasizes the possibility that the God concept resides i specially evolved brain pathways that may have conferred a survival advantag whether they were true or not. For a flavor of this argument see Arnold J. Mandell' "Toward a Psychobiology of Transcendence: God in the Brain," Chapter 14 (pp 379-464) of *The Psychobiology of Consciousness*, ed. Julian M. Davidson an Richard J. Davidson (New York: Plenum Press, 1980).

[32] In fact they can't understand psychological or non-physical metaphors unti just before adolescence. See "The Child is Father to the Metaphor," by Howar Gardner and Elfen Winner, in *Psychology Today* Vol. 12 No. 12 (May '79) pp. 81-91.

[33] For an excellent essay on the secularization of religious functions see Emil Durkheim's "The False Dichotomy of Real and Ideal," pp. 205-210 of *Sociolog and Religion: A Book of Readings*, edited by Norman Birnbaum and Gertru Lenzer (Englewood Cliffs, New Jersey: Prentice-Hall 1969). Pages 209-10 ar especially relevant.

[34] Incidentally, I consider the progression above as entirely consistent wit Loyal Rue's progression from biological adaptivity to brain, from brain t metaphor, from metaphor to "root" metaphor, from root metaphor to broad-base myth, and from myth to culture (p. 31 in *Amythia*). I also hope it is clear that th progression above renders biotheology anything but reductionistic, since eac new stage *adds* something instead of reducing the later stage to the earlier.

[35] See for example Sallie McFague's 1987 book *Models of God: Theology fo an Ecological Nuclear Age*. Note also that while I have not discussed existential-ism, it shares concepts and roots with process theology. For example both it an process theology appear to be rooted in Henri Bergson. See "Henri Bergson Creative Evolution," pp. 167-172 (Chapter 4) in James Bix's *Interpreting Evolu-tion: Darwin & Teilhard de Chardin.*

Part Three: Biotheology in Individual Life

Introduction to Part Three

As we enter Part Three it may be good to review where we have been and where we are going. Part One laid a biological foundation, in four chapters. The first chapter gave an overview of the whole scheme of evolution, underlining four of its aspects, namely variety and commonality, jury-rigging, trade-offs, and niches. The next two chapters began to focus on humans by discussing the roles of thought and language and culture, with particular emphasis on imagination and learning. Chapter Four wove those ideas into a discussion of concepts, noting that they are implanted or formulated in individual neural pathways from early childhood forward.

Part Two related concepts to theology, again in four chapters. Chapter Five examined the history of specific judeo-christian concepts as a product of human theological imagination. Chapter Six discussed scripture, showing how it transmits cultural ideas, and how it supports idolatry when we refuse to modify concepts. Chapter Seven discussed belief, explaining how personal faith grows out of cultural concepts, and how we can evaluate beliefs. Finally, Chapter Eight explored concepts of God, and language's function in expressing whatever God-concepts our theological imaginations can accept. After considering several possible meanings of the word "God," some were rejected as unacceptable to modern theological imaginations, and several meanings were affirmed.

Part Three will now look at additional theological concepts, usually concluding that our traditions accord well with modern biological knowledge. Yet I would guess that you are feeling some discomfort about the direction we are going, even if you accept the reasoning so far. You may be asking whether it is possible to fill the gap left by removing the old supernatural understandings of "God" with naturalistic meanings. I think the answer is yes, but before you can consider whether to make that your answer too, you must understand how Biotheology paints in the background of its God-concept. Just as traditional theology filled out its picture of God with various concepts called doctrines, Biotheology also has subsidiary concepts which support and extend the ideas already presented.

Biotheology's concepts have the same names as the old doctrines—free will, prayer, salvation, etc. The subject matter of each is similar to traditional theology, though there are some significant differences, as you saw in the chapter on God. Part Three will present concepts that are particularly relevant to the fact that we live our lives as individuals. Part Four will present concepts arising from group life. In each part we will be looking for a *payoff*, for practical and useful applications of the principles established so far. Without such applications, theology become a dry and useless philosophy; it merely promises "pie in the sky by and by" or serves as the opiate of the masses.

This notion of payoff tends to blur the classical distinction between the two major categories of religion, namely "natural" religion and "revealed" religion. The theologian George Lindbeck gives a modern form to this distinction by pointing out that one kind of religion describes its underlying beliefs in *propositional* terms, while the other kind describes its beliefs in *experiential* terms.[1] While that division may give some analytical clarity, the payoff concept shows how misleading it can be, at least in thinking about biotheology. That is because biotheology is both; it is a foundational or propositional theology grounded in experience—evolutionary, cultural, and individual. Nobody has seen the interplay of these two components more clearly than the theologian Loyal Rue, who says theology must be grounded in a plausible cosmology out of which a useful morality emerges.[2]

Part Three describes the payoff in our individual lives, beginning in Chapter Nine by explaining how the evolution of judgment produced the striking freedom one finds in humans. Then we will explore some common mistakes and innate tendencies that often lead us astray, in a chapter called "Sin and Morality." Chapter Eleven will recall our earlier discussion of memory pathways, and show how they can be altered in positive ways to confer a basic security—that chapter is called "Salvation." Finally, Chapter Twelve will discuss whether we can extend basic security to produce abundant living and genuine happiness.

That sounds ambitious, but if natural ways of thinking theologically can lead to wholeness and peacefulness and joy, both in experiential terms and in useful propositional terms, these chapters will be worth the effort. Let us begin by asking what free will might be.

Notes

[1] See p. 57 of his *The Nature of Doctrine* (Philadelphia: The Westminster Press, 1984).

[2] See especially pp. 149 and 178 of his 1989 book *Amythia*, cited earlier.

Chapter Nine

Motivation and Free Will

In the years after Darwin, philosophers tended to deny the influence of evolution on spiritual matters,[1] and a few still deny it today. They believe evolution removes any affirmation of spirit. It implies that humans have no free will, and are merely complex robots. Since daily experience persuades them we are not robots, they cannot embrace a philosophy which implies that we are. Other philosophers go to the opposite extreme, arguing that every choice is indeed determined by forces outside our control.

Such arguments are no longer necessary, because a fuller knowledge of biology explains the middle ground between radical free will and absolute determinism, and shows that choice is completely consistent with our evolved nature.[2] And although choices are invariably limited within boundaries, it is not always clear where those boundaries lie; thus it is frequently productive to push against apparent boundaries. Mental pathways are complex enough to support the widest range of choice and freedom imaginable, if it is understood that "free" has never meant absolutely free, either in biology or in classical theology. So even though internal and social pressures push us in certain directions, our mental and emotional makeup allows us to avoid the outside control that characterizes a robot, if we wish to avoid it.

This chapter will establish these principles by presenting free will as a consequence of human biological evolution. It will begin by examining motivational processes, showing how choice and free will grew out of them.

The Evolution of Motivation

Something makes every living thing do the things it does. Something inside an amoeba makes it seek nutritious chemicals, and avoid dangerous ones. A firefly pursues a mate because internal chemistry makes it do so in response to light-blinking patterns. Decreasing temperature or daylength makes birds begin to fret, and finally begin their annual fall migrations.

At these lower levels of organic organization, can there be any doubt that creatures are indeed like robots? The "something" that makes them act is a function of their material construction, itself a result of evolution. Konrad Lorenz has argued that insects, and even birds (except perhaps crows and parrots), are entirely robotic. By that term he means a given stimulus will invariably produce the same response in these creatures.[3]

Such responses developed through normal trial-and-error evolution as a characteristic of particular chemical and genetic arrangements. The "purpose" of such internal motivating processes is to fill ordinary needs of the animal or plant, such as the need for food or shelter or sexual gratification.[4] Any species that evolved genes and neurons which did not propel its members toward such goals, perished. Thus every living being has internal processes which give it "motivation." And even though motivation can be interfered with experimentally or naturally, it is a fundamental characteristic of all life.[5]

These internal forces have been called by various names. "Drive" is often used, but one writer says "draw" would be a better word than drive, because even though the responses are primarily internal, they respond to external enticements in the environment. But whether called drives, draws, goals, or motivation, we now have a better idea of what is actually happening. Evolved internal processes get input from the outside environment, and respond to that input with a redirection of energy. That happens at every level of life, from the cellular to the cognitive. When motivation fails, it is almost always a sign of pathology at one of these levels.

During primate evolution, motivation became more complex as the nervous system became more complex.[6] The motivation to establish alliances, for example, is found far down the chain of life, but only in primates do alliances become very complicated. The same is true of other behaviors like deceit, which seems charming in a killdeer doing its "broken wing" trick, but which can be murderous in chimpanzee or langur monkey troops.[7] And although many animals seek "pure" knowledge by exploring the environment, only humans have a brain complex enough to motivate research programs or general explorations like "the search for truth."

One result of human complexity is that we are conscious of our motivations and goals; we can think about them and work them into our belief systems. Yet we are just beginning to understand two things about human motivation that we didn't know before. One is how much it depends on internal chemistry. A bit too much or little of any number of chemicals, alters our motivations and resulting behaviors, making us depressed or overly aggressive, or even suicidal. Second, our internal processes can be altered by specific experiences, both in early life and later. These insights give the opportunity to modify our own motivations, not only by changing the internal environment medically, but also by manipulating our experiences in positive ways.

The idea of motivation has been used to support a "biology of the spirit," building on the observation that all life is goal-directed.[8] That might be too simple an approach, but by adding the concept of judgment to motivation, we should be in a good position to understand the interplay between biology and spiritual life.

The Evolution of Judgment

Imagine an amoeba in a tank, just getting hungry. Suppose we drop in a bit of nutritious fluid, and watch it swim toward the drop. Now imagine instead that we drop in poison, and watch the amoeba swim away. What would happen if the drop contained only a small amount of poison, and a larger amount of nutrient?

Not having done or read an actual experiment, I don't know the answer, but presumably one proportion would motivate the amoeba to swim toward the drop, and another proportion would cause it to swim away. In between there might be some range that would cause the amoeba something like frustration, but I doubt it. Instead, lower organisms evolved built-in thresholds, on one side of which they respond one way, and on the other side of which they respond another way. With an adequate research grant one could trace the internal chemistry involved in such innate responses.[9]

With more brain power, higher animals have a wider range of responses to middle ranges of stimulus, and they have developed many feedback mechanisms to process subtle clues from those stimuli.[10] Typically these mechanisms interact in predictable ways, and lead their possessors to act in characteristic patterns. To jump over simpler examples, consider a troop of baboons or monkeys raiding crops. Lookouts are posted, and directly transmit information to the raiders about the location of farmers or dogs; the raiders decide whether to keep raiding based on the information conveyed by the lookout. This shows an example of full-blown judgment in a "lower" animal, and gives insight into the evolution of human judgment.[11]

Human feedback mechanisms evolved to become the most elaborate in nature, supported by a brain which is the most highly organized matter known.[12] Like other organisms we get feedback from the senses, but our complex brains enable us to interpret that feedback, to consider it, to learn and remember concepts, to cooperate culturally and otherwise, and in general to exercise judgment to a greater extent than other animals do.

"Judgment" summarizes our complex feedback mechanisms into one word, but it does not completely capture the complexity and range of consideration and decision typically experienced by humans. For one thing it does not include the active seeking we do as part of our innate curiosity, and for another it may connote the "intuitive" parts of decision-making without adequate appreciation of our analytical faculties. Further, it may not reflect the role of culture in decision-making, and neither does it reveal the subtle necessity of choosing between conflicting internal urges within us (more on that in a moment).

"Free will" and "choice" are concepts which attempt to overcome these semantic limitations, and throughout their ragged philosophical history various writers have broadened them to include such mental processes as imagination and rationality and culture. Much of that history shows how inadequate words are for conveying the complexity of mental phenomena, but by focusing on judgment and choice, biology has breathed new understanding into the human experiences referred to by these concepts. By considering free will as part of our evolution, its effects in our individual lives can be seen more clearly.

As judgment and choice evolved along with the complex brain, they conferred greater and greater flexibility in dealing with problems presented by earth's environments, and allowed us to populate more and more of those environments. That is not to say that the evolution of free will made life any simpler, however, or freed us from the internal makeup our history had produced. On the contrary, flexibility slowly worked itself into our very genes[13] in a way that complicates our lives, because it produces internal conflicts in our motivational structure. Let me elaborate on that point, because it is often overlooked:

There is a popular tendency to think of evolution as producing neatly unified organisms, but the facts show otherwise. Instead, our jury-rigged construction produced a large library of potential behaviors within us, just as it evolved a library of physical possibilities like differing eye color or disease-susceptibility. In cases like eye color no choice is required and no conflict arises, but in the case of behavior, internal conflicts often arise which must be resolved by judgment and decision. Moreover, evolution's provision for addressing these conflicts is somewhat awkward, requiring us to establish and maintain certain thought processes. Thus from a biological perspective, the difficulty of making good choices begins deep within our motivational structure, where conflicts reflect the variety of our historical experience.

We have not yet discussed the surprising fact that evolution produces inconsistent tendencies within a single individual, though earlier chapters did reveal some dynamic tensions within groups. In discussing culture, for example, we saw how society promotes both competition and cooperation, and how both exist in dynamic tension within each group. I implied that such conflicts exist within individuals as well as groups, and this section will discuss one such tension to illustrate the reality and power of internal inconsistencies. The example is a sexual one, namely our dual drives toward fidelity and promiscuity.[14]

Both sexual patterns can be observed in the animal kingdom, but within a particular species one pattern or the other is usually dominant. Wolves, for example, are faithful to their mates, whereas domestic dogs are promiscuous. Both patterns are adaptive: fidelity permits the pair's energy to be focused on survival, and not wasted on intraspecies rivalries; promiscuity ensures that every female will have as many offspring as she can, from the strongest suitors. The wolf's harsh historical environment may have put a premium on behavior that saved energy, whereas the lushness of the dog's environment may have favored breeding extravagance. Biologically, each pattern exists as part of individually inherited neural machinery.

Since closely related species often have opposite sexual tendencies,[15] and since there are more individual violations of species patterns than was once thought,[16] those behaviors are apparently altered quite easily at genetic levels. These complexities are illustrated by the greylag goose, and we can get some insight into innate moral conflicts by looking at its life. The goose mates for life and is usually faithful, but the gander occasionally engages in extramarital affairs. When both mate and paramour are present, he feigns ignorance of the paramour, and will not even defend her if she is attacked in his presence (he defends his mate to the death). The goose evidences other sexual practices similar to those found in humans,[17] making it a good candidate for further genetic, neurological, and psychological research.[18]

Geneticists or neurologists may eventually discover the mechanisms that lead to such responses in the goose. In the meantime most biologists would assume the development is in some respect adaptive, gaining some advantages of both fidelity and infidelity. Another possibility is that the extramarital behavior is pathological, a corruption of fidelity which causes harm to geese that engage in "illicit" alliances. Yet another possibility is that nature is conducting another trial-and-error experiment. If so, one behavior will eventually win out in the goose (implying that the behaviors are in a transitional phase), or else both behaviors will be retained in the genetic pool.

One approach that is *not* possible is for the goose to consider whether its behavior is harmful or helpful, or to entertain persuasive arguments against its behavior. Its brain is too simple for that, and it presumably responds to robot-like impulses in yielding to both lawful and illicit attractions. Sometimes human sexual responses seem robot-like too, but since humans also evolved rational abilities, responding to such attractions is more complex for us. Our sexual attractions no doubt have a non-rational genetically-based component. But because we also have rational abilities, we must *decide* whether to have affairs, whether to break them off, and whether to warn young people about their dangers.

In other words genetics may have—probably does have—an influence on our choices, but the primary relevance of genetics for this chapter is that it produces an elegant brain which can consider a variety of factors and make decisions based on them. Indeed, the genes themselves could not contain enough information to determine all the details of our behavior; instead the genes produce the brain, and that organ then considers evidence and makes choices. Thus if one decides to have an affair, it will not do to say "I must just have a genetic tendency in that direction." While all of us are propelled toward our decisions by various forces, we are also subject to the internal processes of choice and judgment. Our available choices usually range over a very wide spectrum, and the one we feel initially inclined toward is only one among many.

Since rational faculties evolved later than more robotic emotional responses, there is a sense in which rational choices don't come as "naturally" to us. Said another way, the nature of rationality is that it takes effort to develop and maintain it. Despite this difficulty, judgment has evolved to become a full partner with emotional responses in humans, and we can only fill our niche by exercising choice and judgment, in response to whatever internal urgings and external circumstances we face.

I do not mean to imply that judgment should always lead us to resist internal urgings. On the contrary, it may be rational to pursue such impulses. All judgment requires is the thoughtful consideration of alternative choices. The next chapter will develop that theme further by considering other conflicts built into human nature, such as the conflict between selfishness and selflessness. There we will explore choices further, to see whether some of them can be called sin. For the purposes of this chapter, however, the sexual example is only meant to underline the reality of choice in our species. For us, these sexual options and many other options evolved along with our brains. We became one of the few species which developed the ability to consider options, and that ability contributed greatly to our success. Yet the ability to choose comes with trade-offs, its most significant disadvantage being the danger of choosing wrongly. That too will be discussed in the next chapter, but for now some remarks must be made about the development of judgment within our individual brains, and not just as a characteristic of our species.

The Neurobiology of Free Will

The paragraphs above did not establish all the details of judgment's evolution,[19] but they should make us aware of how impressive an accomplishment it is for an individual to navigate through life's various decisions. We must each deal with the innate or genetic tendencies we inherit, whether we understand the details of that inheritance or not. Then we must incorporate or reject major and minor aspects of our culture and individual experience. With so much to do, it is amazing that we ever make good judgments, and it is much easier to understand how we manage to make so many bad ones.

Although judgment permits us to incorporate complexity into our lives, evolution also propelled us towards mechanisms which simplify some choices. We gravitate toward group wisdom ("Look before you leap," "Haste makes waste") precisely because it gives us a starting place for making choices.[20] More generally, and as we saw in discussing faith, the ability to learn concepts allows each person to start from a broad cultural base, so that choices do not have to be made entirely from scratch.

Sections of earlier chapters may now be pulled together into a new concept, which may be called a "neurobiology of free will." As we saw in Chapter 7, our brains begin with innate mental pathways which are modified by culture and individual experience to produce a stable framework or worldview, out of which we make life's choices. That worldview (or faith) exists in a physical form in the neural and chemical pathways of the brain.

Our worldview constitutes a significant feedback loop on our free will, however. We do not make our choices in a vacuum, but within the context of our personal histories. Our personal framework is constructed before we arrive at particular choices, which are then made from the foundation provided by that framework. Each choice flows into specific channels as it moves toward decision, so that each choice is partly made before we become aware of it. A choice once made, modifies or strengthens the framework from which future choices will be made.

The impact of pre-existing mental pathways does not mean, however, that they entirely predetermine any choice, because the ability to make choices implies the ability to consider new data and ask additional questions, and to override the tendencies residing in existing pathways. We can train ourselves to the habit of seeking more data, and of considering more than one choice with the same evidence in the same situation. One writer sees the process as one where conscious decisions *supervene* other mental processes, and become causal agents in them.[21]

Moreover, knowing about pre-existing pathways does not take our choices away; it merely removes them one step backward. We must choose beliefs knowing they will influence future choices and behaviors. Indeed, the master-choice of which basic belief-structure to embrace may be the most important decision we make, and is surely one of the most important.

To underline how important the choice of a belief-structure is, consider the probable results of three scenarios. If one accepts the cultural faith with little modification, it may give a great sense of belonging and gain other advantages, but at the cost of corresponding disadvantages. Second, if one chooses an unusual faith, individual integrity might be enhanced, and new insights might be gained, again with associated disadvantages. Finally, if one accepts biotheology's recommended path of beginning with the cultural faith and modifying it in light of evidence, that choice will also have significant advantages and disadvantages. And these three choices are merely suggestive, because the full range of choices is wide enough to include elements of all three of them. One may pull threads from each to construct a tapestry of belief and action more individualized than a fingerprint, though a too highly individualized worldview may offer nothing in the way of sociality. The best choice of all, if it were possible, would be to understand one's complete heritage (genetic, cultural and experiential) and the tendencies it produces, select an existing faith that provides a sense of community and belonging consistent with that heritage, and begin to modify the chosen faith as needed to achieve and maintain personal identity and integrity.

The Limits of Choice

The fact that motivation, judgment, choice, and free will exist as neural processes in each individual, should pave the way for accepting personal responsibility for our choices. Chapters 10-12 will develop that theme further, but first we must modify what has been said about the range of choice available.

That range is indeed wide, and includes not only the choice of a faith nd a whole catalog of moral choices, but also a dozen practical choices day, any one of which may turn out to be important. Most choices though certainly not all) have an effect on what happens to us, just as the hoices of other judgment-exercising animals have an effect on what appens to them. That means we can affect our own destinies to a ignificant extent by making good choices. In humans this fact may not eem obvious, because we often make choices based on difficult judg- nents about long-range consequences. Yet despite this difficulty, our ational faculties let us include the probability of various consequences in ur calculations, so that it is at least *possible* to make choices which roduce the results we want. How to do that will be addressed in Chapter 2, and what constitutes a "good" choice will be discussed in Chapter 10.

But choice is not a panacea. Like language or thought or conceptual bilities or other wonderful aspects of our nature, it is after all only one kind f human behavioral pattern, which evolved because it had more advan- ages than disadvantages, and not because it was limitless. Thus even hough there are significant connections between our choices and the esults we get, it must not be assumed that our choices absolutely or nevitably determine those results. Our freedom is limited by many ynamics, both inside and outside our heads.

I will briefly discuss three kinds of limitations on our choices, related o a) the nature of our brains; b) outside forces which conflict with our esires, including physical impossibility; and c) the large number of vailable choices.

In the first category, we obviously cannot choose to do things or think hings that require a different kind of brain than we have. Thus, if it turns ut that we have mental pathways that incline us toward choosing certain oncepts or certain actions, that limitation must be incorporated into our heology. Exactly where the limit lies would be exceedingly difficult to stablish, but it undoubtedly exists both for the species as a whole and for ndividuals within the species, and I mention this limitation to account for uture research in the field called evolutionary epistemology.[22]

One limitation to our freedom originating in the structure of the brain s that choices are conditioned by "perceptual filters." As language-users ur perceptions are markedly dependent on symbolic and cultural inter- retations.[23] Among other things this means we often misperceive the hoices available to us. In counseling, science, or other problem-solving ituations, it is notoriously difficult to help people see how wide the range f choices is. It is common to hear them say "I have no choice," when they ave many choices. Our later discussion of happiness will consider ways o reduce the limitations of perceptual filtering, by focusing on dynamics that offer a statistical edge in the search for happiness, and keep us from focusing on immediate pressures or attractions.

Besides the limits arising from the nature of our brains, a second set of limits arises from the fact that our choices can conflict with forces external to them. Included in this category are physical constraints, legal constraints, and the choices of other people. Thus the existence of choice does not mean we always get what we want. I have the choice to pursue either A or B, and I make the choice in order to bring about desired results, but other forces besides my choice may intervene to keep the result from being what I want it to be. If I want to buy a lamp at an auction, I can choose to bid a certain amount, but if someone else chooses to bid more, their choice will limit mine.

Physical impossibility is an obvious part of this category. Just as we cannot fly without mechanical aids, neither can we do other magical things. Unfortunately, that does not stop people from choosing concepts which depend on impossibilities. For example people continue to believe in bizarre health practices that shun medical science; or in the concept that planetary positions at conception influence our daily lives; or that numerological analysis of the scriptures can tell when the earth will end; or that fire or snakes can be routinely handled without harm; or that human bodies can spontaneously ignite with no source of fire; or that we have a non-material spirit that can be transported here or there while the body waits on a table; or that long-dead "entities" can talk to us and give us guidance; or unfortunately, a long list of etceteras. My favorite illustration of this tendency comes from Shakespeare, where Glendower says "I can call spirits from the vasty deep." Hotspur responds "Why, so can I, or so can any man; But will they come when you do call for them?"[24]

That choice is limited to the possible should not, however, restrain us from trying to expand the possible. We *can* fly with mechanical aids; we can make reasonable predictions with rational analysis; we can figure out various psychological influences on our personalities and use that information for growth; we can seek healing through expanding medical knowledge; we can watch out for giant meteors or other dangers to the planet and take measures to minimize the danger; we can figure out what causes mysterious death; we can handle snakes and fire by understanding the nature of those things and compensating for their dangers; we can learn from the past through historical and archeological researches; we can contemplate the lives of dead people and learn from them; and fortunately, a long list of etceteras.

Finally, our choices are limited by the very complexity which gave rise to judgment and free will. The relevant facts are frequently so complex as to boggle even our impressive brains. The future is often unpredictable,[2 which means it is sometimes hard to make good choices even if we could do so with more time and information. Life is often difficult.

This chapter's discussion of motivation and free will has yielded a conclusion which supports traditional theology. That conclusion is that we have choice. Traditional theology attributes this to divine design rather than to physical evolution, but it is nonetheless correct in emphasizing this aspect of our natures. A final observation about choice also supports traditional understanding, and that is that some of our choices are bad. By urging people to trust their own choices without reservation, "New Age" faiths ignore the history of our species and of every individual of the species. A more rational theology can instill the same enthusiasm as New Age faiths do, and it too sings the praise of freedom, but a rational theology must also account for our ability to choose attitudes and actions that turn out badly. That is the subject of the next chapter.

Notes

[1] I use "spiritual" the same way M. Scott Peck does, as a synonym for psychological and emotional processes. See p. 11 of *The Road Less Travelled* (New York: Simon and Schuster, 1978). See also the article by Lindon Eaves and Lora Gross entitled "Exploring the Concept of Spirit as a Model for the God-World Relationship in the Age of Genetics," previously cited.

[2] See John Tyler Bonner's *The Evolution of Culture in Animals* (Princeton, New Jersey: Princeton University Press, 1980), especially the chapter entitled "The Evolution of Flexible Responses."

[3] See page 70 in "A consideration of methods of identification of species-specific instinctive behavior patterns in birds" included as pp. 57-100 in *Studies in Animal and Human Behavior*, Vol. I.

[4] The philosophical concept of *teleology* has confused some biology-based philosophy. It technically means "purpose," which at first seems compatible with goal-directedness or internal motivation. However, a conceptual confusion has arisen, and persists whenever no clear distinction is made between normal purpose and "ultimate purpose." If that distinction is kept clear, and if biologists forcefully present their claim that "ultimate purpose" doesn't exist except as a leprechaun-like artifact of culture, then it should not be necessary to coin new words like "teleonomy" to clarify the nature of motivation.

[5] There is always the possibility that some recent genetic mutation still survives in the gene pool which reduces the motivational drives. If it does so severely, it is only a matter of time until that branch of the species becomes extinct, unless further mutations re-motivate the organism in alternate ways.

[6] Though it is somewhat old given its subject matter, Gardner Murphy' *Personality: A Biosocial Approach to Origins and Structure* (New York: Harpe & Brothers, 1947) is still very useful, especially Part One.

[7] For a discussion of murder by non-human primates, see pp. 71-77 of *Socia Evolution*, by Robert Trivers (Menlo Park, California: The Benjamin/Cumming Publishing Company, Inc., 1985), or see Jane Goodall's chilling account at p. 35 of *The Chimpanzees of Gombe: Patterns of Behavior*. For more on the evolutio of deceit, including even self-deception, see pp. 415-420 of the Trivers book an also Loyal Rue's *By the Grace of Guile*.

[8] See *The Biology of Spirit*, by Edmund W. Sinnet (New York: The Vikin Press, 1955).

[9] The psychologist Robert B. Glassman has pointed out (personal communi cation May 17, 1995) that some systems oscillate, and that the usual "approach avoidance" conflict is thus not necessarily resolved simply by a shallow gradier being crossed by a steeper one, as I seem to imply.

[10] See Stuart A. Kauffman's "Self-Organization, Selective Adaptation, and It Limits," in *Evolution at a Crossroads*, ed. David J. Depew and Bruce H. Webe (Cambridge, Massachusetts: The MIT Press, 1985).

[11] Vervets, baboons, and langurs have all been seen doing this, as described a pp. 42-22 of Frank E. Poirier's "The St. Kitts Green Monkey" in *Folia Primatologic* 17:20-55 (1972).

[12] See *Molecular Neurobiology of the Mammalian Brain*, by P.C. McGeer, J.C Eccles, and E.G. McGeer (New York: Plenum Press, 1978).

[13] I do not mean to imply that there is a genetic blueprint for every circuit in ou brains, though that may be an adequate way for non-neurologists to think of it. Se "Hermeneutics and the Analysis of Complex Biological Systems," a very interest ing paper written by Gunther S. Stent and cited earlier.

[14] Irenäus Eibl-Eibesfeldt explores the many tensions within human instinct in his *Ethology: The Biology of Behavior* (New York: Holt, Rinehart and Winston 1970, translated by Erich Klinghammer).

[15] For another example besides the wolf/domestic dog one, see "Anarctica' Well-bred Penguins," by Susan G. Trivelpiece and Wayne Z. Trivelpiece; Vol. 9 (December 1989) of *Natural History*, pp. 28-37.

[16] For some examples, see the sources quoted in footnotes 42 and 43 of th *Science* article quoted in Footnote 18 just below. One of the most rigorous of then is *Sperm Competition and the Evolution of Animal Mating Systems*, ed. Robert L Smith (New York: Academic Press, Inc., 1984), which contains the contributio entitled "Sperm Competition in Apparently Monogamous Birds," by Fran McKinney, Kimberly M. Cheng, and David J. Bruggers" (Chapter 15, pp. 523 545).

[17] Including homosexuality, incidentally. See Konrad Lorenz's *On Aggres sion*, pp. 195-198. See pages 202-205 of the same book for a discussion of adulter in the greylag goose. Pages 29-74 of Lorenz's *The Year of the Greylag Goose* (Nev York: Harcourt Brace Jovanovich, 1978; translated by Robert Martin) recoun

nese same stories and is liberally illustrated with beautiful photographs of geese nd alpine scenery.

[18] For a good summary of the many ways bird research has expanded our nowledge of human nature, see "Contributions of Bird Studies to Biology," by Masakazu Konishi, Stephen T. Emlen, Robert E. Ricklefs, and John C. Wingfield. *Science*, October 27, 1989, pages 465-472.

[19] For more details, see John Tyler Bonner's *The Evolution of Culture in Animals*. See also Robert B. Glassman's "Free Will Has a Neural Substrate" pp. 67-81 in *Zygon: Journal of Religion & Science* Vol. 18 No. 1 (March 1968). For more analysis on the striking errors of judgment humans are capable of as explored by the study of "heuristics" or "rules of thumb," see *Judgment Under Uncertainty: Heuristics and Biases*, ed. Daniel Kahneman, Paul Slovic, and Amos Tversky (Cambridge, England: Cambridge University Press, 1982).

[20] For some entertaining and biologically-informed examples of judgmental shortcuts, and a sophisticated analysis of how that can produce errors in judgment, see Robert B. Cialdini's *Influence: The Psychology of Persuasion* (New York: William Morrow and Company, 1984).

[21] See Roger Sperry's 1969 article entitled "A modified Concept of Consciousness," *Psychological Review*, 76:532-536. Cited with approval by John H. Campbell in "An Organizational Interpretation of Evolution," a paper included in *Evolution at a Crossroads: The New Biology and the New Philosophy of Science*, ed. David J. Depew and Bruce H. Weber. Sperry explains his concept further in "Paradigms of Belief, Theory and Metatheory," pp.245-259 of *Zygon: Journal of Science & Religion* Vol. 27, No. 3 (September 1992).

[22] Bartley, W.W. III, and Radnitzky, G., editors. *Evolutionary Epistemology, Theory of Rationality and the Sociology of Knowledge* (LaSalle, Illinois: Open Court, 1986). See also Chapters 16 ("Evolutionary Epistemology," pp. 393-434) and 17 ("Descriptive Epistemology," pp. 435-486) of Donald T. Campbell's *Methodology and Epistemology for Social Science* (Chicago: The University of Chicago Press, 1988).

[23] The evidence for this assertion is well presented in Richard B. Gregg's *Symbolic Inducement and Knowing: A Study in the Foundations of Rhetoric*. In my opinion, however, modern rhetoricians err by claiming (implicitly or explicitly, depending on the writer) that symbolic experience is the only experience, or that the symbolizing process drastically changes the nature of the experience. They thus relegate a huge portion of our experience (I daresay the larger portion, namely the part we share with non-symbolizing animal cousins) to the realm of the unimportant or even the non-existent. For further discussion see my article entitled "The Precursors of the Eureka Moment as a Common Ground Between Science and Religion."

[24] *Henry IV, Part I*, Act III, Scene i, line 53, Henley edition.

[25] The baseball wag Yogi Berra is given credit for many clever sayings, one of which is appropriate here. Allegedly he said "It is hard to predict, especially the future."

Chapter 10

Sin and Morality

An Outline of the Chapter

A. The Evolution of Morality
B. Morality and the Brain
C. Morality and Choice
D. Morality and Rationality
E. Sin
F. A Note on Sociobiology

Chapter Seven explained the process of constructing a worldview or faith-system. Morality is a component of any worldview, but it is such an important component, and has been part of our nature for so long, that it deserves a separate chapter. This chapter will therefore look more precisely at how moral concepts enter our lives and influence our actions.

In morality as in other aspects of our worldview framework, we start with inherited concepts and modify some of them based on experience. But to avoid repeating the ways concepts develop and work themselves into worldviews, this chapter will concentrate on showing how morality interacts with the dynamic processes already discussed. How does it interact with evolution? With culture? With the complex brain? With choice and rationality? After discussing those questions we will explore the doctrine of sin, which summarizes these interactions in traditional but useful ways. The chapter will end with a discussion of sociobiology, an emerging discipline that addresses some of these same issues.

The Evolution of Morality

The idea that morality evolved, and thus that other animals beside humans may have it, has stimulated great disagreement among moder thinkers, and not a little polarization.[1] One side emphasizes the degree t which animal behavior corresponds to human behavior, concluding th animal morality is essentially the same as human morality, at least i species that exercise the rudiments of judgment. The other side points ou how complex morality is, and how it requires some abilities that animal don't possess; thus they limit true morality to humans.

Biology places the dispute in perspective by reminding us that evolu tion often produces continuums from simple to complex, whether i structures like eyes, or in systems like culture. Sometimes hard evidenc of the transition is lacking; for example how could one hope to find foss evidence of growth in the soft brain cells that underlie mental processes In the case of morality, however, we can get a surprisingly clear notion c the transition. We can do it by considering the evolution of specifi characteristics which had to evolve before there could be such a thing a morality, and by noting which of those we still share with animals.

For example, one pre-requisite of morality is the ability to give an receive sanctions, an ability we share with most mammals and bird Sanctions depend on the existence of aggressiveness, which had evolve even earlier. More subtle behavior like reward and reinforcement evolve too, and may be found in other animals, especially primates. Moreove behaviors like reward and sanctions are magnified by group-identifica tion, already mentioned as having evolved in the social species. Neur pathways underlie all of these behaviors, and the evolutionary steps th produced them can be readily imagined.[2]

While most sanctions in the animal world are external and immediat (for example a threat by a dominant elder may force a desired conduct), th internalization of mores is more complex, producing personal sensation of shame or guilt or pride. But even those components can be found i animals. A dog's body language is our most familiar evidence of that, an if our pets were monkeys and apes we would see it even more profoundl

These simple pre-requisites operate somewhat mechanically in an mals.[3] Some aspects of morality seem so complex, however, that they mu require mental complexity of the human sort. Such characteristics as time awareness, self-consciousness over time, language, and intricate motiva tional structure all depend on complex mental pathways, and each of thos has been called a pre-requisite for morality.[4] Further, some highl regarded writers have persuasively claimed that morality cannot exi without the kind of cognitive skills that humans possess.[5] In other word they conceptualize morality as conduct chosen in conscious awareness of its likely long-term effects.

In trying to understand human nature, these distinctions seem highly significant, even though we must acknowledge the common origin of our morality with animal morality. As we try to understand how our belief-structures and moral choices interact, it becomes especially necessary to acknowledge the role of complex mental operations. It seems appropriate and reasonable to make that acknowledgment by limiting the word "morality" to behavior arising out of such complexity. I will therefore adopt that usage, calling similar behavior in animals "proto-morality."

Using the word "morality" in precisely the right way is not so important, however, as establishing its functions in human life. Of course evolutionary products do not always have specific functions—I have already noted some examples of evolved characteristics that have none. But when we encounter a pattern of behaviors and mental processes as widespread as those represented by morality, and when we are able to find its prototypes in simpler animals, it is usually true that the evolutionary product fulfills an important purpose in our lives. That is easy to establish in the case of morality.

In broad terms that purpose is survival. By mandating or encouraging the development of moral rules, our species avoids energy waste. Morality also helps us know what to expect from one another, and that increases our efficiency in dealing with the environment.

That would be impressive in itself, but morality (like many other evolutionary products) is more luxurious than to offer bare survival. Any characteristic which only resulted in bare survival would eventually encounter some challenge that would overcome it (like the rare freeze that kills a semi-tropical plant growing on the edge of its natural range), but when a species is well within its range and niche, its evolved characteristics usually give plenty of room for error. This cushion not only assures survival on the long run, but also provides the possibility for abundant living in normal circumstances. This observation will be explored further in Chapter 12, but for now let us simply note the claim that morality serves to keep us within the center ranges of our niche.

Morality enhances survival in other ways than by saving energy or by keeping us in the center of our niche. It organizes and solidifies the group, and Chapter Three showed how that enhances survival. Indeed, morality is in many ways an elaboration of this social part of our evolution—it helps to regulate the inevitable conflicts between members of a group, between a group and its own members, and between adjacent groups.

Modern circumstances strain morality's ability to fulfill even these straightforward functions, largely because of the crowding caused by our success. Part Four will suggest strategies for relieving that strain, but such strategies are possible only because of moral choices based in our amazing brains.

Morality and the Brain

Several mammals (notably rodents and primates) evolved a wide-ranging cleverness for responding to environmental challenges. This development was remarkable because it overcame the limits imposed on animals which respond to their environments in strictly programmed ways.

Most species evolved rigid behaviors which succeed or fail because of external circumstances, and not from any judgment by individuals. Behavioral rules are rigidly embedded into such organisms, and in truly robotic fashion each individual responds to challenges in predictable ways. On balance that has been a successful development for those species, even though it produces suffering in many individuals. For example salmon are programmed to swim upstream, and some individuals perish when a bear scoops them up. But because so many of them make it back to the spawning grounds, the programming enhanced salmon life more than it diminished it (at least until concrete dams came along).

For simple animals like salmon the genes can contain enough information to direct virtually all of their behavior, with a minimum of learning.[6] But as I pointed out earlier, genes could not possibly contain enough information to direct all the myriad behaviors and choices humans make. That task requires a brain which can store and process information in flexible ways, and use it to make choices. Thus even though the brain is constructed from a genetic blueprint, it is itself an open-ended computer for solving problems in an endless variety of ways, using whatever resources happen to be available.

Nowhere is this variety of solutions more evident than in our construction of moral systems. Although every known group has a moral system of some sort, and thus reaps the advantages of efficiency and survival that moral systems provide, the details of their moral systems vary widely.[7] Moral systems are thus like language—each human group has one, but the details of any two of them are seldom similar unless they spring from the same historical source.

Theologians and other moralists are often frustrated at the apparent lack of common threads in morality. But over and over again, anthropologists have shown how one society prohibits the same behavior that another society requires. Even the prohibition against incest, often argued to be an absolute moral rule in human and animal societies, has been challenged by anthropologists, because some societies actually encourage it.

What are we to make of this seeming relativism? Does it mean morality is useless, or that there is no reason to adhere to it? Of course not, any more than language is useless because of its relativity. Moreover, the evidence is not all in yet, and we should consider at least four ways that morality might not be so relative as it first appears.

First, it is more probable than ever that some moral rules will turn out to be innate. Just because all of them couldn't be encoded in our genes doesn't mean none of them are. Likewise, just because a particular rule has exceptions does not mean it isn't innate—on the contrary, variety should be expected for any genetically based rule. Furthermore, an apparent exception may simply be a pathology of innate morality, since pathologies can exist in every aspect of our lives, including not only predispositions to cancer but also defects in our neural development. So it is much too early to conclude that no moral absolutes are built into us, however flexible the brain is, and however much those absolutes vary from person to person or group to group.[8]

Second, many moral rules tend to flow from our nature even if not hard-wired in us, and such rules are usually expressed in the moral codes of human groups. For example our nature is somewhat possessive (like that of most carnivores and some primates), and most cultures therefore have rules that regulate our tendency to accumulate possessions. The most general of these is "Thou shalt not steal," but some cultures go further and try to mandate our attitude toward the possessions of others, in rules such as "Thou shalt not covet," or "Thou shalt share."

Third, it is highly likely that there are limits to the flexibility our brains permit. Just as language has deep structures that find their way into each dialect, each moral system probably reflects a deeper syntax than might appear on the surface. Regulating the relation between self-needs and group needs is a feature of every moral system, for example. Moral philosophers aren't always comfortable with relativity, but they are often able to find commonality between moral systems which seem more diverse than they are.[9]

Fourth and finally, psychological development within a given individual operates to produce a stable morality. Although the immature brain is flexible enough to support almost any morality imaginable, the dendrites and pathways connect so as to produce a somewhat cohesive moral structure, based on the individual's cultural and personal experience. Once we have established a moral system, potential new moral rules have to compete with the morality already embedded in our neuronal pathways.

So even though an individual's moral system can be characterized by flexibility and judgment, those processes nonetheless produce firmer moral structures than anthropologists might think, if they fail to consider the biological aspects of individual morality. This is primarily because the memory tracks which hold moral codes and specific rules are highly stable, especially once childhood development is completed. Thereafter one may certainly exercise the freedom to violate or change the moral rules one has adopted, but only at the risk of creating internal dissonances which can cause psychological or physical difficulty. Again, this cannot lead to an absolutist conception of unchanging morality, because other dynamics are operating, especially the dynamics of choice.

Morality and Choice

By acknowledging biological boundaries on moral flexibility, I do not mean to renounce a whit of the chapter on choice. It is entirely possible to modify our moral codes, and still retain respect for biology and early training. Since being human permits—nay requires—judgment and choice, it is quite common for us to rework the moral component of our worldviews, based on our experience and our evaluations of moral rules.

We make those evaluations largely through non-rational or subconscious processes like the gestalt, but we also pause to ask rational and conscious questions about moral choices. Those questions center primarily around the potential *consequences* of each choice; around whether the likely results are what we want. The consequences considered may be short-range ones or long-range ones, but in a world where life is long, it usually behooves us to give particular consideration to longer-term consequences.

One of the basic choices a person makes is whether to accept the moral precepts handed down through family and/or culture. It might be unwise to accept such precepts unconditionally, and they can be incomplete or erroneous, but there is nonetheless a logical reason to give them significant weight. It is that such rules have developed over time and have often been subjected to refinement and modification, which tends to make them more reliable than rules based on the small sample of one or a few individuals. While that refinement may not lead to perfection, it provides a statistical edge over life without such rules.

Seen this way, each moral "rule" compresses human experience into a single statement about the law of averages. "Thou shalt not steal" summarizes the observation by generations of people, that theft causes more harm than it is worth. A modern equivalent is "Thou shalt not smoke." In neither case is it guaranteed that a violation of the commandment will cause harm, but the odds are firmly in that direction. Even when individuals modify morality to fit their own personal experience, the new morality will still partake of this characteristic—it compresses experience into statements that express shorthand judgments about the probable consequences of a course of action.[10]

This view of morality subjects every moral statement to the possibility of investigation and argument, granting significant weight to tradition without seeing it as absolute or permitting it to become idolatrous. Whether theft or adultery or smoking is really harmful, and the reasons for it, can be examined in the light of biology and its sub-disciplines (especially psychology), yielding keen insights and more concrete understanding. I earlier expressed my view, for example, that the adultery commandment will turn out to be based in human biology, but it is for future researches to determine whether that is true.

Seeing morality as a summary of experience also implies that direct connections exist, over time, between factual events and moral rules. Since it has traditionally been argued that this cannot be true, I want to take a moment to discuss a philosophical question called "the *is* and the *ought.*" It is a question that might seem over-technical at first, but it can significantly deepen our understanding of morality.

Ever since Hume, philosophers have claimed that one cannot derive an ought from an is. In other words, the descriptive facts about human nature, such as have supplied the substance of this book, are supposed to be insufficient to help one know how to act. Yet biology's greatest contribution to philosophy has been to show how this is not only possible but necessary, if we are to construct a coherent and useful understanding of human morality. This can be shown in 8 paragraphs, which will summarize the view of morality I am presenting here.

1. Inside each of us, as a function of our muscular and nervous and hormonal systems, is a propulsion toward various kinds of actions. We share that propulsion with animals, and I called it motivation in Chapter Nine. Some actions (along with their underlying thoughts) are aggressive, some are sexual, some are for survival, some are competitive, some are sympathetic, some are alliance-seeking, and so on. These are the "drives" of classical psychology, and although biology sees them in more prosaic ways than early psychology did, nonetheless there seems no dispute that they are part of the "is"; they can be stated as descriptive statements about the nature of humanity.
2. The precise form of response to these motivations, and the ways we abstract them into systems, is a further function of our natures.[11] Our responses depend on innate learning processes interacting with cultural and individual experience, and our resulting worldview systems are thus a function of our natures interacting with experience. This is still part of the "is," and an objective observer (say a man from Mars) would undoubtedly describe it as such.

3. Explicit moral systems result from our individual and cultural tendency to distill and condense experience into pieces of wisdom. Over time we notice that adultery can disrupt society, and we make a rule against it. We make judgments, maybe even incorrect judgments, about the benefits that come to communities when they conquer nature, and those judgments work their way into our moral and legal codes. We judge that honesty creates a more stable community, and thus we penalize perjury. All these judgments are part of the is, by which I do not mean the judgments are correct, but that they may be seen descriptively as arising out of our nature.
4. The next and critical step is to apply if-then statements to these judgments. In computer language, an if-then statement says "If such and such is true, then do such and such." In human morality, an if-then statement is one which says "If one wants such and such result, then one ought to do such and such action or have such and such attitude." The if-then statement is itself a simple observation about the probable consequences of a particular course of conduct or thinking pattern. So if I want a good income, I ought to get an education. And if I want a stable psychological life, I ought not to take myself too seriously.
5. But what makes us *want* a good income or a stable psychological life, or any other product of our choices? The answer has already been given: internal motivations, which are a part of our evolved natures, as influenced by individual and group experience. To put it more broadly, we inevitably and innately "want" to live consistently with our natures, in ways that enhance survival and abundant living[12].
6. In one important sense, therefore, the ought is part of the "is." So long as we exist, we necessarily obey various kinds of internal compulsions toward heeding internal motivations, in various combinations and with various modifications (and subject to various pathologies).
7. Because our evolution produced a brain which is able to respond to the existing variety of personal motivation and experience in a relatively open-ended way, it should be no surprise that moral systems vary. That they can do so is a major function of the flexibility built into our natures. The "oughts" differ from culture to culture and even from person to person, both because the circumstances are different, and because the flexible brain can solve problems in a variety of ways, making different judgments about the consequences of particular behaviors.
8. Some people say they believe in their morality absolutely, and not because of if-then analyses. But they do adopt a moral system, and

they argue in favor of it based on reasons and results.[13] Perhaps they do adopt moral rules without a clear cognitive understanding of the reasons or functions behind those rules, but subconscious processes and acculturation are fully sufficient to explain what is happening inside their heads. Thus the analysis given here is not materially affected; if-then calculations are only removed to realms of sub-consciousness and sub-culture.

This view of morality continues to be resisted, probably because of the specific history of western philosophy. Somewhere we developed the notion that morality should be universal and absolute, and that notion has become the subtle cement of our conceptual constructions. Thus it is not really the jump from the is to the ought that causes the difficulty, but the reluctance to acknowledge the variety inherent in human mental and moral processes. Anthropologists have thoroughly documented the existence of that variety, but philosophers and theologians cling to the hope for universals. Interestingly, biology offers them the opportunity to realize that hope, or a satisfactory substitute for it, if they first abandon the claim that is/ought offers an unjumpable chasm, and accept the exceptions and qualifications that are always necessary to biological observations. We can see this more clearly by looking at the interplay between morality and rationality.

Morality and Rationality

If evolution and rationality worked perfectly, the impulse to morality would slowly move us toward a wonderful world where wise rules are extracted from experience, with occasional modifications where new insights or developments call for them. Life doesn't work like that.

There are many reasons for this. We have other innate mental processes besides those that urge us toward morality, including some that urge us toward self-aggrandizement at the expense of other people. Even where competition is channeled by wise and rational institutions, we analyze situations differently and come to different conclusions. As in all aspects of life and evolution, what we have is the result of trade-offs; it does not produce a smoothly functioning piece of precision equipment, but it does produce moral mechanisms that work, albeit roughly. Eternal vigilance is the price of most worthwhile things, and only through it can we avoid Hitlers and schoolyard bullies that twist our moral impulses to their own ends. And of course we do not entirely avoid Hitlers or schoolyard bullies, because we cannot be that vigilant. We have to relax most of the time and live out of sub-conscious processes and our basic worldview.

Thus the question of how to achieve the good and avoid the bad will remain a difficult one, with which every society and every individual must wrestle. We wrestle with the question by urging greater vigilance, but we also wrestle with it in another way, and that is by the exercise of rationality. That faculty is not necessary for animal-like proto-morality, and even in human life we can live morally without constant thought. But for evaluating whether a given decision or course of action is more good than bad, rationality is our best resource.

How do we tap into this resource? By exercising judgment. But judgment requires thought, thought about actions and attitudes, and about their consequences. We can make judgments privately, and we can make them subconsciously without any commitment to introspection, but often we need help. "Two heads are better than one" makes prediction of consequences more fruitful because community adds thinking power. That is true of any choice, but it is especially true of moral choices because they are usually complex.

Once we think and discuss and decide, our moral choices become components of the larger frameworks of belief discussed in the faith chapter, and we live out of them until something calls them into question. Although our faith-system usually remains stable, it should be clear by now that various experiences can destabilize it. Thus a motivated life frequently faces the challenge of reconsidering (and perhaps reworking) its faith system in light of new experiences and new knowledge.

The demands of rationality and morality must be balanced against the need for relaxation. Later chapters will add that balance, but for now let us continue the discussion above by looking at the traditional doctrine of sin, to see how well it articulates what biology has discovered about morality. The discussion will also point out how easily we can go astray in our effort to live abundantly, thereby emphasizing our need for salvation, and paving the way for Chapter 11.

Sin

The sections above present morality as a set of choices influenced by genetics and culture, but mostly arising from the operation of our flexible brains. If this were a different kind of book, that might be the end of the matter, and we could pass on to a consideration of the practical effects of such a view. However, one of our themes has been that tradition offers an excellent beginning point for the construction of belief-systems and for behavioral choices arising out of them. And in the case of human behavior, intelligent discussions of morality have traditionally been conceptualized in terms of sin.

At least three separate kinds of behaviors are addressed by traditional concepts of sin. First is our tendency to make mistakes, both simple ones and serious ones. Second is our tendency toward selfishness, which ranges over a continuum from healthy self-interest to megalomaniacal obsession. Third are the behaviors which each culture identifies as unacceptable to it. Of course morality is such a many-faceted dynamic of human nature that these three ways of looking at sin cannot exhaust the topic,[14] but by looking at the ranges of possibility within these three kinds of behavior, we should be able to see how the concept of sin has reflected moral thought in the past and still continues to do so today. The goal will not necessarily be to show how to avoid or overcome sin—that will be the task of the next two chapters. Instead I will merely try to show how our biological nature produces the ranges of behavior, and why.

1. Sin as Mistake

These pages have often reminded the reader of how important trial and error is in human life, and indeed in all of life. Evolution works by trial and error, most animals and plants use it routinely, and the modification of concepts depends on it. But as the phrase "trial and error" rolls off our tongues, we tend to concentrate on the successful trials, and take the errors as minor happenstances that have to be endured. We forget the importance of errors, which are actually very important to the process, and very common in human life.

Ordinary errors are necessary to any species with significant learning abilities. The wolf-pup nips at other critters, until a quill teaches it that harassing porcupines is a mistake. Young humans make many mistakes, and if we are wise enough to let them do it, they learn important lessons from their mistakes. But some kinds of mistakes are much more serious, whether made by animals or adolescents or adults. Playing with rattle-snakes or ingesting poisons, for example, can end in death.

It is possible to reduce mistakes through insight, parental training, and the various instructions of culture, but it is impossible to eliminate them entirely. Neither individuals nor culture could possibly construct sufficiently detailed rules for all the complex and ever-changing situations we face. Another reason for mistakes is that we evolved to take risks, including moral risks. Thus a person who resolves to keep every rule without mistake is making another kind of mistake, probably a more serious one, because that resolution can produce too large a tendency toward risk-avoidance. This does not mean we should try to make mistakes, but it means we shouldn't be surprised when we do.

Yet even though mistakes are unavoidable, some are so serious and predictable that we can anticipate them and warn one another about them.

Culture has distilled rules from its experience with these mistakes, and has taught each rule and its reason to its members. Remarkably, we often ignore this wisdom, and risk serious harm to ourselves or others in spite of the warning. In those cases it appears that we are knowingly making mistakes that entail serious risk, or at least we "knew or should have known" the risk. Such behavior is striking, however common it is, and it is appropriate to give it a name. Therefore "sin" is usually defined to include such serious, knowing mistakes.

Calling such serious mistakes "sin" leaves open the question of which actions actually fit into this category, and which mistakes are only ordinary mistakes that are part of the learning process. Yet if we can identify mistakes that really do produce serious consequences, it is entirely consistent with our biological natures as a teaching/learning/language-using species, to conceptualize a special category for them, and call it sin.

The advantages of viewing sin as a kind of mistake are many. First it places sin within the range of behavior regularly seen in our species, and does not consider it a perversion foreign to normal life. Second it implies that we can improve our capacity for avoiding mistake, which will be discussed further in Chapters 11 and 12. And third it depicts sin as growing out of our biological nature as described in earlier chapters, especially that aspect of our nature related to learning and choice. In that sense we certainly have a nature that makes sin inevitable, just as traditional religious doctrine asserts ("We have all sinned, and come short of the glory of God."[15])

2. Sin as selfishness

Some selfish behavior is healthy. One finds it in every living organism, and it is especially necessary for children to act selfishly as part of normal trial and error learning. But in addition to that kind of selfishness, we often choose behaviors which are so selfish as to bring about a disbalance between our nature as individuals and our nature as group-members.

Unfortunately I do not know exactly where the correct balance between those two behaviors is, and I know it must vary with the circumstance.[16] For example frontiers require a strong self-reliance which theoretically stands willing to help others but actually depends on everybody carrying their own load, whereas cities require a strong interdependence even among those that seem isolated from one another. Within an individual life there are times when more or less selfishness is called for. The range exists, and is regularly encountered, precisely because different circumstances favored different responses over evolutionary time.

But even without exactly defining the balance, it is clear that we sometimes upset it. One way we do that is through a kind of self-aggrandizement that takes advantage of social impulses in others, without yielding to our own social impulses. Such selfishness can wreck families or societies, and thus every moral system places some acts in this category, and calls them sin. In American economic life, the robber barons of the 19th century were a notable example of this behavior, and it was necessary to pass laws to control them. Such selfishness easily arises in daily life, and must be controlled by less formal (but sometimes more powerful) mechanisms than law. Bullies are ganged up against; charity is urged; and frivolous lovers are ostracized.

3. Sin as a cultural dynamic

Naturally a culture can make mistakes too, just as an individual can. It can run roughshod over its own subcultures and other cultures, creating pathological conditions that will require treatment later. Since these pathological conditions often begin in serious mistakes or in selfishness, this third category is as much a manifestation of the other two categories of sin as it is a separate category. But because we are such cultural animals, and because culture becomes the primary distiller and enforcer of morals, it is intimately involved in any discussion of sin, and contributes such profound dynamics of its own as to justify a separate listing.

Although individuals face the challenge of identifying serious mistakes and pathological selfishness and other categories of harmful behavior, it is in culture that the identification is preserved and passed down. Cultural processes establish the boundaries between ordinary mistakes and serious mistakes, and between normal selfishness and destructive selfishness. Individual morality begins in cultural precepts. Yet for all its contributions, culture causes a great deal of trouble in the effort to understand morality.

It causes trouble largely because of complexities discussed in Chapter Three. Conflicts between priest and prophet, between new and old concepts, between adjacent groups, and between the group and its individuals or sub-groups—all of these make for highly dynamic and constantly changing interactions. In our culture new claims are always arising by which one group accuses another of undue selfishness, or urges control of individual actions to prevent serious personal or societal mistakes. Essentially we are accusing one another of sin.[17]

This is a proper and quite unavoidable role of culture. Although it may appear to cause a lot of unnecessary pushing and shoving within society, it actually helps us find the middle ground that articulates our nature. This is true whether the pushing and shoving happens in a dignified House of Lords, the intrigues of kingly courts, the gossipy politics of primitive tribes and social cliques, or in the robust deal-making of democratic institutions.

As humans have become more crowded, we have more and more decided what is sinful as an aspect of law, and Part Four will discuss that observation further. Historically, however, and especially when we were not so crowded, these same dynamics have been treated as an aspect of religion. They remain a part of religion today insofar as they are unenforceable by law, or lack the consensus necessary for law.

4. A definition of sin

A concept is useful only insofar as we can articulate it, and yet this chapter shows how complex morality is, and how difficult it would be to define sin in a short sentence. By including some qualifications, however, it might be possible to modify traditional definitions of sin, so as to retain the term for future use.

A classical definition calls sin "a willful transgression of God's law." The doctrinal literature shows that "willful" includes the "know or should have known" concept mentioned above; the definition thus seems broad enough to include both the serious mistakes that one is forewarned about, as well as actions arising out of undue selfishness.[18]

However, this definition arose when God was conceptualized as one who announces rules from on high (like the kings who ruled human society at the time). Therefore it continues the tradition of *declaring* what moral rules should be, rather that searching to *discover* what they are, as part of our natures in the context of specific environments. Declaring morality rather than discovering it keeps us from exploring each rule, and from asking whether it leads to healthy living or not.

Nor does this definition adequately account for variety. It assumes that we are all peas in a pod, and that all moral rules are precisely appropriate to every person. But biology has shown how profound our variety is even at genetic levels, and anthropology has shown how the flexible brain produces additional variety at a cultural level. We need a definition that keeps its focus on willfulness and selfishness, and yet acknowledges variety and discovery.

There is another complication in defining sin, and that is that we can make serious mistakes in more than one way. We can do it through a single big mistake, or we can make a series of small mistakes which ultimately have a big effect. This latter category includes the small cumulative mistakes of habit and destructive attitude, which have traditionally been considered sinful. The scripture that says "the wages of sin is death,"[19] summarizes the observation that small choices made week in and week out, like the work we do for wages, will eventually have just as much effect as a single big choice. Any good definition of sin must include this subtle but powerful reality.

Considering these complexities, I propose the following definition: *Sin is behavior chosen without due regard for the serious risks it holds, or behavior which seeks self without due regard for our nature as a group species, or behavior which willfully disregards culture's distilled wisdom. Sin may be a single act, or a cumulative series of small acts or habits or attitudes engaged in with the same disregard.*

This definition is as specific as it can be and still acknowledge our varying genetic makeups, which might require slightly different rules for different people, just as some of us need slightly different foods. Yet absent persuasive evidence of any claimed variety, it assumes commonality. Thus tradition was reasonable to call things sin that seemed to be perversions of human nature, whereas it may turn out that those behaviors are in fact within normal ranges of variety. Time will tell.

The proposed definition also seems consistent with traditional concepts by limiting sin to behavior made with knowledge. Notice that the scripture is ambivalent on this point. On the one hand it seems to require knowledge in such passages as "To him that knoweth to do good, and doeth it not, to him it is sin."[20] On the other hand are passages which establish a duty to "gain knowledge" or "study to show ourselves approved." Yet insofar as scripture calls behavior sin, it seems to require actual knowledge, and the definition above takes that same approach. This means we must educate people about the probable consequences of their choices, before we have any right to call those choices sin.

I am not entirely comfortable with the definition above, because it somehow affirms our human tendency toward concreteness and simplicity. Morality is so complex as to make suspect any attempt to capture its violations in a short paragraph, especially considering that the concept of sin has been so often used to judge and condemn one another, instead of being used for analysis and growth. But by keeping the focus on human nature, and on the practical role of rules in cultural and individual life, I hope to avoid the grosser consequences of making a definition, and still give meaning to the concept. Moreover, I hope it is clear that the definition would exclude most behaviors that arise from thoughtful consideration, because choices made with "due regard" are not sin under this definition.[21]

Finally, just because we can construct a concept of sin and define it, does not mean the subject is closed. The ongoing dynamics of personal growth and culture, as well as advances in science and philosophy, will undoubtedly keep influencing our view of sin and morality, just as they have in the past. We may want absolute rules without exceptions, but we have evolved as more complex creatures than that. The concept of sin thus changes as our understanding changes. Yet at each stage of that understanding it can help us avoid harmful behaviors, enhancing our security and contributing directly to our salvation, as the next chapter will demonstrate.

A Note on Sociobiology

In 1975 the Harvard entomologist Edward O. Wilson wrote a book that took the intellectual world by storm. The book was entitled *Sociobiology: The New Synthesis.*[22] Entomology (the study of insects) may seem far removed from human sociology, but Wilson's book created a new discipline dedicated to the study and comparison of *all* social species, whether they be insects or fish or wolves or people. Sociobiology claims that we can learn many important lessons relevant to human life by concentrating on the common features of all social life. Naturally this claim provoked severe criticism, but on the other hand some notable scientists and philosophers[23] (and even theologians[24]) have come to Wilson's defense. By now a huge literature exists, and the discipline has begun to mature.

Frankly, I would like to ignore sociobiology. Its concepts are clearly related to biotheology, but its precepts and methods are largely tangential to biotheology's. On the other hand, Wilson and his colleagues have done a masterful job of developing their ideas, and have defined the parameters of current intellectual discussion for any area that cuts across theirs, even tangentially. Therefore I cannot ignore it.

For our purposes it is not necessary to explain sociobiology in detail or to refer to it often.[25] But it is necessary to give an overview, to address some criticisms, and to acknowledge some overlaps. I will do that in three brief sections.

Overview. Sociobiology is the study of the interplay between sociology and biology. It asks what influence biology has on the social aspects of life. In insect societies, the answer is "almost everything." It is inconceivable that insect responses are based on anything other than their biology, regardless of how elaborate their society is. Evolved neural structures in insect brains determine just how they act. That is true even though their neural structures develop in slightly different ways under specific biological conditions. For example a honeybee queen develops if (and only if) royal jelly is fed to a young female.

Sociobiology takes this observation and builds on it. It would be impossible for wolves to develop the social systems found in insects, precisely because a wolf's brain could not support such a system. Yet the wolf's social system is dictated by its biology and brain structure, just as surely as a bee's is. Each of the two species evolved structures appropriate to its respective ecology, as a result of ordinary natural selection. The could not have evolved one another's social structures, because no tract of land could support 10,000 wolves coming to and from a home base, whereas a very small tract can support that many bees.

Despite such practical distinctions, sociobiology has uncovered many similarities between wolves and bees and other social animals, including humans. For example it has given a mathematical basis to altruism, showing the selective advantage of living cooperatively.[26] It also claims to explain why a wide variety of creatures favor their near-kin in these cooperative living arrangements. Most importantly for our purposes, it claims to establish a biological basis for morality.[27]

More details of these claims and contributions will be explored below as we look at some criticisms of sociobiology, and at some overlaps between it and biotheology.

Criticisms of Sociobiology. The criticisms were immediate and strong, and they cut across several disciplines. We can boil them down to four basic criticisms.

1. *Sociobiology's claims are overreaching.* It often makes extravagant claims that cannot be supported. It ignores Lorenz's frequent reminder that we must be cautious when making transfers from one species to another, because closely related species can be very different.[28] Comparisons between distantly-related species are often nothing but disguised analogy—like making comparisons between humans and machines. That may produce some useful insights, but it says nothing about actual evolved relationships.

To give one example of overreaching by sociobiology, consider its argument about altruism. Insects help others because their biology demands it. Mathematical models show that each insect society works in an optimum way to increase survival rates. From this observation, sociobiology claims to understand human altruism. We help others because it enhances our survival.

That much is probably true, at least in general. Moreover it is true (as sociobiology points out) that we are more likely to help near kin. But there is no reason to attribute this fact directly to biology. We may simply help those we associate with, and we are statistically more likely to associate with near kin. Childless aunts frequently help their husband's nephews and nieces, even though they are not close genetically. Perhaps they calculate (subconsciously at least) that it is in their best interest to do so, but it takes a great deal of straining to fit such altruistic acts into a rigid biological schema.[29]

2. *Sociobiology is shallow.* Just as Lorenz criticized Skinner because of the paltry nature of behaviorism's insights,[30] many critics say "So what?" to sociobiology. Of course brains have to have the necessary circuits to support a social arrangement. Of course those circuits evolved.

But in the case of humans, that is a very simplistic point; it doesn't tell us much. Our complex brains allow us to *decide* what social structures and moral principles to adopt, out of a huge library of potential choices. Granted that those choices are limited—it would be hard to believe that humans could successfully choose to live like bees any more than wolves can. But the available choices present a huge number of options, and sociobiology can tell us very little about which option to choose.

3. *Sociobiology is wrong.* Richard Dawkins is a leading sociobiologist who wrote a book entitled *The Selfish Gene.*[31] In it Dawkins argues that selection takes place at the level of genes, and not at the level of individuals or populations or species. Although this is basically a technical and even a semantic argument, it has important implications.

As usual, sociobiology has taken a true statement and expanded it. Mutations at the genetic level begin every new stage of evolution, and mechanisms like natural selection or sexual selection "pick out" the successful changes. But Dawkins concludes from this that we are innately selfish, and cannot be made unselfish without great effort. He ignores the fact that we are also innately cooperative and compassionate.

Sociobiology is wrong on other counts too. The eminent scientists Marshall Sahlins and Richard C. Lewontin have separately counted the ways. In biologist Lewontin's book *Not in Our Genes* he emphasizes that not every evolved characteristic has *any* purpose, much less the "reproductive fitness" purpose that tends to be the only purpose sociobiology recognizes. Anthropologist Sahlins argues that culture influences our choices far more than biology does.[32]

4. *Sociobiology is offensive and perhaps dangerous.* It can be used to support sexism, because it frankly explores behavioral differences that may originate in sexual differences. It could even support racism or nazism if it could be shown that one race had even a slight advantage over another race in any given arena. To sociobiology's credit, it has gone to lengths to avoid such suggestions.[33] It has even been uncharacteristically humble in stating the relevant arguments with reservations, so as to reject even the hint of racist or sexist conclusions.

Overlaps between Sociobiology and Biotheology. Despite these criticisms, sociobiology has stimulated worthwhile discussion about biology's impact on sociology. Shallow or not, basic observations are necessary in order to begin new research programmes. Even overreaching helps determine the limits of a new discipline, and provides enthusiasm for research. In the case of sociobiology, we should not be so sure that its limits have been reached, because it is pursuing several paths that can help us understand our natures better.

Indeed, sociobiology has a certain synergy with biotheology, and some of its less controversial concepts strengthen biotheology. Specifically, it gives further support to the following principles of biotheology:

1. Understanding human nature is easier if we understand biology. This was a theme of early chapters, and although recent chapters have shifted the focus to the flexible brain, there is no doubt that the brain is a biological organ. Neurology and other brain sciences have already made great strides in helping us understand that, and sociobiology uncovers specific manifestations of brain organization in human social life.
2. Culture is an elaboration of biology. Although it is still unclear just how much detail biology imposes on culture, sociobiology serves as further evidence of the conclusions of Chapter 3, which saw culture as a manifestation of human biology.
3. Free will must be exercised within biological constraints. Our brain helps us to cope and to survive, by providing baseline responses and by permitting informed judgments. Sociobiology expands that theme by showing the extent to which judgments are mediated by biology (and by social arrangements which themselves depend on our biology and impact our choices).
4. Moral systems have a biological component. Indeed it is fair to say that moral systems have a biological *foundation.* While such systems are influenced by culture and modified by the judgments of individual brains, culture and judgments merely build on evolved patterns of behavior. Those patterns are fundamentally similar to patterns found in other social species. Thus no action can be seen as "moral" or "immoral" except in the context of the neurology of our species.
5. Our security and even our happiness depend on living consistently with our biological natures. Sociobiology does not overtly make this claim as a scientific conclusion, but there is no reason to take its other claims seriously unless this one lurks in the background. Chapters 11 and 12 deal with security and happiness, and they incorporate this same claim into biotheology.
6. Our social structures—government, education, economic institutions, and religion—work best when they take cognizance of our evolved social natures. Part Four will look further into these social dynamics, placing biotheology squarely in tune with sociobiology's claim that such structures operate better with better knowledge of the biology they are built upon.[34]

Notwithstanding these overlaps between sociobiology and biotheology, it is premature to argue that sociobiology has established its claims. It will doubtless turn out to be wrong in some of its conclusions, and it will be reformulated as it is subjected to greater scrutiny. Eventually it may serve as a beginning place for studying systems of morality and even as a foundation for philosophy. At this point, however, it is still working out its own identity and experiencing growing pains. Therefore it would be a mistake to ground biotheology in sociobiology. Instead we simply acknowledge sociobiology's promising beginning, and return to a broader analysis of biotheology's doctrines and concepts, armed with additional insight into their biological roots.

Notes

[1] For a thoughtful, civilized, and non-polarized review of the issues see several of the articles in *Morality as a Biological Phenomenon*, ed. Gunther Stent. See also "How Did Morality Evolve?" by William Irons, pp. 49-89 in *Zygon*, Vol. 26 no. 1 (March 1991); which includes an extensive bibliography.

[2] For more on the evolution of brain structures, review note 56 of Chapter 2.

[3] They sometimes interact in dynamic ways, however, when individual animals deviate from the usual conduct of their species; for example a murderer was recently found among non-human primates. See Hrdy, B.S. 1974. "Male-male competition and infanticide among the langurs (Presbytis entellus) of Abu, Rajasthan." *Folia primatil.* 22:19-58.

[4] See especially the interesting argument in *The Biology of Moral Systems*, by Richard D. Alexander (New York: Aldine de Gruyter, 1987). See also *On Genes, Gods and Tyrants: The Biological Causation of Morality*, by Camilo J. Cela-Conde (Boston: D. Reidel Publishing Company, 1987).

[5] See several of the contributions of *Evolutionary Ethics*, ed. Matthew H. and Doris V. Nitecki (Albany: State University of New York Press, 1993), especially "The Chimpanzee's Mind: How Noble in Reason? How Absent of Ethics?" by Daniel J. Povinelli and Laurie R. Godfey (pp. 277-324).

[6] Salmon are programmed to return to their birthplace to lay eggs, simply by following the smell of molecules originating there. But they have to remember the smell of those molecules, so "a minimum of learning" is still involved.

[7] For entertaining examples and citations to others, see *Cows, Pigs, Wars and Witches: The Riddles of Culture*, by Marvin Harris (New York: Random House, 1974). Also note that even the differences between male and female may lead to subtle differences in approaching moral issues; see D. Kay Johnston's "Adolescents' Solutions to Dilemmas in Fables: Two Moral Orientations—Two Problem Solving Strategies," pp. 49-72 in *Mapping the Moral Domain*, ed. Carol Gilligan, Janie Victoria Ward, and Jill McLean Taylor [Cambridge, Massachusetts: Center for the Study of Gender, Education & Human Development (at Harvard University Graduate School of Education), 1988].

[8] Two developing insights about innate aspects of morality complicate the concept in interesting ways. First, innate rules may have to be stated in complex

ways; for example, instead of "Thou shalt not commit incest," evidence from the Israeli kibbutzim suggests that the rule should be stated "Thou shall not commit incest with those to whom you have bonded by early psychological processes." Second, the innateness may require completion by experience, and a moral rule may develop in alternative ways depending on whether one has experience "A" or "B." It is possible to teach certain young songbirds one song instead of another, but once learned, that song and only that song becomes fixed in the bird's behavior. Humans are not as subject to this kind of imprinting as birds are, but our early experiences and/or moral training is probably etched into our neurons in some ways, so that modifications in the moral system have to respect this early momentum.

[9] One construction that aims to accomplish both tasks is Mario Bunge's *The Mind-Body Problem* (New York: Pergamon Press, 1980); but note the caution in David O. Hebb's excellent epilogue.

[10] I do not claim that the statements will always be explicit, because we do not always put our principles into language, but with effort a psychologist can usually figure out a person's moral rules. Further, Lawrence Kohlberg argues that moral principles develop in stages, which implies that they are (in some respects at least) independent of language. See his *Moral Stages: Current Formulation and a Response to Critics*, written with Charles Levine and Alexandra Hewer.

[11] The theologian Loyal Rue does not claim to have solved the is/ought problem, but with a minor adjustment at this point I think his schema would do as well as mine. If he inserted "a myth is a set of oughts" in the progression on p. 31 of *Amythia*, he would be postured at p. 46 to bridge the is/ought gap without resorting to the weaker pragmatic argument; he could make a *logical* bridge.

[12] But not so extravagantly as to exceed our psychological and physical carrying capacities—See Part 4.

[13] For an interesting discussion of Protestantism's ambivalence in deciding whether it wants to enter such arguments at the conscious level, see pp. 141-147 of W. W. Bartley III's *The Retreat to Commitment.*

[14] For example I can think of sins that do not fit into these categories neatly. One is conduct knowingly engaged in, which is neither mistaken nor particularly selfish, nor part of the broader societal dynamic, such as snubbing a friend who has a moral lapse. Another is our tendency to distinguish ourselves from family or culture through rebellion too destructive to be called healthy identity processes. I hope the text will be relevant to such subtleties, even though I do not directly address them.

[15] Romans 3:23. Are there exceptions? Might there exist rare persons who make ordinary mistakes without making the dangerous ones that constitute sin? One would expect such sins to vary across the population just as height does. Thus, just as there are athletes that build on their genetic endowment to become the best at their sport, there may be moral giants who commit precious few sins. Indeed, a Christ or a Buddha may live at the extreme end of the continuum, so that it is

poetically accurate to call them perfect. The vast majority of us, however, live somewhere this side of perfection, and make many bad choices.

[16] The tithe gives a hint as to where the line might be drawn. It codifies the observation and belief that we need to give 90% of our attention to our own needs and 10% of our attention to the needs of the group. Or perhaps a range might be more accurate, say from 5% to 15%.

[17] See Peter I. Kaufman's *Redeeming Politics* (Princeton, New Jersey: Princeton University Press, 1990).

[18] Several writers show how willfulness and selfishness arise from deep biological motivations, just as the morality which regulates them does. For example see G.S. Stent's introduction to *Morality as a Biological Phenomenon*, especially pp. 16-17.

[19] Romans 6:23.

[20] James 4:17.

[21] The definition is thus consistent with modern psychoanalysis, as articulated by M. Scott Peck in his book *People of the Lie: The Hope for Healing Human Evil* (New York: Simon and Schuster, 1985). His theme is that evil exists when we refuse to think about our behavior, and consider its causes and products.

[22] Cambridge: Harvard University Press. For an overview and response to critics see Wilson's "The Relation of Science to Theology," pp. 425-434 in Vol. 15 no. 4 (December 1980) of *Zygon.*

[23] especially Michael Ruse. See his *Sociobiology: Sense or Nonsense?* (Dordrecht, The Netherlands: Reidel, 1985).

[24] See the exchange between Ruse and others in Vol. 29 No. 1 (March 1994) of *Zygon.* Ruse's articles are "Evolutionary and Christian Ethics: Are They in Harmony" (pp. 5-24), and "From Belief to Unbelief—and Halfway Back" (pp. 25-35). The theologians in the exchange are Richard P. Busse and Philip Hefner. Busse wrote "From Belief to Unbelief and Back to Belief" (pp. 55-65). Hefner wrote "Entrusting the Life That Has Evolved" (pp. 67-73).

[25] I have already referred to Sociobiology several times (in section B2 of Ch. 1, notes 13 and 29 of Ch. 3, and Subchapter 4 of Ch. 7), and I will refer to it several more times (in section D2 of Ch. 10, note 24 of Ch. 11, C2 of Ch. 13, and D1 of Ch. 16). I hope those references show the basic compatibility between sociobiology and biotheology, despite the criticisms I make below.

[26] For a summary and further references see William Hamilton's "The Evolution of Altruistic Behavior," pp. 31-33 of Section 1 ("Group Benefit or Individual Advantage?") of *Readings in Sociobiology*, ed. T.H. Clutton-Brock and Paul H. Harvey (San Francisco: W.H. Freeman and Company, 1978).

[27] See *On Human Nature*, by Edward O. Wilson (Cambridge: Cambridge University Press, 1978).

[28] Lorenz was the founder of ethology (the comparative study of animal behavior), which was the direct intellectual ancestor of sociobiology.

[29] For a more complete criticism of kin selection see Chapter II, entitled "Critique of the Scientific Sociobiology: Kin Selection," pp. 17-67 of Marshall

Sahlins's *Use and Misuse of Biology: An Anthropological Critique of Sociobiology* (Ann Arbor: University of Michigan Press, 1976).

[30] For example in "The Fashionable Fallacy of Dispensing with Description," included in Richard I. Evans's *Konrad Lorenz: The Man and His Ideas* (New York: Harcourt Brace Jovanovich, 1975). For more on the interaction between behaviorism (a la Skinner) and ethology (a la Lorenz) see two selections in *The Selection of Behavior*, ed. A. Charles Catania and Stevan Harnad (New York: Cambridge University Press, 1988) namely Gordon M. Burghardt's "Ethology and Operant Psychology" (pp. 413-16) and Jack P. Hailman's "Ethology Ignored Skinner to its Detriment" (pp. 431-33).

[31] Oxford: Oxford University Press, 1989.

[32] Lewontin's *Not in Our Genes* was published in New York by Pantheon Books in 1984, and co-authored with Steven Rose and Leon Kamin. Sahlins's *Use and Abuse of Biology* was cited earlier.

[33] For an explanation of the criticism and a defense against it, see Chapter 5 ("Normative Criticisms," pp. 74-101) of Michael Ruse's earlier-cited book *Sociobiology: Sense or Nonsense?*

[34] See Edward O. Wilson's "Biology and the Social Sciences" in Vol. 25, No. 3 (September 1990) of *Zygon*, pp. 245-263. In this article Wilson postures sociobiology as linking two "antidisciplines," namely biology and social science. He suggests that those disciplines can now be studied as nesting hierarchies, like chemistry and physics can. Biotheology basically agrees with this insight, by presenting religion as an element of sociology (see Chapter 16).

Chapter 11

Salvation, The Search for Security

<u>An Outline of the Chapter</u>

A. Love
B. Righteousness
C. Grace
D. Thinking about Death
E. The Biology of Salvation

Establishing security is a primary goal of life. It is a difficult goal, because we are confronted by an astonishing variety of threats. Some of them arise from events over which we have no control, some arise from within human society, and some result from our own choices—we are threatened by weather, accident, disease, insects, economic chaos, crime, aggression, pollution, neurosis and psychosis, sexual dysfunction, alcoholism, and war, to name but a few. If we escape anxiety from these sources, death looms large to remind us of our ultimate insecurity.

We respond to such insecurities in various ways. Where possible we attack them directly, building houses to withstand weather, or establishing occupations to secure an income. We have made great strides in making life more secure over the last 10,000 years, through such inventions as agriculture and medicine. Yet it seems that every advance has brought new problems, leaving us with a great sense of insecurity even today. Since the need for security is so basic, we long for something that will supply it.

In virtually every culture, theological imaginations have constructed a doctrine of salvation to explain how one might find or achieve security. Not surprisingly, each such prescription is tailored to fit whatever threats and insecurities the culture perceives itself as facing, in concepts consistent with the prevailing worldview.[1] Sometimes we are urged to rely on physical strength, sometimes on magic, sometimes on the national group, sometimes on a specific person or god, and sometimes on a doctrine or group of doctrines. Almost always, the search is for an absolute security, something or somebody upon which we can completely depend.

Over time, however, it has become clear that no absolute security exists. Even if we could overcome the threats from outside human life (which we cannot[2]), the ones that arise from inside our species are virtually impossible to eliminate. Indeed, removing the source of those insecurities—by genetic engineering, for example—would probably disturb our positive personality traits as well. Aggression causes war, but it also contributes to the search for new medicines and to the search for mates. Thus we must accept the fact that some insecurity will continue to exist, even while we work to reduce it.

Traditional theology resists this conclusion, going to great lengths to preserve supernatural concepts that would provide an absolute security after all. Modern theologians, however, have come to terms with the inevitability of some insecurity,[3] and have approached that inevitability sensitively and profoundly. Biotheology agrees with these theologians, yet it may be more optimistic than they are about the degree of security possible. While it is indeed impossible to resolve every insecurity, it is possible to find a reasonable security, and even achieve a measure of psychological luxury.

This chapter will begin the search for security by discussing its basic personal aspects, reserving more complex social aspects for later chapters. At the outset, however, we must avoid the naïve conclusion that everybody will be saved. People sometimes have trouble finding even the most basic security. Incompetent parenting or economic chaos can leave us completely lost, and when that lostness is deepened by every additional circumstance, it becomes almost impossible to find security. So our salvation is not assured, and it behooves us to uncover every resource that might aid in the search for security. Especially should we search for ways to overcome the insecurity that results from our own bad choices; in other words we should look for ways to be saved from sin.

Traditional theology is a major resource in that search, because several of its doctrines directly address our need for security, and urge us to make choices that can contribute to our salvation. Three of those concepts are especially compatible with biological understanding, and with minor modification they neatly tie earlier chapters of this book to traditional understanding. So let us spend most of this chapter looking at the concepts of love, righteousness, and grace.

,ove

Like every other human trait, the complex set of behaviors and attitudes ıat we call love, evolved. Its components (such as pair-bonding and lliance-formation) can be found in simple animals. Even blackbirds rely n relationships that parallel our friendships, family relationships, and pposite-sex partnerships.[4] As usual, humans exhibit a particularly com-lex form of the behavior, making easy understanding of its dynamics npossible. We do know, however, that love has always been a dominant ıeme of human life, in one form or another.

Interestingly, it has not always been a dominant theme of human eligion. Fertility gods are quite old, and sexual themes have long been part f religion, but only with the New Testament has love been granted a major nd conscious role in religion, as a feeling that we are exhorted to extend ›ward one another.

One of the modern Christian theologies goes so far as to equate love and iod.[5] Conservative Christian theologies don't go that far, but they give ›ve a central role nonetheless. They say God's love makes Him reach own to us and offer us salvation. Judaism says that too, but Christian onservatives say we can only be saved by accepting the offer. They also ay we evidence our salvation by reaching out to others in love, and on that oint liberal Christians and Jews agree with them.

Biotheology also emphasizes love, seeing it as a dominant human haracteristic. Not only is love necessary to healthy pair-bonding, but it nderlies all healthy interactions within society, including especially amily relations. It impacts on theology precisely because it is a profound art of our lives, broader and deeper than the simple concepts of primitive eligion. That does not mean primitive peoples did not experience love; only means love did not find its way into their concepts in the same way finds its way into ours.

Love is relevant to salvation because it directly enhances security. hapters 4 and 7 discussed early influences on the child's development, ut I have not yet acknowledged the importance of love in that develop-ıent. If little Mary's caretakers love and protect her, she is made more hysically secure. If they communicate their love in calm and confident ays, she is made more psychologically secure, and her identity processes re off to a good start.

Of course love produces security in combination with other processes, nd its interaction with faith is particularly powerful. If little Mary's love-xperience fails to make her feel secure, her belief-structure may feature uspicion and distrust, and she will have difficulty becoming a competent dult. If she does experience the love that instills security, her worldview will be more trustful, and thus connect her to other people in ways tha further increase her security.

Love is as important to adults as it is to children. If I am loved by spous and family and friends, my well-being is enhanced and my identit affirmed. If I am part of a loving community that accepts and supports m even while gently challenging me, I am likely to feel very secure indeed It is especially important for adults to have love as a significant componen of their belief-structure, and to include such love-processes as forgiveness community, edification, and exhortation as sub-units in their framewor of understanding. Those dynamics also hint at grace, which will b discussed in a moment, but their importance for establishing securit through the operation of one's worldview should already be clear.

Love clearly supplies great advantages in meeting life's challenges But human love processes seldom operate flawlessly, nor does love b itself guarantee salvation, because other things can go wrong even if w receive perfect love. This means we cannot depend entirely on love for ou security; we need other processes too.

Righteousness

Early love gives our basic identity a shove in the right direction, but i is difficult to develop or maintain a secure identity without structure Parents who competently discipline their children impart critical psycho logical structures, which will supply as much foundation to the childrens secure identity as love does.[6]

The relationship between righteousness and love is thus easy t understand at one level, because loving discipline implies the gentl imposition of behavioral structures, and we call behavior which is consis tent with that structure righteousness or "right living." Discipline impart family values to growing children and orients them within the broade culture, so that a sense of personal righteousness slowly emerges. Th more loving and competent the discipline, the smoother the transition wil be from imposed discipline to personal righteousness.

Righteousness and love interact in another way, which affects every body to one degree or another. When parental love fails to instill rea security, an adopted righteousness can confer a kind of artificial security Thus a young adult, deprived of an adequate childhood, may gravitate t a sect or philosophy that supplies a sense of security. By acting as behavioral anchor, this adopted righteousness supplies some of the orien tation and identity that parental love may have failed to provide.[8]

Adopted righteousness may be artificial, but it provides a bette foundation for healthy security than can certain other efforts to compen ate for early insecurity, such as drugs or promiscuity. Granted, if the iitial insecurity is too great or if the adopted righteousness is too rigid, the ecurity provided by an adopted righteousness can be quite fragile. But if ie righteousness is not too rigid, and even if it is to some extent artificial, may stabilize one's worldview. That stabilization can then soothe earlier irts by contributing to a more natural love-experience. Thus, adopted ghteousness may allow one to climb out of a pit that was not initially voided, and position one to find salvation after all.

The personal disciplines that grow from childhood structures or dopted righteousness require some conscious attention, but they also epend on subconscious processes which may or may not be programmed onsciously, such as habit. Konrad Lorenz demonstrated the relation of abit to security, even in animals. We (and other animals) do things the ame way over and over because our brains are programmed to make hoices that have proven safe. In fact we often do what has proven safe ven if it otherwise makes no sense, and even if it is harmful to us, provided ie harm is not immediately evident. Habits may thus provide a great deal f temporary security, and save the enormous energy that would be equired if we had to rely entirely on our conscious minds. They let us elax.

Of special interest to theologians is the kind of habit called ritual. By oming together and regularly performing the same acts and words, we nd reassurance and security. I already mentioned Moses's genius in eating rituals for the Hebrews, and we can extend that observation to say l viable societies must employ rituals. Ritual is a "group habit" that applements personal habits to provide baseline security, and avoids the ecessity of constant conscious vigilance.

The idea of group habit brings up another meaning for the word ighteousness." So far this section has used it to denote the structure that ows from discipline or habit. I think that is a fair usage, one which yields isight into the origins of psychological security. Yet it is not the most ommon usage. The most common usage equates righteousness with onformity, conformity to social mores as reflected in religious codes of thics produced and maintained by the group. Conformity is itself a kind f structure and a kind of habit, so let us now look more closely at this kind f righteousness.

People undoubtably carved out a measure of security long before hilosophy or theology existed.[9] But thoughtful people eventually noticed iat security can be enhanced by extracting understanding from experi-nce, and fashioning it into codes of belief and behavior. These codes are ome of culture's most prized products, and it is a primary task of theology nd philosophy to construct and/or maintain them and teach them to the ulture.[10] Historically, some such systems have been rigidly imposed by the force of law, and some have borne only the power of social persuasio or disapproval, but in both cases it was—and is—typically claimed th conformity to the system ("righteousness") is what will save a perso whereas departure from the system will subject a person to great risks.

There is something to be said for such an argument, because we cann easily live without a code of rules as part of our conceptual framework, an we cannot easily construct a personal code that adequately predicts all th challenges we might confront. Especially in times when people lived bri lives and had little opportunity for reflection and study, it was probably good idea to conform to traditional codes and participate in tradition rituals, without expending much energy evaluating whether the traditio was good or bad or somewhere in between.

Over the course of history, however, the concept of righteousness ha changed.[11] As the interaction between cultures increased, a major par digm shift occurred. The notion of personal responsibility was imagine and slowly worked its way into systematic understanding, modifying th requirements of righteousness away from rigid conformity. By the tim of Jesus, Pharisaical rigidity to cultural rules could be seen as the outmode concept it was, at least in areas of personal morality and belief.[12] Th newer paradigm permits us to incorporate traditional components into o worldviews if we choose to do so, but it primarily calls on each of us construct a personal righteousness, and discipline ourselves to live by i

Chapter 7 recommended cultural belief-structures as beginning plac from which to develop a personal faith. Chapter 10 urged cultural moralit as a beginning place from which to develop a personal code of behavio Now it can be seen that the need for security was an underlying reason f suggesting such preliminary conformity. By beginning with conformit one automatically gets a measure of security, both because of the tim tested nature of cultural faith and practice, and because conformity avoi the antagonism provoked by deviant behavior.

But the section on idolatry (in Chapter 6) should show the limits of th kind of righteousness. Unless our belief-structures and behavioral stru tures are dynamic and modifiable, we will be conforming for conformity sake, rather than for the sake of salvation. Our righteousness will be lik dry bones, hard and rigid, and we will miss the joy of personal involveme with our own choices. Our righteousness will be imposed from outsid rather than growing from within. Instead of functioning as a "schoolma ter to salvation," the law will become a confining institution from whic we never graduate, and our security will be as brittle as glass.

But if we adopt a vibrant righteousness instead of a rigid one, it offe a significant increase in security, by giving the advantages of a code whi allowing us to grow from individual experience. Such a code does requi us to reduce our experience to principles, to remember those principles, iscipline ourselves to follow them, and to correct our mistakes. But ortunately, we humans have the capacity to do all those things. Rigidity an always seep back in, but we can greatly reduce the danger of rigidity y adding grace to our baseline fund of love and righteousness.

Grace

The most famous scripture about grace says "For by grace are ye saved rough faith, and that not of yourselves . . .[13]" Yet in its effort to describe race, the New Testament honestly reflects the transitional confusion of n old righteousness confronted by the new paradigm of individual esponsibility that Jesus preached. Some passages seem to accept Jewish galism,[14] but other passages reject it altogether.[15] This raises the question f whether grace is compatible with traditional faith and righteousness, or omething fundamentally different.

I think they are compatible. It must certainly be accurate to say that alvation is through faith, since it seems unlikely that anyone could chieve security without an adequate belief-structure. But although the cripture says salvation is *through* faith, it says salvation is *by* something lse, namely grace. After exploring the original meaning of grace and the ualities it refers to, this section will conclude that the scriptural passage uoted above is quite literally true, and yet quite consistent with the ersonal discipline required by righteousness.

That task is made more difficult, however, because classical theology ade a counterfeit concept of grace. It invented highflown meanings and chnical jargon, largely to fit grace into the straightjacket of ancient upernatural concepts. The resulting conceptualization of grace has hin-ered real grace from operating as a natural part of our lives, and it is also ontrary to the original New Testament meaning of the word. That eaning was really quite elegant, and a bit of word-study should help us place the classical concept with the original one.[16]

The Greek word for grace was *charis*, from which our words charity nd charisma come. *Charis* came from a root meaning "calm and cheerful emeanor."[17] Perhaps the best modern word to capture its meaning is gracefulness." Two athletes might be equally proficient, but one may be uch more graceful, making difficult maneuvers seem easy and fun.[18] imilarly, some people approach life with a grace that vastly overshadows e plodding awkwardness or ordinary competence experienced by the est of us.

According to the scripture, it is this grace that saves us from the nxieties of life, from being lost in the complexity that surrounds us. racefulness is especially powerful for resolving the most serious insecurities we face today, namely those which arise from the difficulties human interaction. By getting and maintaining a peaceful and cheerf demeanor, we may avoid many such problems. Instead of respondir reactively to what happens around us, and instead of constantly trying protect ourselves from one another, we approach life with a calm assuran that resolves most of our interaction problems, when it doesn't preve them from arising in the first place.

That is easy enough to say. The question is, how do we get this grac and how do we maintain it once we get it?

In the best of worlds, grace flows naturally from love and righteousne and faith-development, without much effort on our part—it is truly "n of ourselves." As Chapter 8 would say, the spirit of love—God—com into our lives by the grace of others. Yet it also takes effort to develop ar maintain love and righteousness and faith, and they continue to produ the serene security of grace only if we remain sensitive to each new choi of belief and behavior. Earlier chapters implied this, but those chapte concentrated on cognitive aspects of our faith. I am now suggesting th our worldview must actively embrace the emotional qualities of calmne and cheerfulness, if we wish to live gracefully. Of course the embrace mu be genuine; otherwise the "calm and cheerful demeanor" becomes mere a gladhanding facade which will crumble under stress.

"Calmness and cheerfulness" are not the only attributes of *charis* grace; I merely used them to illustrate the larger array of mental process that make up grace. Let me now complicate the concept a bit by explaini four related attributes which we can adopt to increase grace and augme our security. Those attributes are humility, slow change, acceptance, a mercy.

1. *Humility.* By this word I do not mean any Caspar Milquetoa meekness. Nor do I mean the kind of awe we feel at contemplating t vastness of the universe or the power of God. I mean rather a simp acknowledgment that we might be wrong.

This does not imply wishy-washiness or insecurity with our beliefs. C the contrary, we can live confidently and exuberantly even while knowi that some of our beliefs probably are wrong. If we have worked to ma our beliefs consistent with the spirit of truth wherever it conflicts with t spirit of tradition, and if we are open to change them when persuaded th they are wrong, then we can rely on them in the meantime, relaxed in t knowledge that they can eventually change again. This means a humb person is more likely to be outgoing and enthusiastic, than a meek and m cartoon caricature such as Caspar Milquetoast or his modern equivale Ziggy.

Indeed, humility can even be assertive. Assertiveness is not aggressive insistence on one's beliefs or desires, but it is a calm and dire

aaring of them. A person who is assertive but humble is usually the
)itome of grace. One of our best examples of how to combine those
ıalities is Jesus (at least where he is presented as a struggling model rather
an a master model[19]), and fortunately the combination occurs often
ıough for most of us to know someone who exhibits it.

2. *Slow change.* To maintain a secure and yet humble gracefulness, it
probably necessary to keep some control over changes in beliefs, not in
der to keep change from happening, but to let it happen slowly.
lthough some people have Damascus Road experiences that dramati-
ılly redirect their lives, sudden changes are usually quite stressful. The
ıbconscious requires time to process ideas, and even if a dramatic change
ıs been incubating in the subconscious for a long while, it is a good idea
give conscious processes some time to work also.

Normally, slow change happens naturally through gestalt inklings and
her mental processes, but in dynamic societies like ours, it is necessary
give conscious attention to the principle of slow change. Otherwise we
it from one belief to another, never embracing any faith firmly enough
establish a strong foundation for living. An example will show how one
oup makes slow changes, and although it is a superficial example, it
:monstrates the advantage of slow change.

Most marching bands use a simple style: four persons abreast and
venty-four deep. But the most impressive bands often do it sideways, in
ng lines twenty-four abreast and four deep. It is difficult to keep a long
)rizontal line straight, and that is exactly what makes it so impressive.
he difficulty comes when one end of the line glances down toward the
her end and finds itself a bit behind. If that end compensates by taking
ant steps, and if the other end simultaneously compensates by taking
ıby steps, the line whips down the field in waves, creating a most
nbarrassing caricature of marching.

The secret is to make slow changes. If one end of the line slightly
creases the length of its step, and the other end slightly decreases the
ngth of its step, the change comes in such minute increments that an
)server cannot detect it, and the lines appear impressively straight to
levision viewers and grandstand crowds alike. Turns are especially
aceful, but they use the same technique, requiring only slightly more
actice.

Similarly, if we stay humble and modify our concepts in small
crements, our faith grows steadily, remaining stable through all life's
rns. We gain the benefits of experience and learning while maintaining
stabilizing connection to our previous faith. As we encounter new
ıallenges, we can remember that we have made changes before, and
ılmly begin to make new ones, without feeling we have to do it all
vernight.

3. ***Acceptance.*** Sometimes we must respond quickly. Life can dramatic, and graceful living must include more coping strategies th slow change and humility. Yet by incorporating the attitude of acceptan into our framework of belief, we can respond even to dramatic change a positive way. That is because the concept of acceptance is broad enou to embrace the whole range of possibilities inherent in human life a indeed in all of nature. We must accept the possibility of tornado divorce, economic disaster, personality quirks, and everything else th history has shown can happen. We may increase our understanding those happenstances by reading history and psychology and by observi life around us, but we can only profit from that study by standing ready accept such events in our own lives.

That does not mean we can avoid all anxiety; indeed it would unhealthy to maintain a stoic security in the face of trauma. In a page two we will face the ultimate trauma—death—and it will focus o thinking on the insecurity that comes from really major traumas. But those cases too, a degree of acceptance will help us maintain our grace a security.

At a less intense and yet more pervasive level, there are dozens events we must accept. We have already addressed the probability mistakes which can temporarily disorient us. We may work to minimi them, but Chapter 10 showed how inevitable (and even desirable) they a Thus we cannot live gracefully and stay uptight about our mistakes or t mistakes of others.

Finally, we must each accept our own history, not only our genetic a cultural and family history, but our personal history of experiences a beliefs and actions. That history usually serves as a dependable founc tion, but it inevitably contains cracks and instabilities which could beco dangerous. By accepting that fact, we can avoid the common impote frustration of wanting to have a different history than we actually have, a we can begin to make needed changes slowly and confidently.

4. ***A note on mercy.*** Some theologians say we are saved—that is, avoid the pitfalls of sin and mistake—by the mercy of God.[20] Howev the scripture is clear that it is grace and not mercy that saves Theological literature has got the two all confused, and I want to expl why it is grace that saves us.

Mercy, or the granting of an undeserved reprieve, is certainly example of grace at work. One occasionally hears about a judge w decrees a fine and then pays it for the lawbreaker. In appropriate cases, t is not only merciful but exceedingly graceful, and indeed mercy and gra may be the same thing in those instances. But that is because mercy i kind of grace. All apples are fruits, but not all fruits are apples; similar all mercy may be grace, but not all grace is mercy.

With that in mind, you can see why it is grace and not mercy that saves us. Sometimes we do avoid life's pitfalls and become more secure by the exercise of mercy, either by being merciful ourselves or by receiving mercy. But the more general experience, the one that saves us from even more of life's problems, is gracefulness arising out of an adequate faith-structure.

These four sub-sections of the grace discussion presented the claim that grace saves us, and the three full sections on love, righteousness, and grace explained how the dynamics represented by those concepts interact to produce personal psychological security. But I have delayed one element of traditional salvation doctrines until now, and that is the question of whether we can be saved from death.

Thinking about Death

The ultimate insecurity, I called it above. From primitive times forward it has fascinated us. Salvation doctrines don't seem adequate no matter how much psychological security they give us, unless they also protect us from death. Whether that is accomplished through a heaven, reincarnation, or merger into an oversoul, we apparently feel more secure if someone in authority assures us that individual consciousness will survive the grave. Doctrines such as the duality between matter and spirit support the various afterlife concepts, and doctrines such as hell and purgatory compensate for their problems.[21]

Yet the afterlife concept cannot confer absolute security, unless we can also be convinced that it is certain, or else that faith requires us to believe things in the absence of supporting evidence. Traditional theology inevitably takes one of those two approaches, but some doubt lingers anyway, even among the faithful, and we are left insecure despite the afterlife concept. On the other hand, one is able to feel secure, even without believing in an afterlife, by coming to terms with the cessation of one's individual consciousness. Thus belief in an afterlife is not the only route to security, or even the best. It may even decrease our security, for example where afterlife doctrines make people more willing to kill one another in battle.[22]

Biotheology has no concept of an afterlife, except in the sense that new organisms must have material from which to be constructed, and thus old organisms must die. Even if biological and medical advances allowed for the conquest of death, birth would have to cease too, lest the accumulation of living things exceed the mass of the earth. More fundamental is the fact that life depends on genetic processes, and death is important to biology because it permits genetic vigor in a way that everlasting life could not. By allowing for mutation and modification in response to environmental challenges, death adds to the health and security of the species, even though it makes individual life less secure.

But whether we believe in an after-life or not, our nature as myth-makers (in the positive use of that term) requires the construction of explanations that are consistent with the facts as we judge them. Even if we lay aside our concept of heaven, we still need an acceptable account of what happens when we die. And if the story is to give us psychological security, it must be reasonably optimistic, and it must be consistent with the other concepts in our belief-structure.

Biotheology believes that people die, that that is the end of their physical existence, and that there is no credible evidence of a non-physical realm. After death our bodies return to dust, and re-enter the chain of life through normal processes of decay and reabsorption. No doubt the atoms of our bodies have been re-used millions of times to participate in thousands of life-forms. It is a glorious process, one that we might be just as enthusiastic over as heaven, were we not so worshipful of human consciousness, and so ego-centered in our own individual consciousness. The processes of death and rebirth extend our molecules into the future and give assurance that they will continue to participate fully in life. If one is less than inspired by such a scenario, the realization that death ends consciousness may inspire us in another way, by highlighting the importance of living each day to the fullest, without depending on a fantastic afterlife to give meaning to the life we are given.[23]

Of course there is one way to extend our physical selves into the future, more exactly, while transmitting human consciousness, and that is by having children. That continues our genetic existence to a significant extent, even if our precise combination of genes ceases to exist. Childless persons may even contribute to keeping most of their own genes alive, if that is important to them, by helping their young kinspeople.[24]

But in any case, death cannot terminate a more permanent kind of human consciousness. During our lives we contribute to lives around us or to the culture as a whole, and our contributions certainly survive us. Those contributions can satisfy our need for emotional ties to the next generation, and connect us profoundly to the future of our species. Nor must our contributions be dramatic in order to accomplish this function—the cumulative effect of small individual contributions in the past is profoundly important to our security today, and our individual contributions likewise secure our place in the story of human life, whether or not they are recorded in history books.

For sociological and psychological reasons, this approach to immortality will not satisfy people who feel insecure about their place in life. Th loss of an afterlife will discourage those for whom it compensated for that insecurity. But if the loss makes us redirect our efforts toward enhancing security within the context of this life, and toward finding ways to help people feel good about their contributions to life, then death can become what it ought to be—a natural and unavoidable conclusion to life. Then instead of expending our efforts on the impossible search for absolute security, we may expend them on the search for reasonable security.

The Biology of Salvation

The acceptance of some insecurity is a profound development within theology and related disciplines. By abandoning magical and supernatural doctrines which were necessary to support the concept of absolute security, our imaginations have been freed to address pathological insecurities which we might actually be able to remove. Developmental psychology, for example, has identified specific points at which insecurity can infect the growing child, and cognitive psychology has built on that understanding to develop therapies for correcting the resulting pathologies even in adulthood. Modern theology attempts to reduce our insecurities by urging us to participate in God through love and grace, and to remove sin from our lives so as to avoid the suffering which interferes with that participation.

This more realistic approach to our pathologies is based on the biological principle that physiological events accompany mental processes, acting as feedback loops to add or detract from our sense of security. For example blood pressure is lower when we are calm and graceful; various states of arousal accompany the interactions of love; we are alternately depressed and energized by the disciplines of righteousness or the contemplation of death; etc. Inside the brain, electrochemical processes are working: memory and logic are receiving hormonal messages which can trigger or calm anxiety; faith requires the interaction of conscious and subconscious mental states; programming ourselves for acceptance requires the storage of information about the various events that might happen in our lives; and so on.

The question is, how do we influence our brains and physiology to produce psychological security when we are feeling insecure? Can we go to a brain surgeon, and ask for an operation using tiny tweezers to disconnect one set of dendrites and connect them somewhere else, so that our thinking patterns are surgically altered? No, of course we cannot do that, though we can indeed alter some of our thinking patterns through the application of psychoactive medicines. But those medicines must be reserved for clear instances of well-understood pathologies, lest we take unnecessary risks of interfering with other thought patterns that need no adjustment.[25]

Yet even without medicine or surgery, we have many ways of responding to the world, which do the same thing as the little tweezers would do. By embracing the values of calmness and cheerfulness we literally influence the flow of chemicals and hormones within our nervous system, for example. By adopting a moral code we create memory structures which we access in the midst of life's struggles.

We can influence this physiology in quite superficial and transitory ways (taking drugs or eating ice cream or buying new shoes, for example), but there are also ways that we can enhance our security in more permanent and substantive ways. Just as we can become more physically secure by putting burglar bars on our windows, we can make choices that make us more psychologically secure. We can read and think and discuss options with our friends. We can participate in churches and other groups (see Chapter 16). When those resources fail, a large counselling industry stands ready to suggest dozens of tried-and-proven techniques for changing our mental pathways, entirely without medical intervention. In recent years that industry has expanded its services and competencies far beyond pathological conditions, and it routinely helps normal people change their thinking patterns to achieve a much greater degree of security.

If we can make choices that increase our security by modifying our mental pathways toward greater love, righteousness, and grace, might we go further? Might we make choices in both behavior and thought patterns that provide more than basic security? Might we find happiness and joy? Those questions will be addressed next.

Notes

[1] See Chapter III (pp. 62-71) of *The Idea of God*, by John Fiske and Chapter II (pp. 27-43) of *The Growth of the Idea of God*, by Shailer Mathews.

[2] Through all of history, humans have been able to intentionally eliminate only one organism (the smallpox germ). Even a relatively large organism like the cockroach would probably be impossible to eradicate.

[3] Especially Paul Tillich. See his *The Courage to Be* (New Haven: Yale University Press, 1952).

[4] See Konrad Lorenz's paper "Companions as Factors in the Bird's Environment." In *Studies in Animal and Human Behavior*, Vol. 1.

[5] Based on the assertion that "love" is a predicate nominative rather than a predicate adjective in the scripture "God is love." See # 1 under "Modern concept of God" in Chapter 8.

[6] Parents who want to avoid the harshness of physical discipline must nonetheless supply discipline in gentler (and quite time-consuming) ways. Since

most animals do use physical discipline, a biological approach to psychology tends to affirm the scriptural wisdom "He that spareth his rod hateth his son" (Proverbs 13:24).

[7] For a good discussion of both healthy family love and its substitutes, see Chapters 6 ("The Child in the Family: Loving Comes from being Loved," pp. 162-193) and Chapter 8 ("The Social Control of Love," pp. 238-268) in *The Meaning of Love in Human Experience*, by Reuben Fine (New York: John Wiley & Sons, 1985).

[8] A good example is the "12-Step" program of Alcoholics Anonymous. It is well described and argued for in *The Answer to Addiction: The Path to Recovery from Alcohol, Drug, Food, and Sexual Dependencies*, by John Burns, David Randall, Robert Calhoun, and Samuel Abrahamson (New York: Crossroad Publishing Company, 1990). For two very interesting discussions of cults in American experience see *Heaven on Earth*, by Michael D'Antonio (New York: Crown Publishers, Inc., 1992) and *Cults and New Religious Movements*, ed. Mark Galanter (Washington, D.C.: American Psychiatric Association, 1989).

[9] Hume's case for the "non-philosophical" operation of human dynamics is well presented in Chapter One of Jeffrie G. Murphy's book *Evolution, Morality, and the Meaning of Life*, pp. 17-20 (Totowa, New Jersey: Rowman and Littlefield, 1982).

[10] For a well-written account of this process at work in a non-literate culture (the Navahos) see *The Structure of a Moral Code*, by John Ladd (Cambridge, Massachusetts: Harvard University Press, 1957).

[11] Many good examples of recent changes are found in Gerhard Lenski's *The Religious Factor: A Sociological Study of Religion's Impact on Politics, Economics, and Family Life* (Garden City, New York: Doubleday, 1961). At pp. 270-273 for example, Lenski chronicles the change from valuing straightforward obedience to valuing intellectual autonomy.

[12] Of course there are many areas where conformity remains important, for example in the legal realm. See Part Four.

[13] Ephesians 2:8.

[14] For example "For verily I say unto you, Till heaven and earth pass, one jot or one tittle shall in no wise pass from the law, till all be fulfilled." St. Matthew 5:18.

[15] For example "Ye are not under the law, but under grace." Romans 6:14.

[16] I am not suggesting that original meanings of words are always the most valid or useful, though I am arguing that that is the case here.

[17] See numbers 5485 and 5463 in James Strong's "Greek Dictionary of the New Testament," included in his *The Exhaustive Concordance of The Bible* (Nashville: Abington Press, 1890; 32d printing 1974).

[18] The athletic example is particularly appropriate, because Greek culture was attuned to grace in athletics. And Greek culture was the source of the New Testament's concept of grace.

[19] For example in Chapter 5 ("The Humanity of Christ," pp. 68-86) of *Images of Christ: An Introduction to Christology*, by Glenn F. Chesnut (Minneapolis: The Seabury Press, 1984).

[20] A good example is Richard L. Purtill's "Justice, Mercy, Supererogation, and Atonement," pp. 37-50 in *Christian Philosophy*, ed. Thomas P. Fling (Notre Dame, Indiana: University of Notre Dame Press, 1990).

[21] The abortion controversy is too politically intense to address subtle theological problems just yet, but eventually somebody's theological imagination will have to ask how soon after conception an aborted fetus would go to heaven. Catholic doctrine has sometimes solved such problems with limbo, but conservative protestant theologians will have to compensate for this problem in some other way.

[22] This is the point of Jimmy Carter's excellent book about the Middle East, entitled *The Blood of Abraham* (Boston: Houghton Mifflin, 1985).

[23] John Macquarrie treats all these issues (and more) very sensitively in Chapter XVIII ("Death," pp. 233-242) of his *In Search of Humanity: A Theological & Philosophical Approach* (London: SCM Press Ltd., 1982).

[24] A large literature on "the selfish gene" has accumulated since the publication of Dawkins's book by that name (New York: Oxford University Press, 1976), emphasizing the help typically given to nieces and nephews, for example. But of course one increases the fitness of their genes more broadly when they make contributions that ensure the survival of the human species. So I do not mean to make too much of this point.

[25] For a very thoughtful and provocative essay on the nature and use of psychoactive medicines, see Peter Kramer's 1993 best-seller *Listening to Prozac*, cited earlier.

Chapter 12

Abundant Life: Happiness, Values, and Self-Esteem

In Chapter 10 I juxtaposed the hardships encountered by a plant growing on the edge of its range, over against the luxury and abundance typically enjoyed by organisms living near the center of their niche. I implied that morality offers us "abundant living" if we give attention to this fact. Unlike a plant, humans can make choices about where and how to live, and those choices permit us to intelligently search for the center of our niche and discipline ourselves to live within it.

But the phrase "abundant living" has some unrealistic and visionary connotations, so I want to begin by explaining how I use that phrase and the related word "happiness."

The concept of abundance entered western culture by way of the Bible, and is frequently encountered in both the Old and New Testaments. The Old Testament usage is broader (as one would expect from a document distilled over a longer time period), referring both to physical abundance[1] and psychological abundance.[2] Early Christians knew little of physical abundance, and the New Testament therefore concentrates on psychological abundance.[3] Nonetheless, it promises its readers physical abundance too, if they first concentrate on the kingdom of God.[4]

Whether referring to physical or psychological abundance, biblical abundance always implies luxury—more than enough. And that is how the word is used secularly too. Thus this chapter will use abundance to refer to a condition in which we have plenty of physical resources for our needs and desires, and also the resources for psychological well-being. To be abundant, these resources must not only contain the potential for base-line satisfactions, but for a high sense of harmony with our fellow humans and with the world at large. The operative word is "potential." Even when our resources provide the potential for abundance, we don't always exercise the choices and judgment necessary to reap its advantages.

Abundance is an attractive concept, but it is not usually considered necessary for happiness. That is because "happiness" connotes a lesser degree of satisfaction than "abundance" does. Yet the degree of satisfaction implied by happiness is itself highly desirable, and although the two concepts are sometimes confused, most mature adults would accept simple happiness, without insisting on abundance. Thus one usually feels satisfied if she or he has sufficient physical, emotional, and intellectual resources, without the luxury of abundance. Accordingly, this chapter will first address happiness, recognizing the possibility of abundance once basic happiness exists.

A surprising number of issues can complicate the definition of happiness, and philosophers have asked some interesting questions to clarify just what it is. For example, is a man happy while blissfully unaware that his wife is having an affair? We won't explore this or other interesting questions,[5] but several formal definitions address them. Fortunately, most such definitions try to keep the common-sense meaning of happiness, incorporating minor complications for technical sufficiency.

Of several such definitions, the one I like best is Ruut Veenhoven's.[6] He defines happiness (p. 22) as the degree to which individuals judge the quality of their life-as-a-whole positively. Yet because happiness can describe different mental states, he divides it into three components, namely the affective component, the contentment component, and the overall component. Since "affective" is not a common word, and since Veenhoven uses contentment to refer to the more-or-less rational part of happiness, I will call the three components the emotional component, the rational component, and the overall component.

For our purposes it won't be necessary to constantly distinguish between these three components, and except when looking at subtleties we may think of happiness in its ordinary way, as a level of satisfaction somewhat less than abundance, but more sustaining than temporary joy or physical pleasure. The real question is not definitional anyway, but practical. Can we achieve happiness and/or abundance, and if so how?

The Conditions of Happiness

Let us begin by acknowledging the role of luck. We have no control over when or where we are born or to whom, and those factors can have an enormous impact on our happiness. For example Veenhoven found that people are significantly less happy if their nation loses a war during or just before their lifetimes.[7] And he cites many other factors associated with happiness that depend on luck. Luck is ignored by many philosophers and most theologians, but it is consistent with a biological understanding of life. Nothing is guaranteed, not even the existence of the species, much less the happiness of one individual.

Apart from luck, our happiness is impacted by various group actions or decisions, such as the passage of laws or the spread of social movements. This is part of luck too, except insofar as we may choose to become active agents in persuading society toward actions and decisions that can enhance happiness. Societal dynamics will be discussed further in Chapters 13-16.

But even after setting aside the role of luck and group dynamics, another factor significantly determines our happiness, and that is the choices we make. Regardless of the situation we find ourselves in or the actions of the group, we almost always have a chance to find happiness. Sometimes it takes great effort and insight, but the opportunity is usually there.

Is that true? And if so, what efforts and what insights can increase our chance of finding happiness? Fortunately, many profound thinkers have tried to answer that difficult question down through history, and some philosophers are still bold enough to tackle it today. Indeed, great progress is being made, especially by those philosophers and social scientists who study values. So after establishing some background by looking at four classical theories of happiness, I will present a value-based approach that retains the good features of these classical concepts, and avoids their problems.

1. Classical concepts

a. The Golden Mean. Living within the center of one's niche may sound like a modern biological concept, but it was foreshadowed by the Greek theory of "the golden mean." Greek philosophers formulated it to express their belief that happiness is found near the mid-ranges of human experience: Don't get too skinny or too fat; don't be too passive or too aggressive; don't be too early or too late. Indeed, when the Oracle at Delphi was asked which religion was the true one, she replied "The religion of your own country," reiterating that one should never stray too far from the center.

Under Greek influence, Christianity adopted the golden mean with its passages on moderation. Further, the Christian concept of grace originated in Greek philosophy, which depicted grace as the balance achieved by following the golden mean. Nor is the golden mean limited to Greek and Christian thought, being similar to several other doctrines, such as the Buddhist "middle way."

Although the Golden Mean is a compelling concept, it is not without problems. For one thing it urges us to live in the center of culture, rather than in the center of human nature. If a culture is not itself near the center of human nature, the golden mean cannot lead to a deep happiness consistent with our biological makeup.[8] Nor does the golden mean address the fact that some individuals have a nature which varies from the mean.

The golden mean also forgets that we sometimes need to escape the average. Socrates and Jesus were killed because of their unusual beliefs, and we say they exemplified the highest possibilities in human nature by their willingness to suffer death, even though the golden mean would have advised against that willingness. In fact, the golden mean encourages harmful or sinful activities, if everyone else is doing them.

Despite these problems, the golden mean provides keen insight, because it frankly acknowledges our ignorance about human nature, and suggests that we choose the middle path as a kind of "law of averages" which is unlikely to take us very far astray.[9] By gaining additional insights from other classical approaches, we may be able to compensate for the problems of the golden mean, and modify it toward greater usefulness.

b. The Rule of Law. The rule of law is like the golden mean, because it urges us to accept culture's distilled path to happiness and virtue. But whereas the golden mean approves of individual effort in figuring out the mean, legalistic approaches point out how impossible that is. This reasoning concludes that one should simply trust the law, and discipline oneself to live by it. That idea is well represented in the preacher's despairing conclusion to "fear God, and keep his commandments: for this is the whole duty of man."[10]

While I have already shown the considerable merit of this approach as a beginning point, it has serious defects. Not only can it too easily induce rigidity and idolatry, but it also denies variety and complexity and change. Therefore it cannot act as a guide to happiness without significant modification.

Jesus understood that law becomes rigid if we don't heed aspects of our nature more basic than law, such as emotions and attitudes. Yet his followers quickly returned to legalistic rigidity, and his ideas have yet to find a firm foothold in our culture. Evidently we humans gravitate to rules, and when insecurity threatens, we cling to those rules tenaciously.

c. The Golden Rule. Jesus said something which can help avoid rigidity, and yet give reign to our rule-needing natures. We call his saying the golden rule—"Do unto others as you would have them do unto you."[11] This rule affirms subjectivity and variety, and trusts each person to construct a personal discipline and righteousness.

In terms of a simple standard that each person can apply from his or her own experience, it is hard to beat the golden rule. Provided only that we follow another Greek maxim and "Know ourselves," the golden rule is a biologically sound and ever-present guide, as modern as our natures. If I wouldn't want you to pollute my atmosphere, I shouldn't pollute yours.

Nevertheless, there are two problems with the golden rule. The first is unavoidable and yet seldom noticed. Because of human variety, there are many instances where the golden rule can lead us to treat one another badly. For example, suppose I am a quiet person who values privacy. When I am sick I enjoy visits from close friends, and I enjoy reading. But I do not want mere acquaintances, much less perfect strangers, coming by to cheer me up. You, on the other hand, are an extrovert, and enjoy meeting new people. When sick you would appreciate a constant stream of visitors, whether friends or acquaintances or strangers. Look at what happens when we each apply the golden rule:

When you are sick I will do unto you as I would have done unto me. Assuming we are acquaintances but not good friends, I would not visit, because I wouldn't want you visiting me. That might hurt your feelings, and it would deprive you of one of the visitors you would surely appreciate. If other people interpreted the golden rule the same way and if they were like me, you might lose half your visitors or more.

Similarly, if I were sick, you would apply the golden rule and come visit me, even though I would not appreciate it. In fact I would have many more visitors than I wanted.

Perhaps this way of applying the golden rule would be good, because even though we would not get exactly what we wanted, at least everyone's responses would be genuine. Or perhaps we could avoid these problems by adopting the oriental version of the golden rule, which says "do not do unto others that which you would not want done unto you." This negative version would keep us from directly harming one another, and leave us to use other standards for our positive behavior.

The golden rule has a second limitation, in that it only seeks to tell u how to treat other people. The search for happiness is broader than tha involving both our treatment of ourselves and our interaction with th larger environment. If my happiness is to be found by tilling the soil c typing at a computer, for example, the golden rule gives no hint of it. Th is no criticism of the rule, because it does not claim to be a comprehensiv rule. It guides our human interactions, and does it well where commonalit exists; outside that range we must look to other principles for guidance

Notwithstanding these two problems, the Golden Rule added subjec tivity as an important ingredient in the search for happiness, whic perspective is validated by current research, and which must be incorpo rated into any modern theory.

d. The Rule of Utility—the greatest good for the greatest numbe The three approaches discussed above were articulated 2000 years ago c more. In the ensuing centuries, people still strove for happiness, bu philosophers gave little attention to happiness as a subject of inquiry Perhaps that was because life was so hard during the early Christian er the Dark Ages, and the Reformation. But interest resumed in the 18th an 19th centuries, when Jeremy Bentham and other philosophers searched fc practical ways to increase happiness. They were called utilitarian philoso phers, and their opponents accused them of trying to construct a "happi ness calculation," by which one could figure out just how to maximiz pleasure and minimize pain.

These utilitarian philosophers made great contributions to politic philosophy, and paved the way for the so-called "worldly philosophers, who become the first economists. They and their immediate predecessor were instrumental in the adoption of happiness as a major goal of the U.S Constitution. Yet one particular problem plagued them: the conflic between the individual and the group.

They finally resolved this problem by inventing the concept of "th greatest good for the greatest number." This concept accurately pinpoint a part of our natures, because "the greatest good for the greatest number is built into our genes. We evolved as a species which relies mightily o the group, and therefore we have innate tendencies which enhance th group at the expense of the individual. But do we really find happiness b sacrificing ourselves to the group? Might we not do better to frankl acknowledge an inbuilt competition between ourselves and our ow species?

For example, our innate urges call us to have children, and to hav many of them. That was an excellent strategy when our species needed t keep its numbers up, and it also increased individual security. But heedin that call nowadays puts us at a real disadvantage, and we are clearly bette off by putting ourselves ahead of the species, by having fewer children or none.[12]

Another example of the conflict between the group and the individual concerns the innate urge to believe what leaders say. That urge helps to produce unity, whether leaders are calling us to war or to the cultural faith. Frequently, however, such blind belief harms individuals, who must learn to take "the greatest good for the greatest number" with a grain of salt.

Despite this difficulty with utilitarianism, it focused attention on the fact that our happiness is intertwined with the group—often we do find personal happiness by serving the group. But it is possible to give the group too much allegiance, and we must sometimes put ourselves first. So the utilitarian philosophy of happiness must be modified like the others.

These four theories of happiness do not exhaust the philosophical literature.[13] Moreover, proponents of each approach have "tightened up" their arguments to avoid their problems. Although I think they are basic problems not to be solved by these minor adjustments, it is certainly necessary to incorporate major portions of each approach into whatever theory of happiness we accept. In any case, the discussion above should reveal the main currents of classical literature, and position us to look more intelligently at modern research, and especially at value theory.

2. Modern research

Two researchers have provided great insight into the requirements of happiness. One is the Dutch sociologist Ruut Veenhoven, whom I have already mentioned twice, and the other is the philosopher Robin Barrow.

Hundreds of research projects have focused on happiness, and Veenhoven has performed a monumental service by attempting to correlate them. His book *Conditions of Happiness* is scholarly and practical, it contains a complete bibliography, and it exhaustively examines the research through 1979 (another book is in the works for the research projects since then).

Sociological research typically asks people a wide range of questions. The resulting data provides hard demographic facts, information about health and income, and people's perceptions about such matters as their happiness and contentment. The researcher may correlate one kind of data with another, and discover (for example) whether religious people are more likely to report happiness than non-religious people, or whether third-world citizens are more likely to report happiness than citizens of industrialized states. Naturally there are many difficulties in such research, both in collecting the data and in interpreting it. Any two researchers may use completely different concepts and definitions, an look for completely different correlations.

Yet Veenhoven was able to wade through the work of many othe researchers, and draw some very interesting conclusions. Reviewing ove 1000 studies, he finally selected 245 that could be meaningfully comparec (Previously the most comprehensive survey had compared only 50 stud ies, of which Veenhoven found only 12 useful.) His few conclusions ar modest but firm, and usually hold across time and culture. Among his mos interesting conclusions are the following (p. 349):

There is a very significant correlation (in the 40% range) betwee happiness and

1) Satisfaction with intimate relations, especially marriage;
2) Satisfaction with income; and
3) Satisfaction with oneself.

There is a secondary but still significant correlation (30% range between happiness and

1) Satisfaction with one's leisure;
2) Satisfaction with one's health; and
3) Satisfaction with one's work.

Veenhoven's book is as important for what it disproves as for what proves. For example it disproves both the long-standing notion tha happiness is found in ascetic self-denial; and also the currently popula notion that western civilization is a sink of unhappiness. In fact th contrary is true, largely because of the positive correlation betwee income and happiness. (Note, however, that this correlation has an upwar limit; beyond a basic income, additional money adds little to happiness.[14]

In the pursuit of happiness, it would not be enough to simply rea Veenhoven's book, because its purpose is not to map a path to happiness Whereas this chapter will try to show how an individual person ca actually become happier, Veenhoven simply undertakes to show whic conditions are statistically associated with happiness, from the researc amassed so far. He achieves that purpose very well, and can thus be a grea resource, but he offers no overall framework for working toward happi ness.

The phrase "philosophical research" may seem like a contradiction i terms, because philosophy implies opinion, while research implies facts Some philosophers, however, have done an excellent job of showing ho their opinions arise from factual research, and how the tools of philosoph may be applied to that research. In such cases the phrase is a valid one

In particular, Robin Barrow has elegantly merged philosophy an research in his treatment of happiness.[15] He shows how logic prohibits u from drawing certain conclusions, and how it propels us towards othe conclusions. He uses much of the same research Veenhoven does, but he is able to organize it and reason from it in very persuasive ways.

Barrow ultimately concludes (p. 126) that three things are necessary to happiness, namely (1) Realism; (2) Security; and (3) Self-esteem. You can see that each of Veenhoven's findings will fit into one or more of these categories; for example income has an obvious relation to security, and "satisfaction with oneself" is almost the same thing as self-esteem. The third category, realism, is a basic requirement that underlies all healthy values, as we shall soon see.

3. *Values*

Judgments and choices are not always consciously rational, but they apparently strive towards a kind of sub-conscious rationality. That conclusion arises from a huge literature spanning several disciplines, on the subject of values.

Values themselves come largely from our historical evolutionary experience,[16] which produces some values common to all humans, but also allows some values to emerge from our individual experience, including our experience with culture. Thus an individual value-system emerges in each person which may or may not be similar to the value-systems of other people in the culture.

Does this sound familiar? Since I defined faith quite broadly, to include not only theological belief but one's entire worldview, it is clear that a faith structure and a value system are one and the same thing. Moreover, research demonstrates that values are based on beliefs—beliefs about facts and probabilities, rather than the other way around (which would hold that beliefs are based on values).[17] Even when values are regarded as arising from emotions rather than from rational processes it turns out that values are based on beliefs, and not vice versa.[18] *Attitudes* are also intimately connected to values. Indeed, an attitude may be the same thing as a value,[19] but even if they are different it is clear that attitudes are also based on beliefs about facts. All of this is consistent with the faith chapter, and indeed it strengthens that chapter considerably.

In this chapter I will use the language of values instead of the language of faith, because the values literature is so extensive and so well supported by research. However, there is one important issue in that literature which could be resolved by thinking in terms of the biological and social origins of faith, which I would like to address before looking more closely at the study of values.

That issue is whether values are "agent-relative" or "agent-neutral," which is another way of expressing the problem of relativity in morals. "Agent-relative" refers to values which vary from individual to individual,

whereas "agent-neutral" refers to values which are common to everybod
without considering idiosyncrasies of any particular agent or individua
The question is, must all values be one or the other, or is there room fc
both?

In principle that is an easy question for biologically oriented philosc
phies, because the human species (like any other species) contains bot
commonality and variation. In humans the variety is amplified by cultur
and free will, which guarantees that many values will be agent-relative.
few values, however, are certain to be agent-neutral. Veenhoven'
research tends to emphasize the subjective (or agent-relative) nature c
those values which are correlated with happiness, though some of hi
research would support the existence of some values which are commo
to everybody.[20] (To review the biological approach to commonality an
variety, see the discussion of deep structures and surface structures i
Chapter 3.)

The question of varied (agent-relative) values vs. common (agen
neutral) values is a minor issue for most value theorists, which they do nc
address in depth. I mention it in passing only because I think biologic
analysis avoids a persistent source of confusion. But perhaps the mo
important point made by value studies isn't too greatly infected by th
confusion, and so we can simply make note of it and pass on.

What is the most important point of value studies? It is that happines
can be promoted by one's value system (or faith structure or worldview
That can best be shown by narrowing this section's beginning questior
Instead of asking "How can one find happiness?" we can ask a simila
question, namely "What values can promote happiness?" After discussin
that question at length, the chapter will conclude by recalling som
principles from earlier chapters, which can transform simple happines
into abundant living.

What Values Promote Happiness?

By correlating value studies with the sociological and philosophic
research above, we can clearly see that some values can indeed increas
individual happiness. Specifically, adopting and nurturing values th
increase *security*, *realism*, and *self-esteem*, can greatly enhance our chanc
for happiness.

These three categories, which I take from Barrow, are not mutual
exclusive. Indeed they are highly interrelated. But they do provide a usef
outline for discussing those values and choices which most enhanc
happiness; and I will point out overlaps to avoid any confusion whic
might be caused by the interrelatedness of the categories.

1. Values that Increase Security

Chapter 11 dealt with security, but its emphasis was on the development and maintenance of psychological security, and on the qualities that may protect one from the serious mistakes and immoralities of sin. Physical security was mentioned in passing, but Veenhoven's work shows that it should not be underestimated—for example I have already noted that a decent income is especially correlated with happiness. Several other values researched by Veenhoven also enhance happiness by increasing security, both physical security and emotional security.

In fact, of Veenhoven's six critical correlations extracted above, four of them may be considered as sub-sets of Barrow's security category. They are income satisfaction, intimacy/ marriage satisfaction, health satisfaction, and job satisfaction. Clearly, pursuing these values can increase happiness. If we adopt them, we will try to make decisions which keep income coming in, which nurture our primary relationships, which maintain or restore health, and which find work that is consistent with our other values.

Since I have already given security a great deal of attention, and since I will return to it again later, I will not discuss it further here, except for some additional comments related to the overlap between security and self-esteem.

2. Values that Increase Realism

When Barrow lists realism as a requirement of happiness, he means we cannot be happy unless we overcome our tendency to seek happiness in fantasy and other forms of unreality—we must deal with life as it is rather than as we might like it to be. The values literature greatly expands this concept, and makes realism central to any healthy value system.

This is largely because of the importance of "decentering" in the development of the growing personality. As explained by the celebrated psychologist Piaget, decentering is the process by which we learn to think objectively. Since it would make no sense to define objectivity as outside *anybody's* thinking processes, decentering means learning to consider perspectives other than our own.[21] Thus objective realism refers to the ability to step into someone else's shoes, and understand how those shoes might fit if we were wearing them ourselves.

Indeed, the opposite of decentering is egocentrism, defined as a child-like failure to see beyond one's own narrow needs. It is pathological in that it causes great suffering, both for the individual and for those round about. As with most pathologies, it can occur in varying intensities, and although the basic decentering process usually happens automatically, we must maintain it by consciously de-centering from time to time, lest we get locked into our own perspectives.

Valuing such activities as reading, conversation, education, and democratic dialogue clearly enhances our ability to de-center. Thus if one does not adopt these values or similar ones, there is a real danger of remaining or becoming egocentric, and thus unrealistic. According to Barrow, this will significantly interfere with our chances for happiness.

Inasmuch as values are based on beliefs, a major goal of this book is to urge readers toward realistic beliefs. By understanding and accepting our evolutionary nature, we should be in a better position to avoid concepts that encourage super-natural, fantastic, or unrealistic beliefs about the world and our place in it. Of course imagination is a part of our nature, and we must listen to its (sometimes fantastic) suggestions, but we must test the reality of imagined ideas in objective ways such as those outlined in Chapter 7, if we wish to achieve happiness.

3. Values that Increase Self-esteem

Barrow's third category (and the third of Veenhoven's three most significant findings) is self-esteem. I must immediately note, however, that self-esteem intertwines with security. That is because one usually gets both self-esteem and security from competent parenting. Yet self-esteem may be conceptually distinguished from security, and it may arise even in the absence of competent parenting. As Barrow implies, self-esteem virtually guarantees happiness, provided one also has a realistic worldview and a secure place in the world. On the other hand, it is hard to find happiness without self-esteem. Thus any value-system that enhances happiness must include values that tend to build and maintain self-esteem.

Self-esteem also intertwines with realism. Indeed, the intertwinement of self-esteem and both other categories is so profound that I must make two general observations, before discussing specific values that enhance self-esteem. First, values that increase security will also increase self-esteem. Second, values that increase realism will also increase self-esteem. In fact those two ways of increasing self-esteem may outweigh the additional ones listed in this section.

Those additional esteem-increasing strategies are nonetheless important, and when added to the other two, go a long way toward assuring self-esteem. The three to be discussed here are 1) having consistent values; 2) meeting one's own expectations for oneself; and 3) making good decisions.

Veenhoven thinks value-consistency may be overrated, but the values literature underscores the need for it, at least for one's core values. Gerald Gaus establishes that a reasonable consistency unifies our personality and increases self-esteem. If we don't have that consistency, we will be prone to serious personality disorders if and when we are faced with trauma. This is clear from the effect of the German holocaust on certain of its survivors, and also from other research.[22] Thus if we would increase happiness by increasing self-esteem, we should strive to maintain some consistency in our value structure, and root out inconsistencies which might undermine our other efforts toward happiness.

Veenhoven notes the great correlation between happiness and meeting one's own expectations, and indeed that is the source of his category "satisfaction with oneself." This is a complex topic, however, and offers no neat conclusion such as "adopt a value for being more satisfied with oneself." Simply knowing that such a value is correlated with happiness might program the subconscious to accept it, but we are more likely to find satisfaction with ourselves if we first understand how we come to have self-expectations that we cannot meet.

The literature of perfectionism shows how that happens,[23] and helps therapists correct the perfectionistic messages that often lead to undue self-criticism. This is not a self-help book, and it would take a separate book to show how one can modify unrealistic expectations, but let me list at least a few techniques which can reduce perfectionistic thinking, as a means of explaining the concept further.

a. Listing one's values in writing goes a long way toward making them more realistic. Many different attempts have been made to compile a complete list of human values, and although there is no consensus as of yet, such listings can help one identify one's own values.[24]
b. Ranking these values in some semblance of priority usually reveals conflicts in one's own values. This revelation reiterates the realistic fact that one cannot completely implement all of one's values in every circumstance, and it programs the subconscious to begin resolving those inconsistencies by clarifying the ranking of one's values.
c. One can further clarify one's values by considering the various sources of those values. Indeed, mental health probably depends on the conscious use of at least six sources of values.[25]
d. Direct reduction of self-expectation is usually possible. As simple as it sounds, Benjamin Franklin's advice is wise, and may be applied to more than just income. He said a person may increase satisfaction in either of two ways: by increasing means or by decreasing desires. In theology, Buddhism seems too extreme in its efforts to decrease the desire for achievement and possessions, and yet we probably can increase self-satisfaction significantly by reducing our desires and expectations. In fact the research on aging shows that we all tend to do that as we age, and that may be why

> older people tend to be more satisfied with themselves than the rest of us are.[26]

The matter of meeting one's own expectations is intertwined with the matter of making good decisions, because a "good decision" is one which furthers one's higher values, and avoids the easy mistake of pursuing lower-ranking values while momentarily blind to higher ones. That kind of mistake can undermine one's self-esteem, but it nonetheless happens quite often.

Adopting the value of making good decisions requires a mental commitment, but it also requires learning specific skills for implementing one's values. Those skills should of course acknowledge and consider the entire range of our values, but insofar as we desire happiness, the decision-making process must give special weight to values which enhance security and realism and self-esteem.

Again, it would take a separate book to discuss good decision-making in depth: fortunately many books exist which can help us make good decisions. Some of them are designed for making minor decisions quickly, without much time for gathering additional data.[27] Others require more mathematical rigor, perhaps by prioritizing and quantifying values on a written chart.[28] But each of them teaches skills that will help one develop confidence in one's decision-making process, and that confidence will contribute greatly to the development of self-esteem.

The Possibility of Abundance

This chapter has taken a values approach to increasing happiness. It has argued that we can find happiness by adopting values that enhance realism, security, and self-esteem. For most of us that would be enough. Indeed, if we walked daily in the assurance that we had those qualities, happiness might be indistinguishable from abundance, especially if these qualities are defined to include attributes discussed in earlier chapters.

Thus security would include grace and love and righteousness (see Chapter 11), as well as a good income and meaningful work. Self-esteem would include a set of moral principles and a faith that gave the group its due while satisfying our individual needs; thus we would love our neighbor *as ourselves*. Realism would include an understanding of our place in evolutionary history, a natural concept of God, and an acceptance of such realities as sickness and death.

This would indeed be abundant living, but since we evolved as social creatures, one additional thing would make our luxury complete, and that is a fuller participation in society. Earlier chapters have frequently noted the importance of the group, and Part Four will now examine particular kinds of interaction between us and the group, especially those interactions that have a great bearing on happiness.

Notes

[1] For example "fat cattle and sheep in abundance," "abundance of spices," "iron in abundance," "brass in abundance," "cedar trees in abundance," "victual in abundance," "sheep and camels in abundance," and "money in abundance," to list a few luxurious phrases. For others see any concordance of the Bible.

[2] For example "honour in abundance," "abundance of peace," "abundant in goodness and truth," "abundantly satisfied," "abundantly pardon," and "the abundance of thine enchantments."

[3] For example "abundance of the heart," "abundance of grace," "abundance of their joy," "abundance of the revelations," "his inward affection is more abundant," "abundant mercy," "ministered unto you abundantly," and that most sensuous phrase "the abundance of her delicacies."

[4] For example "Seek ye first the kingdom of God, and His righteousness, and all these things will be added unto you." St. Matthew 6:33.

[5] Robin Barrow discusses several in *Happiness and Schooling* (New York: St. Martin's Press, 1980).

[6] *Conditions of Happiness* (Boston: D. Reidel Publishing Company, 1984).

[7] *Conditions of Happiness*, pp. 170 and 364.

[8] If people are happier in western societies (as Veenhoven claims), it may be because those societies more accurately reflect human nature than other societies do, presumably by finding a balance between cooperation and competition. This balance will be a theme of Part Four.

[9] For a modern defense of the golden mean, see Richard Bosley's dialogue *On Good and Bad, whether happiness is the highest good* (Lanham, Maryland: University Press of America, 1986).

[10] Ecclesiastes 12:13. Note that the preacher had a vivid theological imagination, considering several routes to happiness—money, wisdom, mirth, wine, work, beauty, fame, and power. Yet he examines them one by one, and never tries to harmonize them into a more sophisticated combination. His musings could have led to the golden mean, and his ideas might have helped Jesus conceptualize the golden rule, but all in all the book is best understood as an early and quite simplistic exercise of the theological imagination.

[11] Of course Jesus wasn't the only one to state the rule; various versions are found in other cultures.

[12] Actually, the innate urge is not to have children but to have sex. The "mothering instinct" may be real, but it probably does not arise until a child is born (instigated by vision-mediated and smell-mediated bonding) or at earliest with the hormonal flow during pregnancy. Science and rationality thus allow us to respond

to biological urgings after all, without necessarily producing too many children. Chapter 14 will build on this observation and conclude that both groups and individuals would be better served by limiting the number of children we produce.

[13] For example I omitted the Epicurean philosophy, though it is mostly included in the utilitarian one. Nor have I engaged the "noble savage" view found in Rousseau and (subtly) in "New Age" philosophies. Both seem refuted by the happiness research which follows, except insofar as they recommend adequate relaxation from the rigors of philosophy and religion, which I mean to accomplish in my emphasis on the subconscious. I also omitted Aristotle's notion of eudaimonia, mainly because it is really discussing virtue rather than happiness though obviously the two are related.

[14] For an excellent article on the practical implications of this finding see John W. Kendrick's "Happiness is Personal Productivity Growth" in *Challenge* Vol. 30 no. 2 (May/June 1987), pp. 37-44.

[15] *Happiness and Schooling*. Despite the title, it is really a broad-based analysis, and has relatively little to say about schooling.

[16] See George E. Pugh's *Biological Origin of Human Values* (New York: Basic Books, 1977).

[17] In fact it has been argued that our top 5 beliefs have particular influence on our values, and largely determine many of our decisions. See pp. 223-4 of *Belief, Attitude, Intention, and Behavior* by Martin Fishbein and Icek Ajzen (Reading, Mass.: Addison-Wesley, 1975).

[18] See Gerald F. Gaus's *Value and Justification* (New York: Cambridge University Press, 1990).

[19] See p. 112 of Gaus's *Values and Justification* and authorities cited there.

[20] And of course some writers in the study of values have taken this approach explicitly. Especially see George Pugh's *Biological Origin of Human Values* cited above.

[21] Gaus explains and expands on Piaget's ideas at pp. 200-201 in his *Values and Justification*, cited above.

[22] Gaus, Gerald F. *Value and Justification*, p. 230. The holocaust example comes from Bettelheim, Bruno *The Informed Heart: Autonomy in a Mass Age* (New York: Free Press, 1960), pp. 120-121.

[23] For example see "Perfectionism in the Self and Social Contexts: Conceptions, Assessment, and Association with Psychopathology," by Paul L. Hewitt and Gordon L. Flett, in *Journal of Personality and Social Psychology*, Vol. 60 no. 3 (March 1991), pp. 456-470. This article also contains an extensive bibliography on perfectionism.

[24] See *Choices and Decisions: A Guidebook for Constructing Values*, by Michael Bargo, Jr. (San Diego: University Associates, 1980).

[25] See *A Question of Values*, by Hunter Lewis (San Francisco: Harper & Row, 1990).

[26] See pp. 249-50 and 404 of *Adult Development and Aging*, by Margaret H. Huyek and William J. Hoyer (Belmont, California: Wadsworth Publishing Company, 1982).

[27] For some examples see *No-Fault Marriage*, by Marcia E. Lasswell and Normal M. Lobsenz (Garden City, New York: Doubleday, 1970).

[28] See *A Practical Guide for Making Decisions*, by Daniel D. Wheeler and Irving L. Janis (New York: The Free Press, 1980). For a more philosophical and technical treatment see *Hard Choices*, by Isaac Levi (London: Cambridge University Press, 1986) or the more user-friendly *The Art of Making Choices*, by Ian McGreal (Dallas: Southern Methodist University Press, 1953).

Part Four: Biotheology in Group Life

Introduction to Part Four

Individuality is a wonderful thing: the ability to diverge from the group and solve problems in one's own way, gives "freedom" a deep and profound meaning.

Yet life is almost always lived in the context of a group. We make our choices, we exercise our free wills, we adopt a morality and experiment with it, we process experiences, we examine theories, we construct our faiths, we modify paradigms, we seek happiness and security, and we do all the other things discussed in earlier chapters, as members of a group. This part of the book will examine that aspect of human experience. The individual will not be abandoned, however; each chapter features a section on the individual's role in group life.

Any *theological* discussion about groups must begin with an acknowl edgment of the ancient idea of community, and Loyal Rue's *Amythic* shows just how profound that idea must be for biology-based theologies Moreover, Donald Ortner has argued that since cultural group adaptation are far more dynamic than biological ones, we must encourage social conditions that foster individual creativity to solve our problems—thus we must focus on individuality within the context of society.[1]

We interact with society in dozens of ways, but four ways are crucial and these final chapters will discuss them. Chapter 13 looks at economic life, highlighting biology's recent contributions to the so-called "disma science." Chapter 14 (on politics) presents some biological concepts tha can help solve the most serious problems we face: environmental destruc tion, war, and over-population. Chapter 15 discusses education, and th need for a teaching theory that take advantage of innate learning tenden cies. Finally, Chapter 16 outlines religion's tasks in this Age of Biology some of which are traditional and some of which are new.

Despite the broad subject matter of these four chapters, this part of th book sets forth no idealistic program. Its only claim is a modest one namely that biological concepts offer new ways of looking at traditiona problems, and can lead to better solutions than many current philosophie do. That is because those other philosophies seriously misjudge huma nature, either by presenting a picture that is too idealistic or one that is to pessimistic. Biotheology's vision of the future is built upon more realisti appraisals of our history and nature, and of the behaviors we can reason ably expect from our fellow humans.

Notes

[1] See pp. 127-159 (Chapter 5, "Biocultural Interaction in Human Adaptation" of *How Humans Adapt* (Washington, D.C.: Smithsonian Institution Press, 1983

Chapter 13

Economics: Harvesting the Earth's Bounty

Professional economists quite rightly approach their subject as a complex mathematical model, but we can gain a great deal of insight from a simple classical definition: Economics is *the use and distribution of the earth's resources.* This traditional definition immediately reveals that all of us are intimately involved with economics; we make daily decisions and judgments that become the raw material of statistical econometrics.

Recently, two scholars with expertise in economic anthropology teamed up to show that economic decisions are greatly influenced by a single factor, one that looms larger than any other. They argue that the use and distribution of resources cannot be understood without giving great weight to this one factor—the density of the human population.

In their book *The Evolution of Human Societies: From Foraging Group to Agrarian State,*[1] Allen Johnson and Timothy Earle explain our fascinating procession from a simple small-group existence to the complex social structures we live in today. Their theme is straightforward: As we successfully increased our numbers, it got more and more difficult to make a living in a reasonable number of hours per day. Moreover, our increased density made it necessary to develop technology, institutions, and moral principles which allowed more efficient harvesting of available resources. Without the pressure of numbers, our species could live quite comfortably without working so hard, and without so many institutions and rules.

This is a remarkable thesis, so let us spend a moment exploring th history and developments set out by Johnson and Earle.

There is little doubt that we evolved as a small-group species. Durin our formative eons and right down to 10,000 years ago, we lived an evolved in groups of about 40, typically made up of 5-10 families—i small herds, as it were. Even though we slowly expanded our populatio and geographical range, the economic "system" in effect was apparentl the same everywhere, and featured simple hunting and gathering. W primarily occupied lands that provided excellent human habitat, wher one could conveniently avoid mosquitoes and find good clean water, an collect enough food to live on in a few hours each day. Of course ou existence was not entirely idyllic, any more than was that of other creature we shared space with. We hadn't the foggiest notion of how to cur disease, and occasional famines or hurricanes could devastate particula populations. Such an existence had its charms, but from a biologica viewpoint nothing is static, and change was abrewing from the start.

That is largely because these small foraging groups were successfu breeders. From two hundred thousand years ago forward our species wa probably just as intelligent as it is now, and its "investments" in intelli gence and sociality paid off in population growth. Although certai physiological processes kept the growth rate low for thousands of years, we slowly expanded our range until all the good land was filled.

Other dynamics besides population growth also contributed to th expansion of our range. Normal conflict (over sexual partners, for ex ample) caused small groups to split and migrate into new territory. An since it was usually simpler and less energy-expensive to move on than t do battle, conflicts were probably confined to the individuals directl involved in them. Leadership existed and could theoretically have led t large battles, but primitive group leadership tends to be unconcentrate and varies with the task, so it is unlikely that any leader ever led a larg number of people until much later.

Because so much good habitat was available, population growth di not noticeably affect the structure of economic life for millennium upo millennium. As total population grew, population density stayed about th same, so we could continue meeting our needs through foraging. Yet goo human habitat is sometimes separated by hundreds of miles, and eventu ally we had to start filling secondary lands. Some families moved down th mountain to a swampier site, though they had to contend with mosquitoes Some moved up into the tundra, and had to cope with cold. Some move into the desert, and had to learn to find unusual sources of water.[3]

Once the prime habitat was filled, something very interesting hap pened. In order to support increasing population densities, we turned ou imaginations and intelligence to finding more efficient ways to min available resources. Those mental abilities had evolved to solve other problems, and even this problem had already arisen in local areas, but now it confronted us so profoundly that simple solutions didn't work anymore. Fortunately our inventive minds were up to the task.

Storage was one of the first inventions. Groups in one seasonally harsh environment learned to survive on pine nuts for two months each year, and another group learned to live on pure animal fat for several weeks in winter. With or without storage, trade became important. Coastal populations learned to trade fish for furs. Rock-dwelling groups traded stone axes for clay vessels.

Human imagination not only fashioned such inventions as storage and trade, but also institutionalized them. Feast days were at first spontaneous, a natural impulse to share windfalls such as beached whales or slain mastodons. But population pressures led to the ritualization of feasts and other events. Invited guests were expected to reciprocate when they had a windfall. Proverbs urged everyone to eat all they could before winter—thus even human fat became an institutionalized kind of storage.

These institutions were invented in a more-or-less specific order tied directly to increasing population densities. Marriage rules, formal games, prohibitions against laziness, and other developments can be shown to offer particular efficiencies in using resources. Even war became cost-effective once population densities reached a point where more people could survive by defending (or seizing) valuable resources (salmon runs, for example) than by looking for other ways to survive. Bloodshed was no longer sporadic and personal, and warriors emerged as a special kind of leader. Again, these possibilities had long lain latent in human nature, with violence erupting sporadically in limited conflict, but circumstances now called them to the fore. At the same time our capacities for simple gathering, while not diminished, began to be called on less often.

Finally we got so crowded that these inventions were inadequate to feed everyone, and a fundamental shift became necessary, which we now call the birth of agriculture. No doubt there had been isolated examples of plant and animal domestication, but the permanent invention of agriculture can be traced to about 10,000 years ago.

This shift happened in an opposite cause/effect direction from what one might think. We usually think the excitement of new inventions and concepts beckoned us forward into agriculture, which then released the population explosion. But actually population had been increasing all along; its pressures now forced us to squeeze more resources from the land, and agriculture was the result. Thus agriculture developed as a necessity,[4] not an opportunity. Humans had been intelligent enough to develop agriculture for tens of thousands of years, but there was no need. Why should someone work hard tilling the soil when there was plenty of food for the taking, by working just three hours per day?[5]

It is true, however, that agriculture added new fuel to our growth. A horticulture and animal husbandry spread over large parts of the world they provided the food and social structures that propelled further popu lation growth. More complex institutions came into being that intensifie our use of natural resources—institutions such as tribes and classes an law. Primitive leadership was refined into more elaborate forms, leadin to a generalized "political leader" who tended to replace the specialize leaders that had sufficed before.

For a long time, new technologies and social structures and hard wor permitted us to survive and prosper, even with greater and greate population densities. Eventually, however, our very success began t stretch the earth's resources, until today they are so stretched that eart cannot support us all. We in the west are still relatively isolated from th grosser consequences of growth, but other parts of the world face intens problems caused directly by increased population density, and we ar affected more than we realize. The recent history of starvation wi demonstrate just how close we are to catastrophe.

Starvation

We have good estimates of human population growth over historica time, as summarized in the following chart:

Year			Estimated Population
1,000,000	b.c.	-	125,000
300,000	b.c.	-	1,000,000
25,000	b.c.	-	3,340,000
8,000	b.c.	-	5,320,000
4,000	b.c.	-	86,000,000
1	a.d.	-	133,000,000
350	a.d.	-	254,000,000
1200	a.d.	-	384,000,000
1600	a.d.	-	498,000,000
1750	a.d.	-	791,000,000
1800	a.d.	-	978,000,000
1850	a.d.	-	1,262,000,000
1900	a.d.	-	1,650,000,000
1950	a.d.	-	2,486,000,000
1965	a.d.	-	3,289,000,000
1975	a.d.	-	3,967,000,000
1990	a.d	-	5,000,000,000 [6]

This explosive growth reiterates our ancestors' successes, but it does ot adequately reveal the many instances of hardship they faced. Through-ut our history many individuals and groups have perished because of the avages of disease and weather, and large-scale starvation became a reality nce agriculture arose to support ever greater densities. Our earliest ecords already document frequent famines,[7] but recent centuries make hose early deaths seem insignificant, because we have been starving in ecord numbers for the last few centuries. In fact each century for the last our has set a new record of starvations. This century set its record in a ingle country in four short years (China, 1959-62, when 27-30 million eople starved to death), and it is now almost certain that the 21st Century vill set yet another record.[8]

Here is the numerical record of recent starvations:[9]

Century	Number of People who Starved to Death
17th	2,000,000
18th	10,000,000
19th	25,000,000
20th	27,000,000+ in China alone

It is true that starvation's causes are complex—political conflicts and veather contribute, for instance. But the basic cause underlying more mmediate causes is that we are exceeding the land's carrying capacity. ;very farmer knows an acre of pasture will only support a given number f cows, and yet we have been slow to apply this basic biological fact to ur own existence. One example can stand for a hundred cases, and that s the Nile valley. It could luxuriously support 10 million people, even lespite occasional drought or pestilence, but we have packed its fertile lain with 50 million people. The resulting stresses make abundant living lifficult indeed.

Great hopes have been placed on high-yield crops to prevent starva-ion, but if we continue to reproduce at present rates, it is only a matter of ime before the mass of human flesh outweighs the planet. Absent lunar arming or space travel, it will be impossible to feed everyone.[10] So omething must be done. And that something is to control our population ize.

That is the task of politics and education and religion, and Chapters 14-6 will return to it. It is a long-term solution. In the meantime, we must sk why we can't just return to foraging or simpler living; we must ask what conomic structures can give temporary relief; and we must ask whether moral individual can survive and prosper in today's population-pres-ured world. The rest of this chapter will address those questions.

Why Can't We Just Return to Foraging?

If foraging is so wonderful, letting us survive on three hours' work day, why not just go back to it? Wouldn't it express our essential humanit to interact in small groups as humans were meant to do? Wasn't Roussea right to praise the noble savage, and aren't hermits our highest form humanity?

These suggestions miss two important points. First, foraging was n idyllic. Disease, fear, and other dynamics kept it from being a romant picnic. Second and more important, ongoing change is just as natural a foraging was. The growth described above is normal biology, and norm biology never returns exactly to where it has been before. It is no mo natural to return to foraging than to solve population-caused problems other ways more appropriate to our social natures and to the conditions our lives today.

But even though population growth is normal, it is obvious that contributes to suffering. Contemplating these seemingly inconsiste facts has slowly led to the understanding that the problem is not within o species, but in the mix between our species and its environment. There nothing wrong with an oxygen-breathing species that gets sick when p into an oxygen-depleted environment; the problem is in the mix betwee the two. Likewise, there is nothing wrong with us because we suffer fro overcrowding. The suffering is caused because the earth is of a limite size, and we have expanded beyond its carrying capacity. That means w must control our population, as I have said, but it does not mean there anything wrong with our basic natures. Efforts to somehow change o natures, as by making us learn to love crowdedness or adjust to it unnatural ways, are therefore doomed to failure.

There are also practical problems with a return to foraging. Since goo foraging requires upwards of 100 acres of good land per person, millio would have no place to go. An occasional hermit may live on public lan as a forager, but five-family groups would not be tolerated, and millio of them would not be possible. There isn't enough good foraging habit for that.

Finally, even if foraging could meet our need for food, it would squel other parts of our nature which cry out for satisfaction too. Our curiosi has always been intense, and a certain population density is probab necessary to support that curiosity—I am thinking about cities a universities and taxes to support them. Fortunately we do not have choose between one and the other. We can keep civilization intact and st live in harmony with the planet, without such drastic solutions as a retu to foraging.

Perhaps foraging is too extreme anyway. Could we instead "return to nature" on a smaller scale, say by moving to small tracts in the countryside where we survived by organic farming? That may be a reasonable strategy for some families, but it simply wouldn't work as a general solution to the problems of overcrowding, and for the same reason—there isn't enough good human habitat. Large-scale agriculture is probably the only way we can coax enough food from the earth to keep starvation to a minimum.

No, it is neither a return to foraging nor a return to nature that will solve our problems, but a reduction of population to the level of "appropriate ecology." The remaining chapters will develop that theme further, but in the meantime we must ask whether economics can offer anything to help us cope with densities that already exist.

Economic Systems and Structures

No matter how important the single factor of population density is, other factors must also be considered in thinking about our use of resources. In modern economic theory, that means thinking about such topics as money, tool-manufacturing, public finance, labor, industrial organization, prices, and growth, not to mention international economics and trade agreements.

Biology may seem far removed from these complex cultural interactions, but actually it has the potential to influence economics in fundamental ways, and it has already begun to do so. That is largely because economics must be based on an adequate analysis of human nature operating within its environment, and we are now in a better position than ever to make such an analysis.

This section does not presume to offer economics any new concepts, but by emphasizing the role of biological principles as developed in earlier chapters, it sheds new light on economics, especially on the interplay between small-group economics (micro-economics) and large-group economics (macro-economics).[11]

1. Small-group economics: A question of value

Economics always involves a selection among values, of the kind we discussed in Chapter 12. Although some economic models may obscure this point, first-rate economists already understand how important it is.[12] They explore the values which result in specific choices by families and individuals and small businesses, and they incorporate psychological and biological analysis into economic analysis. Granted, this cross-disciplinary approach complicates economics considerably, but it offers the best hope for understanding how individual and family choices impact on large-scale economics, an interaction that has thus far eluded economists.[13]

Economists usually discuss value in terms of money, but that can trick us into thinking things have more (or less) value than they actually do. Economists have always known this, but there really wasn't a better substitute than price theory for making good analyses, and for most purposes that is still true. It turns out, however, that some segments of the economy may be better understood in terms of energy cost than in terms of dollar cost. It is now known, for example, that it will soon require more energy to produce oil in the continental United States, than can be produced from that oil, regardless of whether labor and equipment are cheap or expensive (indeed we may have already reached that point). Thus it is obvious that producing oil for the purpose of energy-production (as opposed to using it for lubrication, for example) creates a net drain on our economy's energy. We should do that only if the psychological value of oil is strong enough to justify a deficit in energy expenditures. Based on this kind of analysis, it can be clearly shown that the recent woes of the U.S. economy are almost certainly due to an exaggerated psychological investment in oil.[14]

We will return to the matter of values in a moment, when we consider the role of the individual in economic life.

2. Large-scale economics

The second and more philosophical way in which biology must influence economics is to apply concepts about human nature to the study of macro-economics. Of course there have always been writers who claimed (for example) that capitalism is more consistent with human nature than communism is, and other writers who claimed the contrary. But as sociobiology matures it will help us clarify which economic responses are relatively fixed in our natures, and which are more flexible.[15] Especially we need to understand the extent to which self-interest is hard-wired in our natures, and the extent to which it can be modified toward the group's interest.

Hard-wired aspects of human nature should be easier to analyze in principle, but flexible responses are also capable of being understood. That is important, because biologists have begun to realize that most parts of our nature are flexible; we respond differently in different circumstances. Yet it may turn out that such differences are somewhat predictable, so that within certain contexts (population densities, levels of value-satisfaction, levels of security, etc.) one might be able to predict responses quite accurately. Of course economic structures are too complex to permit absolute prediction, but even if we can predict the *direction* in which response might be expected, it can profoundly increase our understanding and direct our planning.

The chapter on morality introduced the idea that our nature is not fixed, and explored some implications. Many of those implications also apply to economics on a broader scale. Above all the theme of human flexibility must be acknowledged, which implies that economic structures must change as circumstances change. To return to the oil example, an honest analysis of petroleum reserves suggests the need for specific modifications in our economic structures to ensure future food supplies, especially in the United States where fertilizers are derived from oil. Fortunately, our particular economic system features flexibility and adaptability, but particular structures can be remarkably resistant to needed changes, and it is the role of government (see next chapter) to bring about those changes through context-relevant policies.

3. Markets

Human imagination has constructed many types of markets, each consistent with human nature, but each appropriate to specific circumstances. Bazaars, fairs, trading companies, guilds, feudal arrangements, and other institutions worked reasonably well (never perfectly, of course) within particular parameters. But as those parameters changed, the existing economies became strained, and new structures had to be invented, though not without stress as old economic paradigms gave way to new ones. Given the persistent population increase and the role of trial and error, it should be no wonder that some economic structures eventually consolidated their grip on society, cutting across large segments of every culture and making the world seem smaller and more compact.

By the middle of the 20th Century we had become so crowded that markets had to include almost the whole world, though there was still room for two dominant economic approaches and several smaller ones. Soon the competition between the two dominant approaches became more and more intense, and communication technologies made each of them acutely aware of the other's strengths and weaknesses. For 50 years we argued about whether communism or capitalism was more consistent with human nature. Is it better to have a centrally planned market, or a chaotic one that depends on the independent judgments of millions of people?

Recent political developments imply that free markets are more consistent with our natures, but that may only be true for the circumstances prevailing in today's world. Given a worldwide economy, central markets simply can't predict all the variables involved in distributing goods fairly or effectively. Free markets can't either, but they come closer, apparently because they send our competitive urges into productive channels instead of trying to squelch them altogether. So even though communism theoretically relies on the cooperative part of our nature, it doesn't reflect the balance in our natures as well as free markets do.

That is not to say free markets have no problems. In fact radical free-market philosophies[16] do not adequately account for social and cooperative urges in our nature, and are thus just as ill-founded as communism is. In the wild, apes and carnivores often gang up on dominant individuals, and human biology urges us toward similar responses, as reflected in our tendency to pull for the underdog. Thus economic regulation is as consistent with human nature as free markets are. For these reasons many economic philosophers believe human nature is best reflected by economic structures and systems that "split the difference" between cooperation and competition, in what is formally called a regulated free market. And not surprisingly, that is the only kind of free market that exists in today's world.

Yet we should be grateful for the communist experiment, because we might need to use some of its principles at some future time. For example a stable world-wide population featuring slow economic growth or equilibrium might require the incorporation of such principles. More likely still, a new conception will arise, which may not look much like either classical capitalism or classical communism.

4. Economic Aid

Another problem with radical free markets is that they would apparently not intervene to keep people from starving, and that is entirely inconsistent with humanitarian urges that well up from deep in the psyche. Massive evidence from human experience and animal research shows how widespread such responses as compassion and altruism are.

This is not to deny that such instincts may have a selfish basis; the hope of reciprocity may be axiomatic in both animal and human altruism. It is certainly clear that compassion and aid increase the overall chance of survival. But whether the deeper basis for altruism is selfish or not, it is built into our natures, and radical individualists do not give enough emphasis to that part of our makeup.

The precise form of such sharing is mediated by culture. Whether surplus is distributed by public feasts, by foreign aid and welfare programs, by charitable organizations or individual samaritans, the result is the same: surplus in one segment of society is shared with others. Economically, the effect is to use earth's resources more efficiently on behalf of the human species.

Economic aid performs another useful function in group life, by keeping resources from accumulating in too few hands. Nobody is likely to support society unless they have a stake in it, and thus revolution always redistributes wealth when "trickle down" theories fail. We can avoid such revolutions through constant exhortations in favor of charity, or we can avoid them through economic schemes that redistribute wealth more rationally. Such programs as the graduated income tax, an efficient welfare system, foreign aid, public education, universal medical care, retirement plans, unemployment funds, and public employment programs all aim to accomplish that. These programs may smack of involuntary altruism by which the poor prey upon the rich, but their ultimate effect is to help the larger society, both by redistributing wealth in an orderly way, and by smoothing over the harshness of unregulated economic competition.[17]

The Individual in Economic Life

So far I have summarized the thesis that population density has a huge impact on economics, and I have listed four ways biology can influence economics short of population control (by discussing microeconomics in terms of value selection, by seeing macroeconomics in terms of flexible responses to external circumstances, by discussing markets in terms of human nature, and by seeing economic aid as arising from deep within our social makeup). Those dynamics are important to the group, but we as individuals also need to locate ourselves within the economic system. Life for us is certainly more complex than foraging, and nowhere near the productive orderliness that might be achieved with a lower population. Can anything be said that might offer a realistic and moral response to the world in which we find ourselves? I think so. Indeed, at least five responses can be listed that are practical, and yet anything but obvious.

1. By understanding that we evolved as a small-group species, we can seek meaning in that context. Even Congress operates in committees, and it is likely that our fulfillment is to be found in small groups too. This does not mean we should all own small businesses, but it does mean powerful inbuilt forces make us more comfortable interacting in small groups. Thus even if we choose to work for large corporations or the government, we may be happier and more effective by focusing our attention on the small groups within it.
2. By understanding that evolved neural patterns produce internal pulls between our personal (or "selfish") needs and the needs of the group, we can more comfortably find the middle ground between the two. There seems little doubt, despite an interesting current dispute among biologists, that we evolved to meet our personal needs first. We cannot "love others as we love ourselves" unless we first love ourselves. Secondarily we look to help the group.

3. There is a partial exception to the previous paragraph, because one group is hard to put in second place, and that is our immediate family. Since we evolved in the context of that particular small group, its needs tend to intertwine with our own. It is therefore no wonder that child-support laws, for example, place restraints on those who put their needs ahead of their family's needs. In rare instances such a strategy might be excusable, but family breakup is frequently an ill-disguised and pathological response to outside economic pressures. If that is true, ancient admonitions to family faithfulness can confer significant economic advantages on those wise enough to heed them.

   That does not necessarily mean one should stay loyal to the *extended* family. Conflicts within that group have always been part of our existence, and distancing oneself from the larger family is itself an ancient response. It isn't often a comfortable response, but we probably have built-in coping mechanisms for the pain involved. Thus "leaving father and mother" to cling to one's spouse and children is good biology and good theology.
4. Lest these observations seem to justify a personal or group-centered selfishness, we must remember the overriding role of rationality discussed earlier. Our brain is not constructed to rely upon basic impulses alone; it is an organ for making judgments, including complex judgments. Since the evidence shows how thoroughly interdependent we are in today's world, we must also give attention to the problems of the larger group, including even global problems. We cannot possibly act purely in our own selfish interest without taking an untenable risk of undermining our own economic well-being and that of our children. Thus the rational and practical pattern must be to seek our personal interest first and our group's interest second, but with due regard for the other levels of human interaction.[18]
5. In implementing this program, there is no reason for excessive guilt feelings as one first provides for self and immediate family. We are not personally responsible for the slow growth of human population that led to our present crisis, and we cannot reverse it except in small ways. Yet those small ways can be important: for example we can keep our own reproductive tendencies in check; we can support ethical efforts at reducing the population; and we can try to ameliorate the crisis through our votes and our charities.

So the conclusion is that we individuals must look out for ourselves and our immediate families, but in ways that recognize the interests of the larger group. We can make humble efforts to reverse the problems caused by over-population, and in fact by being humble, those efforts can help us achieve another goal, that of living gracefully. While it may be frustrating to confront policies and people that seem to perpetuate the problem, patience and humility reduce the need to rail against them, and pave the way for knowledge and progress.

Indeed, there is every reason to hope that grace and knowledge can reverse the dynamics that have caused the problem, through slow changes in the cultural mechanisms of government, education, and religion. These are all ancient institutions through which we express our essential natures as social creatures, and they are the subjects of the remaining chapters.

Notes

[1] Stanford, California: Stanford University Press, 1987.

[2] See a list of such processes at p. 28 of the Johnson and Earle book. See also Steven Shantzis and William Behrens III's "Population Control Mechanisms in a Primitive Agricultural Society," Chapter 9 of *Toward Global Equilibrium: Collected Papers*, ed. Dennis L. Meadows and Donella H. Meadows. See also the last letter in the "forum" section of the November 1992 *National Geographic*, where an M.D. (Mely Lechtich deRévai) reports on the work of Dr. Russell Marker, a chemist who discovered that yams contain a contraceptive chemical.

[3] I have oversimplified in the text, by implying that "good lands" had no problems whatever. Of course there were always predators and disease and other problems, even on lands that were very hospitable to humans.

[4] In support of this thesis, note that agriculture did not develop where it was not necessary. See the examples in *The Evolution of Societies*, cited above. See also "The Earliest Farming: Demography as Cause and Consequence," by Bennet Bronson, pp. 53-78 in *Population, Ecology, and Social Evolution*, ed. Steven Polgar (The Hague: Mouton Publishers, 1975).

[5] Note that although humans came into existence long before the Biblical creation story implies, the timing is about right if we ask when we began earning our living "by the sweat of the brow." Being driven from Eden presents a poetic picture of having to colonize less desirable lands, and the legend may have arisen not long after the shift occurred.

[6] From *Introduction to Population*, Chapter 3, by Judah Matras (Englewood Cliffs, New Jersey: Prentice-Hall, Inc., 1977). See also p. 2214 of *World Population Basic Documents*, Vol. 4, ed. James Avery Joyce (Dobbs Ferry, New York: Ocenia Publications, Inc., 1976).

[7] For biblical accounts of several famines, see Genesis 12:10, Genesis 26:1; Genesis 41:56-47:20; Ruth 1:1; II Samuel 21:1; I Kings 18:2; and II Kings 6:25.

[8] Even an optimist like John Scott believed that by the year 2000 75% of the world (instead of the 50% in 1966) would be seriously malnourished. See p. 73 of his *Hunger* (Time, Inc., 1966). Also see the graphs and discussion in "The

Carrying Capacity of Our Global Environment," by Jorgen Randers and Donella Meadows in *Toward Global Equilibrium*, pp. 315-335.

[9] "Human Ecology," p. 562, by John A. Moore, in *Science As a Way of Knowing*, Vol. 2 (Human Ecology).

[10] For a serious lunar-farming design see "Design of a Controlled Ecological Life Support System," by Steven H. Schwartzkopf in *Bioscience*, Vol. 42, No. 7 (July/August 1992), pp. 526-35.

[11] For more on the biology/economics interaction see Chs. 9 and 10 of *Biology and the Social Sciences*, ed. Thomas Wiegele (Boulder, Colorado: Westview Press, 1982). Chapter 9 (pp. 129-144) is "Economics as a Not Very Biological Science," by Kenneth Boulding; Chapter 10 (pp. 145-184) is "Economics From a Biological Viewpoint" by Jack Hirshleifer. Despite Boulding's title, both selections reveal a great deal of interaction between biology and economics.

[12] For example, see Nobel-laureate Gary Becker's *The Economic Approach to Human Behavior* (University of Chicago Press, 1976) or his *A Treatise on The Family* (Cambridge: Harvard University Press, 1991). See also Kenneth E. Boulding's *Evolutionary Economics* (Beverly Hills: Sage Publications, 1981).

[13] For further insight into the intellectual difficulties facing modern economics, see *The Crisis in Economic Theory*, ed. Daniel Bell and Irving Kristol (New York: Basic Books, Inc. 1982), and *Economics in Disarray*, ed. Peter Wiles and Guy Rath (New York: Basil Blackwell, Inc., 1984). The non-economist who wants a good overview of the current problems in economic theory should especially read the last chapter in the first book (pp. 201-218), entitled "Rationalism in Economics." For a cutting-edge discussion between philosophers and economists, see the papers and comments in *The Quality of Life*, ed. Martha Nussbaum and Amartya Sen (New York: Oxford University Press, 1993).

[14] For a technical and yet readable presentation of these ideas see *Beyond Oil: The Threat to Food and Fuel in the Coming Decades*. The book is written by four members of the Complex Systems Research Center at the University of New Hampshire, namely John Gever, Robert Kaufmann, David Skole, and Charles Vörösmarty (Cambridges: Ballinger Publishing Company, 1986).

[15] See E.O. Wilson's "Biology and the Social Sciences," pp. 245-262 in Vol. 25 No. 1 (September 1990) of *Zygon: Journal of Religion & Science*, especially the section sub-titled "Economics" (pp. 257-258).

[16] Such as Ayn Rand's *The Virtue of Selfishness* (New York: New American Library, 1965).

[17] To see how economics and morality intersect at such issues at these, it is still hard to beat John Stuart Mill's *Principles of Political Economy* (University of Toronto Press, 1965), first published in 1848.

[18] See *The Morality of Scarcity: Limited Resources and Social Policy*, ed. William M. Finn, Jr. and Gerald Alonzo Smith (Baton Rouge, Louisiana: Louisiana State University Press, 1979).

Chapter 14

Politics: A New World Order

Politics is *the process by which leaders organize and motivate us to solve problems cooperatively, especially by getting us to regulate our individual behavior*. Anarchists argue that we could do without politics, but most philosophers and ordinary citizens from Plato forward have considered it necessary. From a biological perspective it certainly seems fundamental to our nature—we inevitably live and operate in leader-centered groups, just as other primate societies do.[1]

At the hunter-gatherer stage, human interactions were relatively simple. Complex political structures arose for the same reason agriculture did—they permit more efficient use of resources. That efficiency is necessary if large groups are to survive, but it introduces many difficulties. It is difficult for leaders to keep large groups motivated over long periods, it is difficult to control the self-serving instincts of the leaders themselves, and it is difficult to create and maintain institutions that adequately implement human values.

This chapter will not try to survey the whole field of government. It will instead discuss three particular problems to illustrate the difficulties and functions of politics in general. But before turning to those problems, the definition above must be slightly modified. It must be modified to make clear that politics can do more than solve problems, to make clear that interactions between leaders and followers can be something less than enthusiastic cooperation, and to make clear that politics does not necessarily imply progress:

First, besides functioning to solve problems, politics can also seek positive opportunities for the increase of human happiness. If the political structure is flexible enough and the society secure enough, seeking positive opportunities may become a major function of leadership. But because societies always live in the tension between tradition and innovation, and because it takes so much energy to maintain existing structures, it is rare to find a political system spending much energy seeking creative new opportunities for happiness. Thus the usual emphasis of politics is on cooperative problem-solving rather than on positive opportunity-seeking.

Second, the "cooperation" involved in politics is by no means straightforward. Group action is often compelled or manipulated by leaders and institutions entirely apart from any genuine enthusiasm on the part of group members. But to the extent the followers "go along" with the initiatives of their leaders and the traditions of their institutions, group action may still be seen as cooperative, and thus this chapter will make no rigid distinction between enthusiastic cooperation and begrudging cooperation.[2]

Third, the definition may imply that group action is always a positive thing, and of course that is not true. The group can make mistakes, it can make judgments that seem right at the time but which have hidden dangers, it may be urged toward action for the ostensible benefit of the group when in fact the action is for the benefit of the leaders or of some sub-group, and so on. Indeed, all of the categories of sin and mistake which were discussed in Chapter 10 can apply to groups and their leaders, just as they can apply to ordinary individuals.

These modifications in the definition above are merely the tip of the iceberg; a complete discussion would require a thick book. But by looking at three particular problems, I hope to limit the chapter to a manageable length, and still illustrate the complications and possibilities inherent in politics. These three problems are critical to life on earth, and if we could solve them we would simultaneously solve many other problems, and take a giant step toward increasing human happiness. It is possible to solve them, but there is no guarantee that politics will be able to motivate us and organize us to do that.[3]

The problems discussed here are 1) Environmental destruction; 2) War; and 3) Over-population. The discussion will demonstrate that leaders can organize and motivate us, and yet it will also show that politics is greatly influenced by dynamics external to it. As we will see, there is only one of these issues that we can do anything about except on the short run, and although politics must give attention to the other two, long-term action should clearly be focused on the one that most impacts the others. After discussing these problems and briefly mentioning some other functions of politics, I will end with a section on the choices we can make as individuals to help bring about the solutions proposed, and I will discuss what we can do to ensure our own happiness while those solutions remain elusive.

Environmental Destruction

Physically and emotionally, the ecology of any species is by far the most important factor in its existence. Unless it has enough air and food and protection from the elements, a species will perish, or at least its quality of life will be greatly diminished. In the case of humans, it is unlikely that we can achieve other goals unless we have a decent environment.

Yet there is little doubt that we are degrading our environment, and at an ever-increasing rate. Although one frequently hears conservative demagogues[4] argue that such statements are only made by naïve cry-babies, it is now established beyond any reasonable scientific doubt that the statement is true. The greenhouse effect is real, air quality and water quality are compromised (indeed the very existence of sufficient water is threatened in some places[5]), and we are losing biological diversity at an alarming rate.

What could be alarming about extinguishing a few species? Hasn't extinction been going on since life began? The answer is yes, and actually there is little danger of the planet dying completely, regardless of what we do. Moreover, there isn't much danger that humans will become extinct anytime soon, given how broadly we range over the earth. The present alarm is not about extinction, but about the quality of life. By losing diversity, we not only miss the chance to harvest cancer cures from obscure plants, but we compromise our existence in other ways too. We are significantly reducing healthy human habitat, and in some cases we are actually poisoning the environment. These actions endanger our physical health, but they also endanger our psychological health, because our psyches evolved in specific environments which we still need.[6]

Despite these dire warnings, there may not be anything we can do about environmental destruction, in and of itself. To show what I mean, let me predict what I think will happen in several instances. These examples are merely illustrative; my prophesy depends on general principles of human life and psychology, and not at all on the details of any one of these examples. Every environmental issue will end the same way, given enough time and given current demographic trends.

The first example is the Spotted Owl, which happens to be at the forefront of the environmental news right now. Loggers in the northwestern United States have slowly destroyed the owl's habitat. They weren't originally antagonistic toward the owl; they simply needed jobs and there was a market for lumber. Conservationists say we should find other jobs for the loggers, because the destruction of owl habitat is also destroying the yew tree, which holds a cure for cancer. They also argue that the destruction is immoral and shortsighted for additional reasons. The politicians are divided, because both sides make good points and the problem is difficult to resolve.

The reason I think the loggers will eventually win is simple. It arises from our biology and our increasing population densities, and it is substantiated by the history of England and China. Loggers will win because worldwide population growth requires more and more intense use of resources to permit the survival of so many humans. In England and eastern China, that has meant the virtual removal of all forests, together with other life-forms dependent on those forests. It will eventually mean the same thing here.[7]

The conservation community hopes to prevent this damage by educating the public about our interdependence with the rest of the planet, and by enforcing laws it helped pass. But those laws could easily be reversed. They were only barely enforced from 1980 to 1992, and then only through the vigilance of conservation groups. All it will take for those groups to lose influence is a long recession, when contributions dry up and conservation workers have to make a living in other occupations. Nobody will have time or energy to speak for the planet.[8]

Nor are the forests of South and Central America likely to be saved. Perhaps some stop-gap measures such as eco-tourism might make it more profitable for those countries to protect their forests than to harvest them, but it is only a question of time before the balance shifts toward use, given a steady increase in the human population.

One organization, The Nature Conservancy, has taken a creative approach that seems to avoid all these problems. It doesn't try to get laws passed or mournfully decry the loss of diversity or natural habitat. Instead it works within the system, acknowledging the power of money and politics by simply buying land that needs protecting, and keeping developers off.

No doubt the medieval Catholic church felt it had worked within the system too. It owned vast reserves and supported many politicians. The business establishment of Northern Europe was tightly bound to it. It could not have been toppled by anything short of a major paradigm shift. But shifts happen, and the church was stripped of its property and influence during the turbulent period we now call the Reformation.

The Nature Conservancy might be secure for now because of capitalism's recent worldwide successes. But even without a retreat from the basic tenets of capitalism, a few simple changes in the law would dramatically reverse the conservancy's achievements. Taxes on contributions or on non-profit corporations, economic-impact requirements that prohibit conservation if it interferes with jobs, prohibitions against taking land out of commerce for perpetuity, laws to pay landowners if they can't use their property as they wish; these and other provisions have already been proposed, and similar ones have been enacted at other points in history when the short-term prosperity (much less the survival) of the masses was threatened.

Perhaps my prophesy will turn out to be wrong, but as individuals and as a species, we are biologically programmed to do what we perceive as best for ourselves. So long as the argument centers around the quality of life, conservationists might continue to have some successes. But with continued population growth, the argument will eventually come down to survival, and it is already perceived in those terms in many sub-groups, even in prosperous countries. Abstract principles of conservation cannot stand in the face of profound insecurity.

In a moment we will discuss how we might have both security and conservation, both survival and biological diversity. But the politics of the matter is so dependent on economics and human nature as to require a very broad solution, one that goes to the heart of all the problems we are discussing. The confrontational politics of loggers vs. owls, or even the subtle aggressiveness of buying up all the land, just won't work in the long run.

Notwithstanding the need for a long-term solution, politics must also address problems on the short run, and we desperately need the time they can buy. Excellent proposals have been placed on the table by scientists and activists and politicians, which offer hope for protecting the environment even in the face of growth. But those proposals must not delude us into thinking they will do more than buy time, and they must not deflect our attention from the only solution that can ultimately solve the environmental problem. That solution is to reduce our population, and I will get to it after discussing the second major challenge we face.

War

I said above that there is little danger of the planet dying completely. Even if a "nuclear winter" destroyed higher plants and animals, simple life-forms would survive to begin another round of evolution. Yet few of us are so lacking in affection for higher lifeforms as to relish that prospect, and it is sobering to realize that we could indeed destroy human life.

While the fragmentation of the USSR and the thawing of international relations may seem to reduce the threat of nuclear winter, we still have the same nature and technology that we had during the cold war, and there is no reason to relax our concerns about the potential for destruction. There is every reason to believe smaller wars will continue, any of which could escalate into a world-wide conflagration. And even if they don't, war is itself a great impediment to human happiness, as we saw when looking at the happiness research in Chapter 12.

But isn't war inevitable? Isn't it innate and therefore unavoidable? At first blush it appears to be. Despite the protests and bemoanings of moralists, wars and rumors of wars continue. And in the words of an old song, "it's been going on 10,000 years."[9] Given the persistence of war, it is hard to dodge the conclusion that it is indeed innate, a human response impossible to eradicate. Wouldn't politicians do better to try to control it instead of trying to eliminate it, avoiding its grosser consequences by outlawing nuclear weapons, for example?

Perhaps so. While moralists can persuasively point to the eradication of slavery to show that we can rise above our history, war seems even more deeply ingrained in our psyches than slavery was. Moreover, our nature as symbolic creatures implies that we need a "team" to root for. Our need for identity feeds on the existence of nations and parties and ethnic groups, and physical battle is an all too natural way of expressing and consolidating our commitment to such groups. So again, the implication is that political efforts should be focused on moderating war rather than eradicating it.

All this is only at first blush, however. On further analysis war is seen to be like anything else: it has causes. While its pervasiveness certainly makes one think that it rises from causes within the human psyche, there is also evidence that war is caused as much by circumstances as by human nature. Said another way, it appears that war might be an innate human response *to certain circumstances.* If that is true, we might be able to eradicate or greatly diminish war by addressing the circumstances that give rise to it, rather than by attempting the impossible task of changing human nature (which might have undesirable consequences even if it were possible).

What is the evidence that war might be caused by circumstances? There are actually two lines of such evidence, though they boil down to the same thing.

First, recent research on the causes of war has reduced the number of explanatory theories from about 16 in 1973 to about 8 today. Those theories may be divided into three categories, namely the "ultimate" causes of war, the "immediate" causes of war, and the causes that originate in earlier wars. Of the three, we are most interested in research about ultimate causes. And that research suggests that while some aggression is innate, it doesn't necessarily lead to war absent the presence of certain circumstances. What are those circumstances?

Insecurity is the foremost circumstance leading to war. For example, the *fear* of food shortages has been shown to be positively correlated with war, even more than food shortages themselves. Competition for scarce resources, mistrust of neighbors, and meteorological disasters like floods or droughts have also been positively correlated with war.[10]

The second line of research shows that warfare did not emerge until 10,000 years ago. Since humans existed long before that, this fact suggests that some new circumstance arose at that time, which interacted with our nature to cause war. That new circumstance was the one we explored in Chapter 13: increased population density. And since density is directly related to food shortages, competition for scarce resources, and other kinds of insecurity, it is no over-simplification to say that increased population density is a direct and primary cause of war.

Indeed, there is no reason to believe we can prevent war without first addressing over-population. Certainly we must make efforts to reduce war and eliminate nuclear weapons, and I am not saying those efforts are bound to fail. But given a natural aggressiveness embedded in our nature because it served us well over eons of evolution, and given the stresses caused by over-population which aggravate that aggressiveness, there is no current theory that offers hope for eliminating war, or much hope for preventing its worst consequences.

If this section and the preceding one are correct in concluding that we can only solve the problems of environmental degradation and war by first solving the problem of over-population, then political action should certainly be focused there. So let us turn to a fuller discussion of that problem.

Over-Population

The concept of carrying capacity was originally an agricultural concept, but it is rapidly becoming the guiding principle for intelligent politicians who aspire to genuine nonpartisan statesmanship. Those politicians are aided by sophisticated academic studies, by computer models and statistical analyses, and by the average person's growing awareness of the dangers presented by increasing densities. Indeed, almost everybody knows that we need to do something about the population problem. The questions are what and how.

Persuasion is part of the answer, and we will return to it in the final chapter, but we humans are programmed to seek our self-interest. It is hard to persuade people to have fewer children merely to prevent future world-wide disaster. Therefore persuasion must be aided by policy.

Fortunately policy seems to work. China is attempting to reduce its population by implementing policies to encourage "one child per family." Initial indications are positive, and since China is both a leader of third-world economies and the home of 1/4 of the human population, its policies could have profound consequences. Some moralists decry those policies because they tend to encourage abortion, but China sees the choice as one between abortion and starvation, and abortion seems preferable. (Remember that almost thirty *million* Chinese people starved to death in the four short years from 1959 through 1962.)

Denmark has also instituted policies that control population, and although they arise from a different history, they have the same effect, and without relying so heavily on abortion. A review of those policies gives great insight into the options available to politicians, and indeed the most sophisticated computer models validate Denmark's strategy, even though it arose as an accidental by-product of another policy.[11]

Denmark's food supplies were nearly cut off in each of the world wars. In response it sought to diminish its dependence on imported food, by instituting strict land laws. Those laws made it nearly impossible to convert farmland around Copenhagen to industrial or residential use. An unforeseen result of the policy was to make it difficult to get an apartment. That meant young people stayed at home longer, waited longer to get married, and had fewer children. People began to plan their families carefully, using birth control to ensure that they only had children they could house. The birth rate declined.

Of course Denmark's experience may not have much relevance to other countries, where large families are perceived as the only route to old-age security even though the actual effect is exactly the opposite. In those cases politicians must institute other policies to reduce population. Above all, we must find ways to provide people a sense of security without forcing them to depend on their own offspring.

What other policies might work to reduce population? I will list two general policies first, and then get more specific.

1. We can provide research money. Great progress has been made toward understanding population dynamics, but much more work is needed, especially as we strive to determine just which policies might effectively reduce the population in a variety of social systems.
2. We can avoid imposing our ideology on cultures that already have workable mechanisms for maintaining population equilibrium. For

example the Marin Tsembaga of New Guinea manage to control their population, and although they do it in ways unpalatable to Western minds, we should tread softly in disturbing such equilibrium until we understand it better.[12] Indeed, it is clear that global equilibrium will eventually come, either through rational planning or biological disaster, and therefore we should be closely observing primitive cultures to see what we can learn that might be relevant to our own future.

One way we can avoid imposing our ideology is to reject policies and treaties that too easily urge third world countries into "modern" life, especially if our ideological evangelism unleashes population growth that can only end badly. I am not talking so much about religious evangelism (that will be addressed in the last chapter), as about economic evangelism. If we allow our corporations to rush into third-world countries to reap the advantages of cheap labor, we may unwittingly urge those countries to increase their populations. Whether that is true will vary with the country and the culture, and deciding which countries to allow our business people to enter simply underlines the need for more research. Even preliminary research, however, can offer great insights if policy makers understand its implications for population growth.

3. In a more direct vein, we can pass laws which take cognizance of the ways capital-formation impacts on population growth. The Denmark example shows how such laws and policies can have unexpected results, but ongoing research is beginning to suggest that such laws may offer the best hope of reducing population.[13]
4. As cruel and counter-intuitive as it sounds, we may have to reduce humanitarian aid to all countries except those that are trying to reduce their own populations. Private charities are already grappling with the uncomfortable realization that assistance increases population, and they are beginning to target countries where relief might do the most long-term good. In practice that means aid goes to countries that are trying to reduce population. And although such a severe policy fails to prevent human suffering on the short term, we can mitigate that suffering by providing food and medical aid in true crisis situations like famine or epidemic.

   This regrettable policy will become even more necessary as our own resources become strained from the effects of increasing world population. It is estimated that U.S. food surpluses will disappear as world oil supplies diminish, and that will probably happen within this generation.[14]
5. Speaking of oil, we are using it up at a frightening rate, and some proposed policies aim to conserve what is left while we develop

alternative energy sources. In light of Denmark's experience, such policies would probably discourage population growth as an added bonus. For example if we stopped giving tax breaks for agricultural practices that depend on oil for cheap transportation and cheap fertilizers, we would incidentally stop contributing to the explosion in third-world populations.[15] Many proposed policies for conserving oil and developing alternative energy sources would also delay environmental destruction, buying even more time to work on the population problem.

6. Insofar as we can do so without overtaxing natural resources, we should adopt policies that tend to increase the standard of living in third-world countries. That is because increased standards of living are clearly correlated with the decision to have fewer children.[16] If we were able to buy great chunks of time, as by new energy sources like fusion or hydrogen or solar power, then this policy might be a sound one (and make no mistake, such breakthroughs would not solve the problem; they would only buy time, and not much at that). But given the damage already caused by resource depletion, it may be too late to increase standards of living directly, without first reducing the number of people that share the available resources.
7. We can support moral methods of birth control, and teach people to use them. Although it is not certain which methods are moral, it is clear that many moral objections are ill-founded, if long-range effects are considered. For example it is foolhardy to limit birth control to rhythm methods, and it is to be fervently hoped that the Catholic Church will abandon its prohibition of other methods. In the meantime forward-thinking politicians must find strength to oppose such short-sighted religious doctrines, by voting for policies that accept a wider range of birth control.
8. As an important adjunct to birth control, we must support lifestyles for women besides child-rearing. Most women will continue to choose child-rearing as a first career, but with long lives and fewer children they need a second career too. We can adopt policies to support this need, for example by giving loans and grants for occupational education. And because population reduction is so important, we must also encourage couples that choose childlessness, as by giving them the same tax breaks larger families enjoy.[17]

As I said, there is no guarantee that we can solve the population problem, but nature will solve it if we don't; even yeast dies back dramatically when it expands beyond its food resources. If we are smarter than yeast, we will find rational ways to reduce our population. If successful our rewards will be great, because population reduction will lso diminish our assault on the environment, help prevent wars, and solve ther problems. Perhaps we can partially accomplish those goals in other ays than by reducing the population, but from a biological perspective, nd also from a theological one (as the final chapter will argue), reduction f population is the soundest way to solve the political problems facing us.

* * *

Politics has other functions besides solving problems like those dis- ussed above. It must spend enormous energy maintaining a system of aws and basic services. It must regulate the factions that inevitably spring p within society. It must complement other group functions such as conomics and education and religion. It has significant responsibilities or public health. Again, I have not addressed these issues.[18] Yet if the roblems above can be solved, and especially the problem of growing opulation, then these other tasks should be manageable. The planet has esources and humans have imagination to use those resources, if only we an avoid the population densities that stretch both resources and creativity eyond the breaking point.

In the meantime, what responses are available to individuals in the urrent political climate? Can we find effective and moral ways to live raceful and happy lives in a dynamic culture that features severe lisagreement about political decisions?

Individual Responses to Politics

Politics is an unavoidable part of life. We owe society some allegiance y virtue of our nature as a member of a social species. Our neurons propel is toward basic cooperation, and we slip into pathology if we ignore their irges. On the other hand there has always been a tension between the group and its members, and we must decide when to pursue our interests even if they are contrary to the group's interests. Further, modern densities nake the group much more intrusive than it was over evolutionary time, and that underscores the need for seeking our own legitimate interests even while trying to accommodate the group.

This section will explore some suggestions for striking the balance between society's demands and our personal interests. In classical Chris- ian terms, we are trying to discover how we might love ourselves, and how we might love others as ourselves.

The balancing would be easier were it not for a quirk of evolution—not a quirk really, so much as a commonly encountered characteristic of evolution not generally understood. It is that evolution pushes us in certain directions without supplying any brakes. That may sound odd, so let me present a simple example of the phenomenon, and then examine some implications.

Along with other primates and many insects, we humans have an innate taste for sugar, which was a useful adaptation because it motivated us to seek fruit and other sources of energy and nutrition. In pre-historic environments there was little chance that we would eat too much sugar, and thus there was no advantage to developing brakes on our appetite for it. But horticulture and refinement technology made it possible to eat too many sweets. Fortunately, our rational abilities helped us figure that out, but rational abilities often conflict with our older, non-rational urge to eat all the sugar we can get.

This example hints at broader political problems caused by a lack of inbuilt brakes. We each have an urge to use resources for our personal betterment, but that urge was developed when resources were more plentiful than our ability to use them. In an age when over-population requires conscious consideration of future supplies, we must use our rational facilities to control our "wasteful" urges. I put *wasteful* in quotes because our urges would not be wasteful if it were not for over-population. Ecological policies are required precisely because our urge to use resources conflicts with the needs of a large population.

Our aggressive urges are similar. In this case evolution has provided a set of brakes,[19] but it doesn't work very well in modern life. For example, bullies are not well tolerated in small groups, but it takes great vigilance to control them in large groups. The emergence of an occasional Hitler shows that vigilance isn't easy, and necessitates the establishment of institutions and laws to keep aggression in bounds.

How do these insights impact individuals? In the case of sugar (or fat, which has a similar history), they help us understand a specific temptation. We can discipline ourselves to avoid obesity, even in the face of the group's success at producing all the sugar (or fats) we want. We must listen to our rational thoughts, and modify our habits accordingly.

Similarly, we can control our tendencies toward wastefulness and aggression, realizing that they evolved without the needed brakes. We may also gracefully insist that other members of society curtail their own aggressive and wasteful urges. We can do that in small personal ways, and we can ask politicians and institutions to require cooperation from us and our neighbors.

Let's pursue the discussion about aggression a bit further. If war emerged as a response to insecurity, especially the insecurity caused by crowding, we can each strive to avoid the insecurity that stimulates aggression. We can do that partly in psychological ways, as by learning the skills of grace. But we can also do it in economic ways, by providing for ourselves and our families with due sensitivity to broader needs, but without undue guilt over our inability to meet those broader needs.

Moreover, understanding the wellsprings of war may modify our enthusiasm for the excesses of our own countries. When stresses come and aggressive responses build, politicians get swept along in a patriotic fervor, and few of them (or us) can withstand the deep social pressures that produce inter-group hatred. We send our young people to war, and reduce population in the worst possible way. By understanding such dynamics, the country would be in a better position to avoid war, but even if the country doesn't do that, we as individuals can. In extreme cases individuals must decide whether to leave their country and seek a saner life, or stay and try to persuade their country to resolve international tensions through diplomacy instead of war.

Obviously we must not go so far as to squelch our aggressive tendencies altogether, any more than we should avoid all sugars or fats. Our taste for them evolved to fill legitimate needs, and it is only the lack of brakes that we must compensate for.

Apart from larger issues like war and waste, there are dozens of ways individuals can respond to the coming political changes caused by increased population densities. Some of them are socially responsible in that they urge the group toward sound solutions; but some are frankly self-protective, to compensate for the failure of politicians to find solutions. Some responses depend on having the economic means to protect ourselves, some depend only on the exercise of discipline, and some depend on attitudes that we can adopt. I will list seven different coping strategies, knowing full well that some of them are outside the capabilities or interests of most people, but confident that everybody can profit from at least one of these proposals.

1. First, we must learn to use resources more efficiently. Depending on our circumstances, this may mean using less energy, using public transportation instead of private automobiles, or insulating our houses better. These practices are already common in Europe, and in many large cities of the United States. Such measures may also reduce environmental stress, but I list them as ways of ensuring personal comfort and economy in difficult times.
2. To protect our families on the long run, we should vote for politicians that distinguish between harvest and exploitation: Harvest focuses on sustainable yields of food and other produce, whereas exploitation relies heavily on capital depletion, either by using huge shares of non-renewable resources, or by using renewable resources so fast that usage outstrips renewal.
3. People who can afford it can take more extreme measures. Buying land might not be necessary, but that would offer some security if

it were an otherwise attractive purchase. People with investments should invest with a view toward oil depletion and other inevitable results of continued population increases.

4. Even if one does not have money to make such investments, everyone should get more education, either formally or informally. That is because individual knowledge can help us adapt to whatever changes come. If possible, the education should be cross-disciplinary and should include some practical skills, in case the changes turn out to be cataclysmic.
5. We can participate in organizations that work toward the goals of this chapter. Indeed, that might be an excellent way to maintain individuality within the context of a small group which adds security and meaning to one's life. Chapter 16 will suggest that churches can fulfill this function, but many other organizations also offer the opportunity for such participation.
6. A particular kind of organization was popular in the past, and may become popular again, and that is the fraternal organization. These groups frankly seek member interests, in the belief that groups can accomplish more than individuals can. Other co-operative ventures can also serve as economic and political vehicles to reach similar goals.
7. Finally, we can adopt attitudes of grace as discussed in Chapter 11. That means we pursue our own interests with a cooperative attitude, without running roughshod over others. We can obey the law, even while working to change parts of it we don't like. At the least, graceful attitudes propel us toward acceptance of change, even when we are not adequately prepared for it.

This chapter has also listed other things we can do to prepare for change, including 8) controlling our own aggression; 9) having fewer children; 10) targeting gifts to countries that are working to reduce population; and 11) asking government to encourage population reduction. Obviously these ideas are only illustrative. The more general thing we can do is contemplate change, and decide what to do about it. Some individuals may decide to do nothing, some may decide to make small changes, and some may decide to contribute large amounts of time and energy to causes they consider relevant to the concerns of this chapter. In any case, each person must respond to the political and social climate he or she lives in, and any one of the above proposals could turn out to be very important for particular individuals.

* * *

History teaches one primary concept—change. That is true whether we are talking about the long history of human evolution, the shorter history of human civilization, or the brief history of a particular country. Over and over again, either because of ecological changes or because of developments within the culture, we have had to make adjustments.

For us to think change has stopped would be ludicrous. Some changes are easy because they happen slowly, but some are fast and traumatic. Slow change is not always easier than traumatic change, however, because it can induce anxieties that are harder to deal with than traumas are. That is especially true when a transition is both slow and subtle, deluding us into thinking things are the same when they aren't, and worrying us at a subconscious level. It is only when individuals and nations see change coming that adjustments can be made in advance.

In our case it is easy to see change coming. It will come because of two dynamic trends that will intersect before the year 2050 a.d. One is that the population will double, and perhaps double again if our policies and decisions do not intervene. That means there will be 4 times as many people on the earth then as there are now (in the last decade of the twentieth century). That cannot happen without major readjustment.

The second factor is the depletion of resources. In the continental United States it already takes more energy to take oil out of the ground than we can extract from that oil.[20] Worldwide oil supplies will be virtually zero before 2050. Other resources, including even some renewable resources, will be substantially depleted by then.

I do not presume to know what adjustments will be required, or how to prepare for the coming changes, except in three ways. First we must adopt political strategies for reducing world population. Second, we must fashion individual strategies for coping with the difficulties we will face if government is unable to reduce population. And third, non-political institutions must contribute to those endeavors.

Foremost among those institutions are our educational institutions and our religious institutions. If they do what they can during the coming transition, it might be possible for us to avoid disaster. Let us therefore look next at the role of education, and then end with a discussion of religion.

Notes

[1] See *Primate Politics*, ed. Glendon Schubert and Roger D. Masters (Carbondale, Illinois: Southern Illinois University Press, 1991).

[2] For more on interactions between leaders and ordinary people, see Thomas Wiegele's *Biopolitics*, (Boulder: Westview, 1979), pp. 40-44, 90-92, and 102-106.

[3] Paul Kennedy's *Preparing for the Twenty-First Century* (New York: Ran dom House, 1993) elaborates this chapter's themes in a most persuasive way, an the interested reader is encouraged to read it.

[4] It is ironic that people who are against conserving the environment are calle conservatives. Radio entertainers like Rush Limbaugh are taken seriously onl because we are creatures of language and habit and hyperbole. If a thing is sai often enough and strongly enough, we tend to believe it. Thus Rush should adop the exclamation point, and not the ditto, as his identifying punctuation mark.

[5] See Sandra Postel's *Last Oasis: Facing Water Scarcity* (Washington, D.C. Worldwatch Institute, 1992).

[6] See Holmes Rolston III's "Rights and Responsibilities on the Home Planet pp. 425-439 in Vol. 28 No. 4 (December 1993) of *Zygon.*

[7] On March 25, 1992, which was after I wrote this prediction but befor publication, the Supreme Court handed down a decision favoring the loggers. Th battle is far from over, but my prediction is already beginning to come true. Se *Robertson vs. Seattle Audubon Society et. al.*, 118 L. Ed. 2nd. 73.

[8] I wrote this paragraph in early 1992, before it became clear that George Bus might lose the American presidency. Although President Clinton will probabl delay the prophesy given, the underlying analysis remains the same. Even majo environmental organizations acknowledge the likelihood of this result; for ex ample see Tom Horton's "The Endangered Species Act: Too tough, Too weak, o Too late?" in *Audubon*, Volume 94, Number 2 (March/April 1992), pp. 68-74 (The paragraph was also written before the conservative victories of 1994.)

[9] Peter, Paul, and Mary (Peter Yarrow, Noel Paul Stookey, and Mary Travers song written by Peter Yarrow), "The Great Mandala (The Wheel of Life)," fron *Album 1700* (Burbank, California: Warner Bros. Records, Inc., Album # 1700)

[10] For further discussion and citations, see "Gauging the Winds of War," b Bruce Bower, in *Science News*, Vol. 139, no. 6 (February 9, 1991), pp. 88-9. Als see John Horgan's report on the Correlates of War project in the April 1991 issu of *Scientific American*, pp. 24-5.

[11] See Jay Forrester's "Counter-Intuitive Behavior of Social Systems" i *Toward Global Equilibrium*, ed. Dennis Meadows and Donella Meadows. Th report on Denmark's policy is on p. 29, but the entire article is very informative

[12] See "Population Control Mechanisms in a Primitive Agricultural Society, an analysis of the findings of Roy A. Rappaport by Steven B. Shantzis and Willian W. Behrens III, included as Chapter 9 (pp. 257-288) in *Toward Global Equilib rium*, ed. Dennis L. Meadows and Donella H. Meadows.

[13] For a summary of the policy alternatives see Jay Forrester's "Counterintuitiv Behavior of Social Systems." For more on global modelling see *Groping in th Dark*, by Donella Meadows, John Richardson, and Gerhart Bruckmann (Nev York: John Wiley & Sons, 1982) and *Beyond the Limits of Growth*, by Donell Meadows, Dennis Meadows, and Jorgen Randers (Post Mills, Vermont: Chelse Green Publishing, 1994).

[14] See pp. 177-215 of *Beyond Oil: The Threat to Food and Fuel in the Coming ecades*, by John Gever, Robert Kaufmann, David Skole, and Charles Vörösmarty. lthough this book acknowledges the ultimate need for global equilibrium, its cus is on one country, namely the United States. In keeping with this focus, it oes so far as to propose limitations on immigration as a way of conserving energy. have not listed this idea as a proposed policy for reducing population because it trikes me as counter-productive and short-sighted. Isolationist policies have been dopted throughout history to reduce stresses caused by growth, but the problem as outgrown all boundaries. It must be solved comprehensively.

[15] See pp. 243-246 of *Beyond Oil*. Note the additional policy initiatives roposed there; although they address energy supplies instead of population, most f them would also act to stimulate a reduction of population.

[16] Randers, Jorgen and Meadows, Donella. "The Carrying Capacity of the lobal Environment." Chapter 12 of *Toward Global Equilibrium* (p.323).

[17] See p. 238 of *Beyond Oil*, by John Gever and others (cited earlier).

[18] For a review of the modern and historical functions of politics see *The rowth of Government*, by Geoffrey Fry (London: Frank Cass and Company, Ltd., 979). For a biologically oriented treatment with an emphasis on the future see Valter Truett Anderson's *To Govern Evolution: Further Adventures of the olitical Animal* (Boston: Harcourt, Brace, Jovanovich; 1987). For some fascinat-ng new ideas on the interplay between government and economics see the iterature of equilibrium theory, especially Robert Kuenne's *General Equilibrium conomics, Part I* (New York: New York University Press, 1992); Edmund helps's "Synthesis of the Global-Economy Theory," Ch. 14 (pp. 230-242) in his *tructural Slumps* (Cambridge, Massachusetts: Harvard University Press, 1994); ohn Shoven and John Whalley's *Applying Global Equilibrium* (Cambridge Jniversity Press, 1992); Cristina Bicchieri's *Rationality and Coordination* (Cam-ridge University Press, 1993); and pp. 202-3 of Michio Morishima's *Capital and redit* (Cambridge University Press, 1992).

[19] In *On Aggression* Konrad Lorenz suggested that the smell of blood inhibits ggression. If that is true, primitive hand-to-hand conflict may come with inbuilt rakes. However, technology has produced "aggression-at-a-distance," where uch inbuilt brakes never have a chance to operate.

[20] On March 25, 1992, shortly after I wrote the sentence in the text, Mobil Oil nnounced that it would stop drilling in the continental U.S. It will be interesting nd sobering to chronicle similar announcements by other companies.

Chapter 15

Learning and Teaching: Education from a Biological Perspective

<u>An Outline of the Chapter</u>

A. The Biology of Education
B. Pedagogy: Learning How to Help Learners Learn
C. Beyond Schooling: Individual Learning
D. The Content of Education
E. Teaching Morality and Grace
F. Teaching Social Responsibility

Chapters 2 and 7 discussed learning as if it were an isolated individual process. The more complex truth is that learning takes place in a social context, and it takes place continuously. Indeed many writers have pointed out the necessity for constantly relearning what we have already learned; only in that way does society reaffirm what counts as "knowledge," and only in that way do individuals shape and retain particular concepts from all the ideas that pass through their brains.

On the other hand, it is true that social knowledge is simply a combination of individual knowledge, and it is also true that knowledge can exist in the absence of social structures and even in the absence of human beings. This chapter will show how such seemingly inconsistent things can be true, primarily because of the biology that underlies our social nature.

The Biology of Education

Like other higher animals, we collect data with our senses, and ou brains sort it into categories. Everything we "know" resides in the brai as memory traces, memory pathways, modified cell structures, chemica compounds, or other biological forms.

Despite the size and sophistication of the brain, we cannot possibl store every insignificant piece of data that passes through our senses Therefore the question arises as to how our brains decide what data t extract and retain. We have already seen how the subconscious uses tria and error to make that decision, primarily using gestalt processes. Othe processes are also involved in the subconscious retention decision, fo example habituation, which accustoms us to nightly train whistles or othe repetitive stimuli so that we do not have to think about them. And of cours rational processes play a part, allowing us to consciously consider whethe a piece of data is important.

The field of evolutionary epistemology explores the biological histor of such knowledge processes. That label is a good one, since "evolution ary" implies a historical process within the biology of the species, an "epistemology" refers to the nature of knowledge. Although it is jus gaining momentum as a separate discipline, evolutionary epistemolog has already helped us realize that the impetus for learning comes fror evolved internal processes interacting with the outside world, usin various feedback loops and information branches so that we often hav several chances to recognize important data.[1] This "schoolmarm in ou heads," directs our attention and helps us figure out how to learn even i we don't start off knowing very much.

The social component of this basic process can be better understood b exploring two concepts that are addressed by evolutionary epistemolo gists. One is called *hermeneutical pre-knowledge* and the other can b labelled *the sociology of knowledge*. An explanation of how they wor together (and in conjunction with other knowledge processes) will help t show why education is so critical to human experience.

We discussed hermeneutical pre-knowledge to some extent in Chapte Seven. You may remember that Hermes was a mythological figure wh transmitted and interpreted messages to humans from the gods.[2] Thu "hermeneutics" came to mean interpretation, and refers to the fact that w not only ***have*** experience; we also ***interpret*** it. Our interpretations depen largely on whatever knowledge and concepts we bring to each experienc so that there is no such thing as "pure" knowledge. It is always mediate or influenced by what we already know. Thus the knowledge that exist before each new interpretation of experience is called hermeneutical pre knowledge.

Evolutionary epistemology establishes that some knowledge is already
n our heads at birth, having got there through normal trial-and-error
volution over millions of years.[3] Thus a baby knows to seek the breast
ithout training, and is pre-programmed in many other ways. Most
nportantly, it knows how to learn from other people, and that brings up
e social aspects of knowledge.

Humans are not the only creatures that learn from one another. Indeed,
is kind of learning is quite common in the animal world. Chapter 2
cited several examples, such as the birds that taught one another to
move lids, and the baboon that taught her peers to winnow grain. A
cent experiment shows that even octopuses can learn from one another;
ey learned to retrieve a light-colored ball instead of a dark-colored one,
imply by watching a trained octopus do the same thing.[4]

Yet humans rely on social knowledge more than any other animal does.
Ve learn our earliest concepts from our group, embedding them in
emory pathways and interpreting subsequent experience in light of those
oncepts. Moreover, biology provides a long maturation period during
hich we learn from our elders. Language allows—perhaps compels—
e accumulation of social knowledge, so that we can teach one another
ithout having to duplicate one another's experiences. Reading even
llows us to learn without a teacher present. This societal knowledge
erves as a crucial part of our hermeneutical pre-knowledge, once it gets
corporated into our mental pathways.

Of course we are not "locked in" to hermeneutical pre-knowledge or
social knowledge—Chapter 7 explored how they are modified by
xperience and evaluation. Yet hermeneutical pre-knowledge and social
nowledge profoundly affect the interpretation of experience; therefore it
important to consider both the methods we use to teach, and the content
f what we teach. This chapter will do that in two sections called
Pedagogy" (the theory of education) and "The Content of Education." In
etween those sections we will briefly discuss a form of adult education.

edagogy: Learning How to Help Learners Learn

Jean Piaget and his disciples brilliantly explored the operation of innate
earning processes in children. Children interact with their environments
n relatively predictable ways, because of specific developments in their
rain structures at each new stage of growth. At the earliest stage they are
nwardly focused, but they soon learn to control their crying and smiling
nstincts to optimize the attention they get from others. Next they become
nore aware of objects and people, though they cannot conceive of the
ontinued existence of objects or people that are hidden (under a blanket,

for example). Once they learn language, concrete logical operation become possible, and finally children start learning to use abstract logic At each stage they make delightful mistakes, but soon they correct thos mistakes and start seeing the world basically like adults do.[5]

Piaget's work implies that children must be taught things in a certai order and at a certain time, because their brain development determine what they can learn. Thus there really isn't much point in trying to teac a 10 year old to understand abstract principles of government (fo example), whereas a 14 year old is ready to learn such things. This concep is called "readiness" in educational theory. Before a child can lear something, he or she must be prepared, not only in terms of age and menta maturation, but also by mastering materials that are prerequisite to th current material.

It is not clear just how far the readiness concept can be stretched. Ou formal educational system is built on it, so that we set up pathways o learning from the simplest concepts to the most advanced. Yet beyond th predictable developments of early childhood, learning is so complex, s individualistic, and so involved with social components, that it is difficul to construct a simple system even in elementary education. By junior hig school, different children are ready to learn different things, some childre are highly resistant to schooling, teachers become frustrated, and th public argues endlessly about whether its educational dollars are well spent

In an interesting book entitled *Learning from our Mistakes: A Reinterpretation of Twentieth-Century Educational Theory*, the educato Henry Perkinson argues that we could solve these problems by takin advantage of one aspect of human nature.[7] Building on evolutionar principles, he argues that humans have an innate desire to resolve incon sistencies, using trial and error. Successful teachers like Madam Montessori, A.S. Neill, and Carl Rogers intuitively understood tha students want to resolve inconsistencies, and they developed techniques t help students see their inconsistencies and correct them. All good teacher can use the same techniques these great teachers did.

How can teachers do that? Perkinson examines the great teachers c this century and concludes that it can be done by creating educationa environments that are 1) free 2) supportive and 3) responsive.

Free doesn't refer to a financing scheme, nor does it mean the student run wild (though Neill seems to have given them that freedom at hi Summerhill school[8]). It means rather that students are encouraged t freely reveal their present knowledge, instead of listening to teacher reveal theirs. The teacher is non-judgmental and does not seek to correc errors. Given such freedom, the natural exhibitionism and expressivenes of students can be relied upon to ensure that they will share whateve knowledge they have.

Second, the teacher must be supportive. That mostly means accepting mistakes, not just because they are inevitable but also because they are productive. The biggest impediment to education is the intimidation most students feel when they make mistakes. Teachers must foster the idea that we are in school and in life to make mistakes, to enjoy our mistakes, and to correct them. Enjoying our mistakes does not mean laughing at one another, but rather delighting in the discovery of error, so that it may be corrected and knowledge can be increased.

Finally, the educational environment must be responsive. It must respond to the student's curiosity and display of knowledge, and it must do so in honest and realistic ways. The teacher should do this by presenting information, but it should be made clear that the information itself may be full of mistakes. Since knowledge is fashioned by human beings, and since all human beings are constantly making mistakes, the student should be encouraged to criticize what is taught. But the criticism must focus on improving the knowledge and correcting its errors through solid reasoning from facts and experience. This will usually lead students to correct their own errors, and to accept well-established knowledge as presented by the teacher.

Anyone bold enough to present a theory of education is bound to sound naïve, and Perkinson is no exception. Yet his theory sows the seeds of its own improvement, by candidly acknowledging the need for criticism. In my opinion at least three points of criticism are called for.

First, the theory does not adequately account for discipline. Even though it calls itself Darwinian, it ignores the lesson of nature, where discipline (and sometimes harsh discipline) is the rule. Although college students may be mature enough for Rogerian methods, children need the moral and social discipline imposed by Montessori. Thus Perkinson is mistaken when he sees moral and social discipline as a flaw in Madame Montessori's method.[9]

Second, the theory does not give enough weight to the transmission of knowledge. Indeed it denies that transmission takes place at all, since students learn only by correcting their own mistakes, and not by simple acceptance of knowledge from the teacher. It is true that lectures can be ineffective when presented in an authoritarian way. But it would be highly inefficient to expect every student to re-trace the learning path, to do all the experiments that led to our present knowledge, or to re-invent the light bulb. Instead, students instinctively trust most passed-down knowledge, and they usually accept it unless it is poorly presented or there are other reasons to challenge it. Perkinson's warnings about authoritarianism are well taken, but even a small increase in student-centered techniques pays big dividends. It is not necessary to completely abandon the benefits of transmission.[10]

Third, Perkinson's theory flies in the face of practicality. It woul undoubtably work if the student-teacher ratio was only 10 students pe teacher. But current ratios are 30 to 1. In such a large classroom it i impossible to give the individual attention Perkinson's theory require and some trade-off in favor of authoritarianism is necessary.

There may be a way to do what Perkinson recommends, however, eve with large ratios. Despite his criticism of behavioristic teaching machine computers may be needed to implement his ideas. After all, students ar willing to spend long hours in front of video games, and those games ar already being modified to teach traditional subject matter in interestin ways. But how can impersonal computers take the place of traditiona teachers?

The team-teaching concept was popular in the 1970's, but it fell fror favor because teachers didn't like to coordinate so closely with thei colleagues. Perhaps we could get the benefits of three worlds (classic teacher, team-teaching, and modern technology) by letting a human team teach with a computer. Suppose a class of 30 students, composed of 2 private carrels (monitored from above by TV cameras for disciplinar infractions) and a small circle of 11 chairs. The teacher spends 1/3 of th time with each rotation of 10 students, and the other students learn vi interesting computer programs. The time with the teacher is sper discussing the computer material, discovering and resolving apparer inconsistencies between its knowledge and the students' prior knowledg At the computer, students learn new facts and apply them to probler situations, progressing at their own pace and going into whatever dept they desire. They could also be graded by the computer on most tasks

Such technology may not be necessary at the college level, and indee Perkinson's vision is already being realized there, largely because studen can be required to learn outside the classroom. Granted, many professo do not adequately criticize the knowledge within their field, much les encourage their students to do so. And granted that there will always b a need for re-assessing teaching techniques (for example see the compe ling argument in favor of changing the way we teach introductory colleg biology in *Science as a Way of Knowing*[11]). But overall, it appears th college teachers do try to lead students in a discovery process that they ar also participating in, using sophisticated techniques of trial and error.[12]

Beyond Schooling: Individual Learning

Few of us really begin to see conceptual flaws in what we have learne until after our student years are over. That may be because students are nc trained to ask questions or explore inconsistencies, but it is also becaus hey lack experience, and because they have not yet accumulated a fund f knowledge from which to criticize what is taught. Once people have ome experience and knowledge, they are better able to evaluate and uestion the information and theories presented to them.

There is a built-in disadvantage to experience, however, which grows ut of hermeneutical pre-knowledge: *our knowledge puts blinders on us*. s Popper has shown, we enter every new situation with expectations, ased on our prior knowledge and experience.[13] We even begin life with nnate expectations based on millennia of genetic evolution, and it takes ard work for our brains to modify those expectations where they are rong. Even as we gain knowledge and concepts, we continue to expect he world to conform to our new framework of understanding. When that xpectation is not met, it causes an internal disharmony that sparks further hought and conceptualizing. Theoretically this should mean we undertand the world better as we grow older and wiser. But several things keep hat ideal from being realized, some of which are as follows:

- society may not possess the knowledge that would resolve disharmonies;
- a given person may not know how to access knowledge even when it exists in society;
- life's practical problems are usually too pressing to permit a comprehensive search for answers;
- we may be loyal to dogmas that are hostile to available knowledge, and thus we may never seek it out even if dimly aware that it exists;
- we may not have the skills for realizing that a disharmony exists in the first place, though it influences our choices and confuses our belief structure.

These problems could perhaps be addressed institutionally, for exmple through adult education. But few participate in such formal rograms, and thus continued learning is largely up to individuals. Each erson must decide whether further learning is worth the effort. To be orth the effort, adult learning must be perceived as productive and fficient.

To help learning be reasonably efficient, we need a good concept of earning itself. Perkinson and Popper and Piaget agree that efficient earning depends on discovering our errors as soon as possible. Unfortuately our errors tend to become habits; thus we must give conscious ttention to principles and compensatory habits for overcoming that endency. We can do that in three ways, namely by becoming wellnformed, by concentrating on questions instead of answers, and by eeking breadth instead of narrowness. Let me briefly expand on each of hese ideas.

1) *Becoming well-informed.* Popular media make information more readily accessible than ever before, but deadlines and profits water down the benefit, and longer treatments must be sought out if one is really to be well-informed. Well-written journals and books provide much more in-depth information than television can. Full texts of judicial opinions reveal much more subtlety than newspaper accounts can dream of. Both sides of contemporary moral issues can be thoughtfully discussed in long essays or in good fiction. And it really does not take much more time or effort to be informed at that deeper lever than it does to read a newspaper or sit in front of the "tube" night after night.[14]

2) *Concentrating on questions.* Questions are crucial to helping individuals construct their own ideas or concepts, because questions make it more likely that all sides of any issue will be considered. Socrates demonstrated how important questions are to the growth of knowledge, and Popper's emphasis on particular methods of discovering error underscores Socrates's insight. Another advantage of books and other longer treatments is that they typically explore underlying questions in depth, whereas popular media can only mention the questions as an introduction to sketchy answers. Notice how scripture summarizes ancient wisdom on this point by saying "Ask, and ye shall receive; seek and ye shall find; knock, and it shall be opened unto you."[15]

3) *Seeking breadth instead of narrowness.* In our occupations we usually need precise knowledge, but resolving disharmony in our concepts requires broader knowledge. The more we study disciplines other than our own, and the more we subject ourselves to the opinions of others, the more integrated our concepts and worldviews are likely to be. That is because broad study and experience supply the raw material from which subconscious gestalt processes manufacture insightful connections and new concepts. Konrad Lorenz makes this abundantly clear in his earlier-cited book *The Waning of Humanness*.

* * *

So far this chapter has discussed the *theory* of learning at various ages. Equally important to biotheology, however, is the *content* of education, including especially the teaching of moral and social responsibility.

The Content of Education

Ever since Plato's *Republic*, it has been clear that we need to teach four things, namely *skills, knowledge, ideas, and responsibility*.

Skills includes the general abilities needed to navigate through life, as well as specific occupational skills. Schools seldom have the time and resources to teach more subtle skills, such as the skills of interpersonal interaction and grace, but people need to learn those skills too. It is presently up to individuals to learn those things on their own, but as crowdedness continues to erode cultural structures it may become necessary to teach them formally, perhaps through our churches.

Knowledge is the present focus of our educational system, and although religion usually supports the teaching of knowledge, certain scriptures are suspicious of knowledge,[16] and some fundamentalists are hostile to any knowledge that challenges their doctrines. On the other hand, scripture says knowledge casts out fear. It is easy to see that this is true, because knowledge increases security, and a more secure person is less fearful. A basic aim of education is to provide the knowledge that can make society more secure, and mainline religion (as opposed to fundamentalism) usually supports education's efforts to spread knowledge, even when it challenges belief.

Thus knowledge seeps over into the realm of *ideas*. Schools teach ideas as part of the content of each subject, and a few subjects lend themselves to teaching ideas directly (history and government, for example), even without formal philosophy. Religion also teaches ideas quite directly, even to very young children, but it usually fails to incorporate any critical evaluation of the ideas taught, with consequences explored in earlier chapters. Thus it is up to families to make sure their children can competently evaluate ideas gleaned from church and school, and it is up to each individual to organize and develop ideas into a cohesive framework of belief, as Chapter 7 said.

When it comes to teaching *responsibility* and the values associated with responsibility, schools in the United States have been in a state of confusion for decades, but they may be emerging from that confusion. It is still considered unethical to evangelize students toward particular dogmas in public schools, but teachers now recognize that it is impossible to teach any subject without imparting values. In fact schools have always taught values, if only in confused ways, because values arise from beliefs about facts (as Chapter 12 said). Schools strive to teach facts, and that requires a selection among the many facts that could be taught. Moreover, it seems clear that families and churches are not adequately teaching values, and that fact makes it more likely that schools will have to spend more time teaching values and responsibility.[17]

Values and responsibility are broad topics, and overlap many of the ideas we have already discussed, including morality (Chapter 10), faith (Chapter 7), free will (Chapter 9), culture (Chapter 3), salvation (Chapter 11), and happiness (Chapter 12), not to mention the social responsibility which is the subject of these last four chapters. Those concepts incorporate not only the responsibility one has to the small group (family and neighbors), but also the responsibility one has to one's world and one's self. Thus we must expand the topic into two final sections of this chapter, which address in more detail the teaching of morality and grace, and the teaching of social responsibility.

Teaching Morality and Grace

One can conceive of learning morality by trial and error, and errors are effective moral teachers. Moreover, we tend to over-emphasize the danger of moral mistakes, probably because some of them can be so devastating. Unfortunately, the resulting anxiety spreads to our children and reinforces them to keep moral mistakes secret. Indeed, we have trouble acknowledging our own moral mistakes, largely because our religious concepts have condemned them.

That is why A.S. Neill allowed his Summerhill students to construct their own morality.[18] Certainly they made mistakes, but the result was a strong morality that they felt committed to. Interestingly, St. Paul's radical permissiveness had the same result.[19] Both Neill's students and Paul's followers constructed a strong morality notwithstanding the permissiveness of their mentors. That observation combined with the fact that even atheistic societies adopt and enforce moral principles, reminds us that morality is innate in humans, just as Chapter 10 said.

Yet innate impulses must be trained and shaped. There is no inconsistency in that, because training and being trained are themselves innate behaviors. Our impulses result from eons of evolution interacting with dozens of environmental situations, but we did not merely evolve kneejerk reactions to those situations; we also evolved a brain. An important function of the brain is to consider which impulses are appropriate to the actual circumstances of our lives, to distill moral rules from our experience, and to teach what we learn to others.

Training and shaping for morality is not simple. There are at least three prerequisites to it, in addition to trial and error. They are 1) discipline, 2) example, and 3) discussion. In pre-adolescent childhood, discipline is the key ingredient, just as for other mammals. After that, discussion is the key. Example underlies the whole process, and all the processes grow out of basic principles of trial and error.

For example, discipline is technically a form of trial and error. Our job as adults is to make sure the trials of children are met with firm responses, so the children will gain confidence in their role models. When the trials are in a direction we believe productive (based on our own experience and concepts) we try to give encouragement and reinforcement. When in an unproductive direction, we aim for consistent and appropriate punishment or other negative reinforcement. By trial and error the child learns which behaviors bring about desirable responses, and parents learn how to elicit socially acceptable behaviors. A delightful elaboration of this kind of interactive behavioral modification is presented in Karen Pryor's popular book *Don't Shoot the Dog,*[20] though it may go too far in recommending the same tactics for training one's fellow adults as one uses to train dogs or children.

Discipline is thus a good and natural thing, and its use in teaching morality is biologically sound. We rely on it too much, however, when we try to impose it all the way into late adolescence. By that time young people are ready to start making their own decisions, and they don't respond well to external discipline (as opposed to self-discipline they adopt, or discipline they agree to as part of negotiated contracts). At that point we only have two good choices, namely 1) trusting early discipline and example, and 2) making adult-to-adult agreements. From our conservative adult perspective, it will often appear as if our children are on the edge of disaster, but if we have imparted psychological and physical security to them, they are very likely to adopt the same morality they were taught, with some modifications based on their own experience.

Teaching morality is still very possible in adolescence, however, provided we understand and respect the new stage of growth it represents. At this stage discussion and explanation and other verbal strategies are necessary, much as among adults. Teenagers are particularly interested in moral speculations, and teachers should pay attention to that interest. For example they can select literature which displays hypothetical results of moral choices, which teenagers are motivated to discuss; history is stranger than fiction, and produces results that are anything but hypothetical; psychology and philosophy explore the reasons for adopting one or another moral precept. All of these things can be taught in normal educational settings, and wise parents can learn to discuss them at home with open minds.

Preaching can also play a role in teaching morality to adolescents, but it can undermine morality if preachers don't understand the psychobiology of human development. Threats of punishment in hell or messages that say "Don't do that because it is wrong," are not likely to be heard by adolescents, unless they are accompanied by good explanations of *why* something is wrong. *Mad Magazine* appeals to adolescents because it supports its preaching with persuasive reasons; it says "Don't do that because it is stupid" rather than "Don't do it because it is wrong." For example a *Mad* cartoon might show three hilarious frames of an outrageous character smoking cigarette after cigarette, and end with a final frame showing his lungs rotted out. The message is graphic to the point of grossness, but the point is made. If we want to teach morality, we must ground our opinions in reasons and in foreseeable results, and not in magically revealed truth that is immune from discussion or from trial and error.[21]

When it comes to grace, it is harder to see how any of these dynamics are involved. Isn't grace a personal quality that one either has or doesn't have? And is there any clear reason that we need it? Certainly example has something to do with those questions, and Chapter 11 explained discipline's role in grace. But is there any role for education, for studying and explaining the techniques and theory of grace?

Recall Chapter 12 again. It said that our attitudes and values are based on beliefs, especially our beliefs about facts. If we can be persuaded that graceful attributes can improve our lives as a matter of fact, that belief will in turn influence our attitudes and values toward greater gracefulness. We will become more humble, more accepting, more calm and cheerful, and so on.

But how is one persuaded of the fact that grace improves life? First, we must remember that grace is but one aspect of our natures. It does not operate in a vacuum of holiness, but rather interacts with other parts of our nature. Ungraceful behaviors like aggressiveness and selfishness evolved because they had survival value, and grace merely moderates them. On the other hand naked aggression and raw selfishness are recognized as pathological, and we have learned to address those pathologies through medicine and psychological therapy (or through punishment if the pathology is not too great). In the home, parents can prevent a great deal of developmental pathology by learning to provide a foundation of love and security that will allow their children to grow in grace, again as Chapter 11 said.

Grace can also be taught in our institutions. Formal education can teach grace like it teaches other things: through literature and through real-life examples that present themselves in the school routine. Churches can teach grace through homilies and sermons and discussions about specific scriptural passages, extrapolating them to the requirements and opportunities of daily living. Grace can be taught in the workplace through formal training sessions which emphasize a mixture of assertiveness and cooperation, and by providing counseling and additional education when tempers flare.

These notes on teaching morality and grace have been sketchy, but I hope they supplement other chapters to show how we can foster the growth of those qualities. Now let us see how we can teach responsible attitudes toward serious contemporary problems.

Teaching Social Responsibility

Our society faces at least three serious challenges as described in Chapter 14—environmental destruction, war, and population growth. Those challenges could be overcome by concerted action based on accurate knowledge. Human brains are receptive to such knowledge, and to attitude shifts based on that knowledge. Thus it seems obvious that education can play a role in teaching people social responsibility.

This does not mean that we have entered a brave new world where Big Brother government brainwashes us in school. It does mean we take responsibility for correcting the problems we face, and that we do it by teaching people about them. Provided our teaching remains humble and permits criticism of the materials, it cannot become mere propaganda. Indeed the person who coined the phrase "groupthink" pointed out that we can avoid propaganda's dangers only by critical evaluation.[22]

Schools can provide much of the needed education from elementary school forward. Dilemmas can be presented and discussed at quite young ages. Civics and government classes can discuss societal problems, and other courses can include units relevant to them.[23] History can discuss the philosophy of war; biology cannot be biology without discussing ecological and environmental issues; geography can discuss population density and its effects; sex education can teach the responsible use of one's reproductive urges.

Schools can do an excellent job with these issues, and teachers are enthusiastic about teaching them. The main impediment is that some citizens don't want them discussed, and school boards are often disproportionately attentive to those citizens. Fundamentalists are especially insecure with these issues, apparently because they are committed to an unchanging faith, and these issues seem to require flexibility and change. They cite Jesus's metaphor that one cannot put new wine in old wineskins lest they break.

Fundamentalists are not merely opposed to new wine in old wineskins, however; they are also opposed to new wine in new wineskins. They only want old wine in old wineskins, because they have found it safe. Yet Jesus was not making an argument in favor of conservatism. On the contrary, he was saying that new wine must be put in new wineskins. We can only provide new wineskins by letting our children learn about the problems we face, and by letting them learn new ways of dealing with those problems. Only in that way can we encourage them to conceptualize newer and better solutions for the problems that have become increasingly evident over the last few decades.

As to educating the larger public about serious contemporary problems, there are several things we can do. But first let us acknowledge what a good job is already being done. Journalism conscientiously presents the problems and enthusiastically reports on promising solutions; non-profit corporations raise vast sums of money from concerned citizens, and marshall volunteers to tell the rest of us about their work; government finds ways to regulate the various competing interests, and at the very least our political campaigns provide an opportunity to focus attention on the issues; international agencies work to uncover relevant facts and to hammer out practical agreements.

Notwithstanding these efforts, the three problems listed in Chapter 14 have not gone away. That is not so terrible; we have really only got a conceptual handle on them within the last few years, and they are all very difficult problems that will take time to resolve, if indeed they can be resolved at all. Other than educating students about them, we can also educate the electorate about them in hopes that they will vote for candidates and policies that can solve the problems. And beyond education, we can ourselves be involved in ways already mentioned.

Indeed, there is cause for optimism. Not only is it true (as Chapter 13 said) that a basic solution to the population problem would also solve many subsidiary problems, but the open-ended nature of learning offers additional hope. New concepts are yet to be imagined, new discovery can take place, new facts will emerge, new devices can be invented, and in the meantime, individuals can continue to learn more about the issues.

There is one other component to educating one another about the serious problems we face. That is to construct an adequate theology, implemented by renewed religious institutions. Only in that way can our churches become agents of change instead of agents of obstruction, helping us meet the need for balance between a sound doctrine and a vibrant practice. The next chapter, the last one, will address this all-important need.

Notes

[1] Even without knowing about evolution, Plato was able to see that some kind of "external fire" mixes with some kind of "internal fire" to produce knowledge in our heads. In the centuries since Plato, philosophers have argued about whether knowledge comes from within or without, and it is only with evolutionary epistemology that we are able to answer that question confidently. For a book on evolutionary epistemology not previously cited, see Konrad Lorenz's *The Waning of Humanness* (Boston: Little, Brown and Company, 1987).

[2] Incidentally, St. Paul was called Hermes by the citizens of Lystra, because he was the chief messenger or speaker for God; see Acts 14:12 in The Revised Standard Version of the Bible.

[3] See Gunther Stent's "Hermeneutics and the Analysis of Complex Biological Systems," pp. 209-225 in *Evolution at a Crossroads*, ed. David J. Depew and Bruce H. Weber.

[4] See Boris Weintraub's report of the experiment in the "Geographica" section (page not numbered) of Vol. 182, No. 6 (December 1992) of *National Geographic*. See also Boesch, Christophe, "Teaching Among Wild Chimpanzees;" *Animal Behavior* for March, 1991, pp 530-2.

[5] See Piaget's *The Principles of Genetic Epistemology* (New York: Basic Books, Inc. 1972; translated from the French by Wolfe Mays). To demonstrate a common mistake at the pre-operational stage (ages 2-7), suppose you have ten brown marbles and a white one. The child knows there are more brown marbles than white ones. But even if you explain that all the marbles are glass, she or he will not understand that there are more glass marbles than brown ones.

[6] For more on the directions learning can take see *Developmental Plasticity*, ed. Eugene S. Gollin.

[7] Westport, Connecticut: Greenwood Press, 1984.

[8] For a sensitive argument against too much freedom, see Holmes Rolston III's "Science Education and Moral Education," pp. 347-355 of Vol. 23 no. 2 (September 1988) of *Zygon.*

[9] In the United States, discipline has been lax since racial integration began. That is probably because nobody wanted to be guilty of racism, and black children tended to cause more discipline problems, as the poorest children almost always do. But now black parents are beginning to demand discipline too, requiring only that it be fairly administered. Their demand offers a great hope for the rehabilitation of public education.

[10] The most sophisticated treatment of transmission I have seen is in a book called *Complexity of the Self*, by Vittorio F. Guidano (New York and London: The Guilford Press, 1987). For a down-to-earth and influential article by a veteran teacher with similar ideas, see "Reinventing Teaching," by Deborah Meier, in Vol. 93 No. 4 (Summer 1992) of *Teachers College Record.*

[11] Volume 1 of the series by that name is entitled *Evolutionary Biology*, and the lead article by John A. Moore outlines what must be done to improve the teaching of biology. See also Weingartner, Rudolph H., *Undergraduate Education: Goals and Means* (Phoenix, Arizona: Oryx Press, 1993), especially p. 87.

[12] For comments on changing University classrooms to reap the benefits of cognitive learning theory see Chapter 7 ("Active Work and Creative Thought in University Classrooms," by Howard E. Gruber and Lucien Richard) in *Promoting Cognitive Growth Over the Life Span*, ed. Milton Schwebel, Charles A. Maher, and Nancy S. Fagley (Hillsdale, New Jersey: Lawrence Erlbaum Associates, 1990). See also Howard R. Bowen's *Investment in Learning: the Individual and Social Value of American Higher Education* (San Francisco: Jossey-Bass Publishers, 1978). And of course nobody can clearly think about the current state of university education without considering the many provocative ideas in Allan Bloom's *The Closing of the American Mind* (New York: Simon and Schuster, 1987).

[13] Without quibbling over the profundity of Popper's insight, I will point out that it is subject to an important criticism which does not affect the argument in the text. That criticism is articulated by Konrad Lorenz, a champion of inductive method in the theory of knowledge. Popper says we cannot justify the statement "all swans are white" on the basis of seeing a million white swans, because that "knowledge" can be contradicted by one black swan. Lorenz, on the other hand, argues that we evolved to rely on inductive experience—knowledge begins in hypotheses arising from induction. Indeed, he says evolution itself is the same kind of process. (See Lorenz's footnote 73 of "Gestalt Perception as a Source of Scientific Knowledge," included in *Studies in Animal and Human Behavior*, Vol.2, cited earlier.) Thus Lorenz would affirm the process of induction, even though he believes we have to test the induction when it begins to look suspicious. Of course Popper would agree that the black swan does not significantly change the *probability* that the swans we encounter will be white; see his *A World of Propensities* (Bristol, England: Thoemmes, 1990).

[14] There is nothing inherent in television or newspapers that prevents deeper treatment, and they do sometimes treat issues in depth. As multi-media develops, it could help keep us well-informed; but it is just as likely that "information highway" will turn out to be a euphemism for "entertainment highway."

[15] St. Matthew 7:7.

[16] For example "For the wisdom of this world is foolishness with God." (I Corinthians 3:19.)

[17] Perhaps the most widely used text for doing this is one which builds on the work of Lawrence Kohlberg, namely *Promoting Moral Growth*, by Joseph Reimer, Diana Pritchard Paolitto, and Richard H. Hersh (Prospect Heights, Illinois: Waveland Press, Inc., 1983). Another approach widely discussed is the hands-on Foxfire approach; see *Sometimes a Shining Moment: The Foxfire Experience*, by Eliot Wiggington (Garden City, New York: Anchor Press/ Doubleday, 1985). In *The Waning of Humanness* Lorenz neatly combines knowledge, ideas, and responsibility; he says we need to supply children hundreds (even thousands) of facts, especially facts from nature, in order to supply data from which the gestalt can extract both truth and a sense of responsibility. Indeed, his development of this idea provides an excellent biological foundation for Perkinson's principles of freedom and support, and shows why the Foxfire approach is so effective.

[18] Neill founded the school in 1921, and it is still in operation. The basic premise is that children are trustworthy, and they are given a great deal of freedom to govern themselves and the school. The results have been good, judging by the later success and testimony of students. See Neill's *Summerhill School* (New York: St. Martin's Press, 1992).

[19] It may seem strange to think of St. Paul as permissive, but see pages 14 and 115 of *According to Paul*, by Joseph A. Fitzmyer (New York: Paulist Press, 1993) and also *Pauline Theology* by the same author (Englewood Cliffs, New Jersey: Prentice Hall, Inc., 1967). An even more thorough and unrelenting treatment is

Paul's Ethic of Freedom, by Peter Richardson (Philadelphia: The Westminster Press, 1979). For a sampling of scriptures that support the claim see Romans 7:6, 8:2 and 21; I Corinthians 8:9 and 10:29; II Corinthians 3:17; and Galatians 2:4, 5:1 and 5:13.

[20] New York: Bantam Books, Inc., 1985.

[21] For a delightful elaboration of this point see Vernard Eller's "The 'Mad' Morality: An Exposé," at pp. 1647-9 of the December 27, 1967 issue (Vol. 84) of *Christian Century* magazine.

[22] See pp. 260-276 of Irving L. Janis's *Groupthink: Psychological Studies of Policy Decisions and Fiascoes*, 2d edition (Boston: Houghton Mifflin Company, 1983).

[23] Based on a personal communication from Dr. Susann Dorman dated June 12, 1993, I believe that many humanities professors are undertaking to teach social responsibility in college classes, in addition to the "official" subject matter of the class. One text that is widely used for this purpose is *Habits of the Heart: Individualism and Commitment in American Life*, by Robert N. Bellah, Richard Madsen, William M. Sullivan, Ann Swidler, and Steven M. Tipton (New York: Harper & Row Perennial Library, 1985).

Chapter 16

Religion in the Age of Biology

One can think about religion in many ways. For example one can think in terms of what religion says about truth, or in terms of the functions it plays in society, or in terms of its impact on individual psychology. One can strive to see if there are common threads that run through all religions, or one can explore the distinguishing characteristics of a specific religion. One can ask whether religion's overall impact is good or bad.

All of these approaches (and others) have been taken by honest scholars trying to understand a very broad subject. Although each approach offers some insight, the one that makes most sense to me is called "the sociology of religion." It is the study of how religious concepts interact with social life. It is a sound approach because it begins in the study of group dynamics that preceded religion, and then asks how particular religious concepts and practices affect the group and are affected by the group. Even within the sociology of religion there are several different approaches, but I consider those that focus on the *functions* of religion most helpful.[1]

Functions can themselves be looked at in several ways. One can take either a historical or an analytical approach, for example. This chapter will soon analyze the functions which religion can perform in the next century, but the historical approach is also useful, because past functions give insight into the potential for religion's future functions.

What a historical overview shows most clearly is that religion has performed almost every societal function imaginable. At one time or another, it has been in charge of government, health, philosophy, education, and science, and in fact each of those functions began from within religion. Even morality, though more ancient than religion, was incorporated and refined by religion, and became a major component of religious practice and theory.

Eventually, most of these disciplines matured beyond the walls of religion, like a child that grows up and moves away from home.[2] The separation was not always pleasant, however; sometimes mother religion was too insecure to let her children go gracefully, and the children were often ungrateful for the help they got at home. There are still many family-like interactions between religion and its offspring: for example charitable groups rely on religion for help when their own resources fail; other groups urge government to revive its lost religious values; even philosophy, sometimes seen as thoroughly secularized, frequently dips back into religion for concepts and methods. By and large, however, separate disciplines have emerged to organize and implement all of the above functions except perhaps morality, and it is unlikely the others will move back home anytime soon.

As each discipline became separate from religion, there was still plenty for religion to do. Once it was divested of government, it could turn its attention to education; and once secular society assumed responsibility for Harvard and Princeton and hundreds of other educational institutions, it could turn its attention to hospitals. Today religion is addressing new challenges, such as third-world agriculture, racial and ethnic conflict, and family disintegration. In fact it sometimes seems as if churches constantly send out tentacles into society, ready to serve any function not being filled by somebody else.[3]

I do not mean to say religion always acts for the good. History shows that it sometimes performs functions harmful to society. It has directly supported slavery, witch-burning, wars, and despotism. In some cultures it has supported prostitution, gambling, and other crude vices. Today's churches often support bigotry and social injustice and an assortment of charlatans. I have said before that eternal vigilance is the price of a great many things, and we must be ever watchful lest our religious institutions do more harm than good.

That is especially true at this juncture in history. As earlier chapters have argued, we are on the verge of a great paradigm shift. This change is coming because science has helped normal biology create a population crisis, and because we are running out of various natural resources. As we strive to meet the resulting challenges over the next century or so, what functions might religion fill? Are there any functions it should abandon?

Are there new functions it should take on, or old functions it should reacquire?

This chapter will propose answers to those questions. It will start by reviewing ancient functions of religion as already discussed in earlier chapters. Then it will ask what particular functions religion seems to be performing now. And finally it will suggest some directions for religion in the new world order.

Some Ancient Functions of Religion

Chapter Five noted that religion constructed the first systematic explanations for both ordinary and mysterious events. Yet even before it began to construct systematic explanations, religion had organized society through ritual. Both functions—explanatory and ritualistic—are still part of religion today.

Ritual's contribution is powerful and poorly understood. Suffice it to say that ritual resonates with rhythms deep in our natures, and religion uses those rhythms to convey its message. In some religious groups the rhythm is as profound as the message itself, to such an extent that one might say "the medium *is* the message."[4] It is unclear whether modern rituals are more powerful or less powerful than those simpler rituals of dance and drum, or whether each is a separate but equal "mask of God."[5] In other words it is not entirely clear whether religion has made progress in its ritual, or merely changed its forms.[6]

But in its explanatory aspect, religion has grown steadily from its earliest beginnings. It developed concepts and then reworked them, as conflicts arose between old and new ideas, between the group and its members, or between the group and adjacent groups. On balance it was conservative, and it focused more on cultural uniformity than on diversity. Nonetheless its concepts grew and changed in response to new ideas produced by cultural interaction and by the theological imagination. Today one can still approach religion as a primitive experience without "studying to show oneself approved," but a sophisticated array of concepts also exists within religion, from which one can construct a modern and cohesive faith that is vibrant and useful.

From the first, religion also performed educational functions, carefully teaching its latest concepts to each succeeding generation. Informal education doubtless preceded religion, as humans taught their children rudimentary skills like hunting and gathering,[7] but apart from such simple teachings, formal education was begun by religion. After writing was invented, concepts could even be preserved for generations yet unborn. Above all, religion became the primary agent for teaching morality, and yet its role in teaching philosophy and cosmology were almost as important.

It is unlikely that pre-historic religion ever imagined a cohesive and comprehensive plan for individual salvation, but it nonetheless increased security by solidifying the group. It also developed strategies for salvation from evil, such as magical incantations, superstitions, sacrifice, and afterlife concepts. Later religion realized how complex salvation can be, and how dependent it is upon individual belief and action. Thus it stopped promising either temporal or ultimate security except to those who did what was necessary to achieve it. That included not only righteous actions and the adoption of an adequate belief structure, but also a continued participation in ritual.

As Chapter 11 showed, Jesus simplified and re-interpreted these salvation doctrines to emphasize the *attitudes* underlying righteousness and belief, attitudes like grace and love, mercy and justice, acceptance and brotherhood. On rare occasions religion has gone even further, preaching that a sound faith can provide happiness as well as salvation.

Some Functions of Religion Today

Chapters 13 and 14 may seem to have abandoned any interest in religion with their discussions of economics and politics, but the sociology of religion has thoroughly documented the influence of religion in both spheres, and vice versa.[8] That is largely because religion has always been involved with the same subject matter as the other two disciplines, namely the use and distribution of resources and the development of policy. Throughout history religion has usually supported the economic and political establishment, and for the most part the establishment has supported it in turn. On the other hand, critics of the establishment have also used religion, for example to argue for a more just distribution of resources or to organize civil rights protests.[9]

Today, government and economics operate without thinking of themselves as having much interaction with religion, except insofar as politicians pander to religious groups. Yet when one thinks of all the functions religion performs in modern life, it is clear that the relationship is quite intimate. Indeed, given the pervasiveness of religion in society today, and given the hundreds of religious organizations, it is safe to say that religion has a profound impact.

What are the functions religion performs today? Of course it still performs some of those mentioned above, but it performs some of them differently than it did in the past. A listing of three or four contemporary functions will show what a significant role religion plays, even though some functions are secularized.

1. Our century has been characterized by changes in family life. Because we are having smaller families, and because our economic system features geographic mobility, many people lack the support of an extended family. Divorce further fragments the family, and other cultural changes drive wedges between the generations.

   Churches have responded to these changes in various ways. Initially they tried to stop the changes, but now they usually accept them as inevitable. They do try to keep marriages together through a greater commitment to counselling ministries, for example, but they also minister to social changes through childcare centers, activities for single adults, and youth recreation programs. They also provide various charities to compensate for economic hardships resulting from fragmented family life. More fundamentally, however, they provide a substitute for the extended family by having constant meetings, which have increased in direct proportion to the fragmentation of family life. While these meetings may appear to be primarily for worship and ritual, they also function to bring people together. That provides the opportunity for friendship, for occupational networking, for watchcare over one another, for romance, and for mutual exhortation and edification. Especially among our senior citizens, the church functions to give support and friendship that might otherwise be lacking.

2. Another special function of religion today is the legitimization of emotion. Although plenty of emotion is expressed in our society (for example by popular music and sports and television soap operas), those outlets are often considered frivolous. "Serious" worldviews tend to de-emphasize emotion, leaving a vacuum that religion fills by affirming emotional expression. In academic circles literature and art also perform this function, but they are relatively unattractive to ordinary people.

   The trade-off is that religion can lead us to pursue emotional fantasies. As Chapter 12 showed, that is not the path to abundant life. But insofar as religion distinguishes healthy core emotions from superficial or unhealthy ones, and insofar as it is able to help us balance our rational responses with healthy emotions, it performs an important function in contemporary life.

   On the other hand, religion does *not* function as well as it could on the rational side of life. This is primarily because of the residual effects of magical and supernatural concepts, including the lingering idea that devils and other immaterial spirits exist.[10] Nonetheless, a few philosophers and pastors have sifted through religious concepts, salvaged some and modified others, and moulded them into rational and productive worldviews. Contributing to that sifting and salvaging and reshaping is of course a main purpose of this book.[11u]

3. We live in a legalistic culture. What is right and wrong, approved and disapproved, allowed and disallowed, has become almost exclusively the province of law and government. Yet everybody knows that morality is more complex than that; some things are wrong that are not illegal, and some things should be permissible even if the law says they aren't. It is of utmost importance that we preserve mechanisms and institutions for persuading one another about such matters outside the stuffy halls of government.

   Religion performs this persuasive function. It does not do it alone, because journalism also does it, and so do informal conversations of various kinds. Indeed, with the maturation of television, journalism may eventually develop a medium that performs this function more comprehensively than religion does. Nonetheless, mutual discourse and persuasion is a monumental task, and religion still contributes greatly to the moulding of public opinion. Its long heritage renders its moral arguments especially worth listening to.
4. Even where other institutions are performing functions once performed by religion, it is often important to have parallel structures—even seemingly redundant structures—for some functions. One wonders whether the former Soviet Union might not have come unglued precisely because it didn't have the parallel structures of religion for performing some functions. If churches had been encouraged, for example, might not they have tapped into international sources of aid? Might not that have eased the roughest economic hardships, and allowed the state to survive? More importantly, might not churches have functioned to develop and organize public opinion toward gradual changes in the social structure, so that cataclysmic change might not have been necessary? The church could have functioned as salt to season the otherwise bland bureaucracy, and (to change the metaphor) to harmonize the underlying discord in society. It has certainly performed these functions in western societies. For example America's black churches have consistently influenced society toward social justice, liberal white churches served as forums for questioning the Vietnam War, and fundamentalist churches call us to examine our attitudes about abortion.

These are four of the functions religion seems to be performing today, and they all seem exceedingly worthwhile. The remaining question is whether there are special functions that religion will be called on to perform in the next hundred years or so.

Religion in the Age of Biology

Throughout earlier chapters I noted various ways that we should modify our theological concepts to conform to biological or scientific understanding. In this section I will pull all those notes together into a list of ways that religion should change its concepts, if it is to be a vibrant force in the next century. I will organize and discuss the list under three headings, namely 1) Morality and values; 2) Philosophy and theology; and 3) Action.

Morality and values

For the most part religion is still in charge of morality. Even if people don't act morally, they usually look to religion for the ultimate word on what is moral and what is not. But unless religion modifies its traditional approach to morality, it is in danger of losing its credibility in this arena.

That is because most religious groups still tend toward a rigid support of traditional wisdom. This leads them to *declare* what morality is instead of trying to *discover* what it is, whereas biologists and anthropologists are working hard to discover what it is and how it operates. I am not claiming that our biological urges are the same thing as morality, or that primitive tribes are more moral than civilized nations—far from it. But unless we understand underlying biological and psychosocial dynamics, there is little hope that we can build a consistent and persuasive moral philosophy.

Since every religious college has a biology department and a psychology department and a sociology department, the framework exists for cross-disciplinary moral researches. Religion's strong moral heritage qualifies its educational institutions to sponsor such researches. But so long as those institutions spend their energies trying to support a narrow tradition, especially by avoiding evolution and its implications, just that long will moral leadership continue to slip away from religion. Already the lead has been seized by secular institutions that are sponsoring studies in sociobiology, but it is not too late for religion to reassert its leadership.

The questions that might be researched are endless, but almost all of them are important. What are the biological underpinnings of homosexuality? Would it be moral for parents to plan their families so as to avoid having a homosexual child? What methods of family planning are moral, especially in a world condemned to ever-increasing starvation absent several kinds of approved birth-control? What are the biological roots, if any, of fidelity and promiscuity?[12] Of honesty? Of aggression? What are the implications and variations of our small-group nature? Does one indeed enjoy a statistical edge by heeding the advice of traditional moral systems, and if so why?[13]

These researches are ideally fitted to religious colleges, but greater moral challenges are faced by our churches. The coming strains on society could easily push churches into a position of reactionary self-righteousness. They could slip back into their pre-Reformation role of supporting economic and political parties in the name of morality, instead of supplying true moral leadership. In fact one must acknowledge that scenario as a likely one, for all but the most forward-thinking churches.

One writer, representing a prestigious group of futurists, has suggested how the church can avoid this trap. He suggests that the over-arching function of religion must be to focus on the *long-range* values of society, as opposed to the shorter-range values that occupy governments and other institutions.[14] The Ten Commandments show this focus, because they represent values that are believed to be valid over very long time periods, relevant to more than contemporary concerns. That is why they have kept their power, whereas the lesser jots and tittles of Leviticus and the other "thousand points of law" have not.

Indeed, whenever religion strays from its role in establishing and maintaining long-term values, it becomes merely an arm of secular society, subject to all the short-sightedness and tunnel-vision that politics and other secular institutions are capable of. This is not necessarily a criticism of politics and secular institutions. Somebody needs to be working on short-range problems. But modern countries are not organized or inclined to think very far ahead. The United States is organized to think in four-year increments, and thus it is no wonder that it typically addresses problems in a crisis mode. The United Nations is theoretically structured to think in long-range terms, but it is so responsive to the moods of its constituent governments that it shares their short-sightedness.

Religion, on the other hand, is ideally situated to think in extremely long time periods, if it can avoid the temptation of secular attractions, including especially the seductive appeal of power and prestige. Those are very strong attractions, and religion has not often avoided them. But scriptural and other traditions call the church away from secular temptation, urging it toward longer-term values.

The question is, which values are long-term? Which values are so embedded in our natures, or so important to our ultimate survival, as to deserve the support of organized religion? And which of those values is so fragile as to need the support?

These questions return us to academic institutions, who are most able to formulate the broad conceptual schemes necessary to answering such questions. However, earlier chapters have already suggested several long-term values that are nonetheless fragile: security, morality, family life, happiness, love, righteousness, and grace. Although these values are somewhat abstract, we have also noted more specific values in passing: sexual fidelity, developing a personal faith, securing an income, and the need for regulating individual behaviors. I will mention some of those values yet again, but this section aims to concentrate on specific opportunities available to the church in the 21st century. If the church seizes those opportunities, it stands in an excellent position to encourage critically important long-term values. If it does not, and if no other segment of society picks up the slack, important human values will go unfulfilled.

In deciding which values are long-term ones, religious and secular thinkers alike must focus on the evolutionary foundations of those values. For example the value for chocolate is certainly not very important, but if we had co-evolved with the cocoa bean, chocolate might have been crucial to health. In that case the value for chocolate would be an important long-term value. The biophilosopher George E. Pugh has written a famous book entitled *The Biological Origin of Human Values*, in which he examines long-term values and divides them into our selfish values, our social values, and our intellectual values. Religion could profit greatly by adopting Pugh's schema, using it to organize and expand its own traditional concepts.

I will not explore all the implications of such an approach here, but one example will show how fruitful it could be, and that example concerns the need for rest. The Ten Commandments set aside one day of seven for rest, and although we often forget that commandment today, rest is as important as it ever was for the physical and psychological needs of human beings.

It is by no means obvious that we need weekly rest. From a strictly biological viewpoint, there is nothing magic about seven days—at least nothing that has been discovered so far—and it might seem that nightly rest would suffice. At the most, young people might need extra rest only every 10 days or so, and old people might need an extra rest every afternoon. Nonetheless, the writers of the Old Testament understood that society must be organized around discrete time periods, and seven days serves as a rough average for the entire society. They distilled from long experience the principle that the Sabbath is good for us; it is in our selfish interest, both as individuals and as a society.

Jesus's insistence on looking for the principles underlying the Sabbath led him to point out that "the Sabbath was made for man, and not man for the Sabbath." After Jesus the church promptly returned to a rigid interpretation of the Sabbath, but religion has recently picked up Jesus's idea again, and is now beginning to figure out how to apply the Sabbath to modern needs. For example a professor of religion and personality has

recently pointed out that we need "Sabbathing" (rest and contemplation
in proportion to the hectic pace at which we live.[15] While a literal day o
rest was certainly appropriate to agricultural life, and while one day i
seven still seems a good average for middle-aged people, it may be that th
church will have to find new ways to implement the underlying reasons fo
the Sabbath. Those reasons clearly include physical rest, but they als
include time for emotional rest, contemplation, and solitude.

If religion can explore the social and biological wellsprings of mora
institutions and principles like the Sabbath, it may yet help rehabilitat
traditional wisdom, by showing just which values are most important in th
long term. But when it insists on a rigid Sabbath or on rigid adherence t
moral rules without trying to understand the dynamics underlying them
religion loses credibility and psychological power.

Philosophy and theology

People need a structure of belief, and indeed some structure is inevi
table, even if it is only a hazy subconscious structure that constantl
reorganizes and modifies innate or cultural values based on individua
experience. That minimal structure is likely to be somewhat loose
however, and can easily lead to confused responses in our complex world
Philosophy is the attempt to "tighten up" our belief structures to make then
more comprehensive, more cohesive, and more useful.[16]

Religion has become wary of secular philosophy in the last few
centuries, primarily because scientific discoveries relied on by philosophy
seem to challenge religion's doctrines. But if religion could modify those
doctrines toward greater compatibility with science, it would actuall
increase its credibility, by offering a belief structure that keeps the bes
components of the old philosophy while incorporating the new.

Making that transition will primarily require religious philosophy tc
abandon supernaturalism. That seems like a tall order, because supernaturalism has been part of religion since its inception. Yet it might not be as
impossible a task as it seems, and indeed it has already begun to be
accomplished, as we have already seen. It has begun to be accomplished
because sensitive theologians have learned to distinguish between literal
truth and symbolic truth, and have reinterpreted old doctrines in light of
new knowledge. (See especially Chapters 7 and 8.)

The real challenge will not be the transition from supernatural tc
natural explanations—that will be accomplished one way or another. The
real challenge will be to do it without unduly disrupting individual lives.
If churches try to deny the truth of scientific discovery, they will put their
members in the position of having to choose between believing various
fantasies and rejecting religion altogether. Such a course runs a grave

anger of decreasing human happiness, because fantasies undermine appiness, as we saw in Chapter 12.

Given the choice between believing fantasy and rejecting religion ltogether, our best young minds will choose the latter. But there is another hoice. We can show them that humans think and communicate symboli-ally, and that we need adequate symbols in order to understand the world nd our place in it. Religion has a good inventory of symbols, and it has ne opportunity to rework and maintain them so that they can continue to ct as foundation-stones in our worldview structures.[17] Such refurbished ymbols can inspire us toward good choices and reasonable beliefs, and nat is the essence of a sound philosophy.

Biotheology is a particular philosophy that reinterprets traditional eligious symbols, as the last section of Chapter 8 explained. It argues that ur religious symbols must be modified in light of modern biology, and it rgues that ongoing modification will always be required as our knowl-dge grows.

ction

Religion can do more than fashion an understanding of morality and alues. It can even do more than to construct a philosophy which is iologically sound and yet modifiable. It can also motivate action.

The needed actions are often unclear, and the prescriptions for meeting hem can be as confusing as a doctor's handwritten instructions. But in our entury the foremost challenges are quite clear, and the needed actions are egible—the handwriting is on the wall. The challenges are the same ones iscussed in Chapter 14: ecological degradation, war, and the population xplosion. Religion has an important role in confronting each of these roblems, which we can immediately summarize, as follows:

The great judeo-christian heritage is founded on a stewardship para-ligm; early in its scripture we are admonished to be good stewards of the arth. Nothing could be more biologically sound than this call to ecologi-al husbandry, and therefore Jews and Christians can validate a fundamen-al premise of their culture by helping to protect the environment. Other eligions also stress our responsibilities toward nature, often more power-ully than the judeo-christian heritage does. For example, Buddhism and Iinduism have always emphasized harmony with nature, and so do the eligions of more primitive peoples.[18]

How might religions help protect the environment? Through sermons, hrough literature aimed at sins of environmental insensitivity and waste, hrough condemnation of special interest groups that squander our re-ources, through encouragement of modest lifestyles, and through the levelopment of specific stewardship doctrines relevant to modern life.

That is the persuasive function, and it is probably where religion can contribute most. But religion can also require its institutions and missionaries to practice and demonstrate good stewardship, through such measures as energy conservation. (As it is, mission headquarters are often among the most energy-wasteful operations in third-world countries, and we should immediately abandon such bad examples, unless we are trying to teach consumerism instead of moral commitment.) Religion could also sponsor researches, at least on a small scale, into ethical environmental practices. If it were willing to transfer funds from doctrinal evangelism it could even afford to sponsor such researches on a large scale. And it can support laws that protect the environment.

Religion's contribution to reducing the dangers of war is less clear, if only because war and religion and other cultural dynamics are so thoroughly intertwined. Yet it might be possible for religion to avoid being caught up in patriotic fervor, by serving as a moderating force instead of an incendiary one whenever conflicts loom. Its traditional doctrines emphasize peace, and it must function to shape public opinion in that direction. Even when patriotic fervor sweeps up the whole culture religious people could privately support young people who see the futility of war, affirming their misgivings rather than trying to somehow persuade them toward a warlike patriotism.

Racial and ethnic strife within a country are similar to war, but religion's role in reducing such tensions can be even more effective than its role in international conflict. It can perform that role in many ways, for example by understanding the underdog's viewpoint and arguing for social justice. On the other hand, it can explain to complaining segments of society why violence usually does not resolve tensions, even when the impulse to violence seems justified. More directly, it can organize and provide aid to people who need it, and it can provide effective mechanisms for self-help within each group, both by marshalling available resources and by demanding equal opportunities from the larger group. Above all religion can teach tolerance to all sides, helping us celebrate diversity and creativity rather than trying to squelch them.

Religious contributions to environmental soundness, and to the reduction of war and social tensions, are very important. But ultimately, religion can make its longest-range contribution by confronting the dynamics of our population explosion. At first blush this might seem contrary to one of the first commandments, which urged us to be fruitful and multiply. Yet nobody could argue that we have failed to fulfill that commandment indeed it may be the only one we have fulfilled. Therefore our modern task is to harmonize the stewardship commandment with the growth commandment. And the only way we can do that is to live within the earth's carrying capacity.

Although it is arguable what the carrying capacity is, it is inarguable that we are presently exceeding it, given the ways we organize our agricultural and transportation technologies. And even if the earth is theoretically capable of carrying more people than now reside on it, we are plainly hurtling toward a population that cannot possibly be supported without widespread starvation (See Chapter 13). Certainly religion cannot ethically support such a course of action, if it has any prophetic vision at all.

What is it to do? For one thing it must abandon rigid notions of birth control. Fortunately all religions except Roman Catholicism already permit a wide range of birth control, and many practicing Catholics ignore their church's narrow commands.[19] Yet that is not enough. Religion needs to actively encourage birth control, and not merely permit it.

Another thing religion can do is to harness some of its traditional doctrines to the task of reducing population. For example, urging family faithfulness would be easier if religion also encouraged small families, primarily because economics works better in small families. If families are under less financial stress, religion's admonitions for fidelity and its promise of security are more likely to be kept. To deny such economic impacts on morality and family values in our world is to keep one's head in the sand.

In its third-world operations religion can also contribute to population reduction. It is surprising—even shocking—to find missionaries with large families, and mission boards should prohibit that sexual misconduct among their employees as vigorously as they prohibit other forms of it.

Finally, religion can help reduce population by supporting social security systems that do not depend so greatly on having children, especially lots of children. If supported by religion, many different kinds of laws could contribute to social security, such as laws against monopolies, laws permitting labor unions, laws that guarantee education, and laws that provide adequate health care. Naturally such laws must be balanced against other laws that support individual responsibility, as they strive to "split the difference" between a radical socialism and a radical individualism. But when religious conservatives preach against the first kind of laws, they are conveniently forgetting such biblical concepts as the Year of Jubilee, which completely redistributed wealth every 50 years.[20] They are also forgetting Jesus's condemnation of those who forgot to help the poor in the name of religion.

Individual Responses to Religion

If religion were able to accomplish all the things suggested above, individuals could simply relax and participate in the changes proposed, thrilled at the prospect of growth and yet grateful for the firm foundation provided by churches. However, the likelihood of religion making those adjustments anytime soon is slim. Therefore we must ask what an individual can do to avoid religion's short-sightedness, and yet maintain a healthy connection to the group. I propose three things:

1. Develop a consistent religious philosophy.
2. Participate in forward-thinking religious groups.
3. Don't depend entirely on religion for meaning.

In developing a consistent religious philosophy, one must give great attention to logical and cognitive considerations, with an emphasis on cohesiveness, comprehensiveness, and realism. One should be able to explain and support his or her beliefs, and stand ready to change them when the evidence is persuasive. That is a formidable challenge, and will ordinarily require a fair amount of reading, conversation, and conscious thinking.

But since I have already given a great deal of attention to developing a cohesive worldview (also known as a faith or value-structure), I will not concentrate further on the cognitive aspects of developing a consistent religious philosophy. There are two non-cognitive components of such a philosophy, however, which should be mentioned.

The first concerns the importance of sub-conscious processes. Chapter Seven discussed those processes in relation to faith, but it did not strongly identify them with any traditional doctrine. It is clear, however, that sub-conscious processes have much to do with traditional concepts of prayer. One may tap into those processes in specific ways, and reap enormous benefits. For example we can "take our burdens to the subconscious and leave them there," confident that our brains will search our experience and society's experience for principles that might offer relief. If we learn to access this resource, potential answers to our problems will rise to consciousness for further processing and action. Likewise, we can program our sub-conscious brains with grateful meditations about the joys and blessings of life, and this programming will incline us in positive directions and away from destruction, all without our conscious awareness and without too much fretting. In other words we must learn to pray without ceasing.

Second, a sound religious philosophy should also feature commitment. Life is so much richer when one confidently makes choices based on a strong belief structure, and commitment enables us to do that. It keeps us from being swayed against our own true beliefs, and it urges us toward outgoing enthusiasm in sharing those beliefs. Of course commitment can easily develop into an unhealthy dogmatism, so one must also cultivate the attributes of grace, such as slow change and humility.

Dogmatism is not necessary to commitment, however, and we have the ption of trying to develop a graceful spirit that features a calm and humble ommitment. One way to develop commitment without sinking into ogmatism is to participate in forward-thinking religious groups. Such roups also provide other advantages, such as like-minded friends and a afe forum for testing one's ideas even while standing ready to change iem.

As to what constitutes a forward-thinking religious group, there can be o absolute criterion, but a simple litmus test would probably serve in oday's intellectual and social climate. You will not be surprised to hear ; it is whether a group argues against evolution. If so, it must be suspected f not being forward-thinking. I am not so optimistic as to think there are ery many religious groups that yet consider evolution a foundation of ieir doctrine, but there are some who do not argue against it. In fact many iinisters and lay leaders are sympathetic to scientific ways of thinking, nd one could do worse than to cast one's lot with them.

Of course other indicators can serve instead. If a group meeting nearby ; silent on evolution but preaches a strong environmental message, it is bviously more forward-thinking than one that does not. Depending on ne's interests and commitments, such a group could provide an excellent pportunity for growth and participation.

For persons of particularly strong faith and commitment to the prin-iples of biotheology, another choice is available. That is to serve as non-ogmatic missionaries to traditional churches, urging them to modify their eligious philosophies toward biotheology.[21] That can be especially ewarding in groups containing intelligent people who simply have not yet iought very much about biology's relevance to theology. In Bible tudies, for example, one may often point out the relevance of current cientific knowledge to the passages discussed. Or one may develop riendships with pastors and provide books for them to read, such as those have cited in footnotes. This sort of work often flushes out deep riendships from the open-minded participants in a group, and that is a velcome plus for any missionary.

Not everybody is inclined to be a missionary, and not every geographic rea is sure to provide forward-thinking religious groups or even groups vith a few open-minded members. During the transition to more modern vays of thinking, it may take awhile for such groups to form. In those cases may be necessary to look for one's primary meaning in groups other than eligious groups. Fortunately, most cities have purely secular groups that re interested in peace, population control, environmental action, and arious charities. If such groups don't already exist, there is usually nough interest to start one.

Although these groups may not be overtly religious, they are clear
pursuing principles that are consistent with biotheology, and with the spir
of truth and knowledge which was identified with God in Chapter 8. Th
deep emotional satisfaction offered by such groups may be greater tha
that offered in nearby churches, especially if those churches actively striv
to obstruct truth and progress. Thus the spirit we call God may be mo
fully found in those groups than in churches.

In case of doubt, however, church is often a better group to choose f
one's main focus, for several reasons. First, whatever spirit it does exhib
is likely to have a broader base than that contained within the narro
agenda of a specialized group. Second, it is likely to meet social nee
better, because churches tend to attract a wide variety of people, who als
give a broader base from which one's spiritual and emotional needs can b
met. Third, a church's familiarity with doctrines of grace and moralit
creates a reservoir of love and community that can be deeply meaningfu
even if the church does not understand the biological wellsprings of thos
doctrines. Fourth, the traditional wisdom and organizational heritage
churches often gives them a stability and longevity lacking in secul
groups. It can be discouraging to participate in a group that flames wit
passion but soon burns itself out.

Moreover, there are good reasons to hope religion will eventuall
adopt the challenges and philosophies revealed by sophisticated biologic
knowledge. It eventually accepted Copernicus and Pasteur, and th
justifies the hope that it will ultimately accept biotheology, using it as
foundation for helping to solve the challenges that face us. A church th
is moving in that direction, however slowly or clumsily, can keep th
momentum going, and it can also provide a place of personal service an
meaning. Thus it is not yet time to abandon our churches, except in ve
limited geographical situations.

* * *

This book has argued that the challenge of the future is primaril
biological. Government, education, and economics must each address th
difficulties caused by population densities, and religion must provide th
worldview that underlies those efforts. This last function is probably th
most important of all, but religion can only fulfill that function b
modifying its theology to accord with biological insights. That will in tur
help us live in greater harmony with the planet, with each other, and wit
the underlying reality traditionally called God.

otes

[1] For a brief (but comprehensive) description of the functionalist approach, see hapter 1 of *The Sociology of Religion*, 2d edition, by Thomas F. O'Dea and Janet 'Dea Aviad (Englewood Cliffs, New Jersey: Prentice-Hall, Inc, 1983).

[2] For a discussion of the secularization of the various disciplines see pp. 220-5 of *Sociology and Religion*, ed. Norman Birnbaum and Gertrud Lenzer. The ction is found in Henri Desroche's contribution to the book, entitled "Socialism d the Sociology of Christianity."

[3] Religion is certainly more than institutions, as pointed out by Peter Berger and homas Luckmann in "Sociology of Religion and Sociology of Knowledge" p.410-418 in *Sociology and Religion*, ed. Norman Birnbaum and Gertrud enzer). I concentrate on the institutions because they are the "group voice" of ligion.

[4] The phrase and the concept come from Marshall McLuhan's *Understanding edia: The Extensions of Man* (New York: McGraw-Hill, 1964).

[5] This phrase and idea come from Joseph Campbell's *The Masks of God* (New ork: Viking Press, 1964).

[6] Konrad Lorenz speculates that the function of ritual in humans is much the me as in the animals he observed, namely an increase in security. His hand-ared goose turned around on the stairs when let inside each night, and was visibly erturbed one evening when it came in late and forgot to perform this ritual. It came ack down the stairs, did its turn, and then went up. Apparently it reasoned (at some dimentary level, perhaps the same rudimentary level we reason from when erforming rituals) that things had gone OK whenever the ritual was performed, d forgetting the ritual made it feel very insecure. Similarly, Malinowski's robriand islanders (see next note) did not know whether the garden would fail if ey neglected magical rituals, but they were not about to find out. After all, a ritual es no harm, one might say.

[7] Bronislow Malinowski, in his studies of the Trobriand islanders, observed at they knew a lot about fishing and gardening, their means of subsistence. They lied on magic for things they couldn't control, but they relied on themselves for ings within their control. If a fence needed repairing they repaired it. See his 954 book *Magic, Science and Religion*, pp. 28-29 (Glencoe, Illinois: The Free ress).

[8] Troeltsch, Ernst. "Religion, Economics, and Society." In *Sociology and eligion*, ed. by Norman Birnbaum and Gertrud Lenzer (Englewood Cliffs, N.J.: rentice-Hall, Inc. 1969).

[9] O'Dea and Aviad call this the *prophetic* function of religion. See pp. 14-15 f their earlier cited book for a list of six functions they ascribe to religion (namely e support function, the priestly function, the social control function, the ophetic function, the identity function, and the maturation function).

[10] For a tantalizing discussion of the interplay among magic, religion, and ience see the selection by James George Frazer entitled "Magic and Religion,"

pp. 31-39 of *Sociology and Religion*, ed. Norman Birnbaum and Gertrud Lenze

[11] For an excellent example of a pastor wrestling with the issues see Lind Jarchow Jones's "Wildflowers and Wonder: A Pastor's Wanderings in th Religion-Science Wilderness," in *Zygon* Vol. 29 no. 1 (March 1994), pp. 115-12

[12] For example, it would be a fascinating study to follow up on Konrad Lorenz' work with the greylag goose, in which fidelity to a later mate is interfered with b early sexual experience. See his *On Aggression*, pp. 202-205, cited earlier. Suc research could check for similar findings in monkeys, apes, and humans.

[13] Incidentally, a recent study says Nevada is in terrible physical and psycho logical shape compared to Utah, and immorality is the reason. See "Nevadans ar Gambling with their Health," by Jennifer Warren, at p. A-1 of the *Los Angele Times* for Sunday, August 2,1992.

[14] See "Churches at the Transition Between Growth and World Equilibrium, a fascinating paper by Jay W. Forrester presented to the National Council o Churches and included as Chapter 13 (pp. 337-353) of *Toward Global Equilib rium*, ed. Dennis L. Meadows and Donella H. Meadows.

[15] See James B. Ashbrook's "Making Sense of Soul and Sabbath," cited earlie

[16] The psychologist Erik Erickson says we need ideological orientation a much as we need food. See p. 176 of "Religion and the Crisis of Identity," (pp 173-183 in *Sociology and Religion*, ed. Norman Birnbaum and Gertrud Lenzer

[17] Reinhold Niebuhr sensitively and profoundly analyzes the workings of tria and error in the conflicts among various worldviews, and the necessity of suc conflicts, in his essay "The Commitment of the Self and the Freedom of the Mind, pp. 55-64 of *Religion and Freedom of Thought*, by Perry Miller, Robert L Calhoun, Nathan M. Pusey, and Reinhold Niebuhr (NY: Doubleday and Com pany, 1954).

[18] Other religions will continue to outdo Christianity in this arena unless redefines its message in more ecologically sound terms, according to Lynn Whit Jr.'s "The Historical Roots of our Ecologic Crisis" in *Science* Vol. 155 (March 1 1967) pp. 1203-7. James M. Gustafson gives a slightly different perspectiv acknowledging that our western philosophical heritage contributed to the prob lem, and yet showing how its focus on individual dignity offers a basis for hop see his "Ethical Issues in the Human Future," pp. 491-516 (Ch. 17) in *How Human Adapt: A Biocultural Odyssey*, ed. Donald J. Ortner and cited earlier.

[19] For an interesting perspective on the Catholic opposition to birth control, se *Population, Evolution, and Birth Control*, 2d Edition, assembled by Garre Hardin (San Francisco: W.H. Freeman Co., 1969).

[20] See Leviticus 25-27.

[21] Nowhere is the need for "getting back in church" by science-literate peop more persuasively presented than in the closing pages of Loyal Rue's *Amythic* cited earlier. I am inclined to quote long passages from his clarion call, but instea I will urge the interested reader to read it (and indeed the entire book), especiall if you are even slightly attracted to the idea presented in the text, of being missionary to church.

Bibliography and Index of Authors

Adams, Gerald R.; Montemayor, Raymond; and Gullota, Thomas P., eds. *Biology of Adolescent Behavior and Development.* Newbury Park, California: Sage Publications, 1989. [Ch. 4, note 19]

Albright, William Foxwell. *From the Stone Age to Christianity: Monotheism and the Historical Process.* Doubleday: New York 1957. [Ch. 5, notes 44 and 45]

Aldred, Cyril. *Egypt: The Armana Period and the End of the Eighteenth Dynasty* (Volumes I & II of the Cambridge Ancient History). Cambridge University Press, 1971. [Ch. 5, note 43]

Alexander, Hubert G. *Language and Thinking.* Princeton: D. Van Nostrand Company, 1967. [Ch. 2, note 36]

Alexander, Richard D. *The Biology of Moral Systems.* New York: Aldine DeGruyter, 1987. [Ch. 7, note 4; Ch. 10, note 4]

Alt, Albrecht. *Essays on Old Testament History and Religion.* Sheffield, England: Sheffield Academic Press, 1989. [Ch. 5, notes 26 and 46]

Altizer, Thomas J.J. and Hamilton, William. *Radical Theology and The Death of God.* New York: The Bobbs-Merrill Company, Inc., 1966. [Ch. 8, note 30]

Anderson, Walter Truett Anderson. *To Govern Evolution: Further Adventures of the Political Animal.* Boston: Harcourt, Brace, Jovanovich; 1987. [Ch. 14, note 18]

Arbib, Michael; Caplan, David; and Marshall, John; eds. *Neural Models of Language Processes.* New York: Academic Press, 1982. [Ch. 2, note 12]

Armstrong, Este and Falk, Dean, eds. *Primate Brain Evolution.* New York: Plenum Press, 1982. [Ch. 2, notes 3 and 4]

Ashbrook, James B., "Making Sense of Soul and Sabbath: Brain Processes and the Making of Meaning," in *Zygon: Journal of Religion and Science* Vol. 27 No. 1 (March 1992), pp. 31-49. [Ch. 4, note 21; Ch. 7, note 34; Ch. 16, note 15]

Asimov, Isaac. *The Relativity of Wrong.* New York: Windsor, 1989. [Ch. 7, note 25]

Atlanta *Constitution* for May 14, 1988, p. A-1. [Ch. 2, note 34]

Axelrod, Robert. *The Evolution of Cooperation.* New York: Basic Books, Inc., 1984. [Ch. 3, note 21]

Balaban, Miriam, ed. *Biological Foundations and Human Nature.* New York: Academic Press, 1983. [Ch. 3, note 14]

Balton, Neil. *Concept Formation.* Oxford and New York: Pergamon Press, 1977. [Ch. 4, note 1]

Barbour, Ian. *Religion in an Age of Science: The Gifford Lectures, 1989-91 Vol. 1*. San Francisco: Harper and Row, 1990. [Ch. 8, note 25]

_____. "Science, Religion and the Counterculture" pp. 380-397 in Vol. 10 No.4 (December 1975) of *Zygon: Journal of Religion & Science*. [Ch 2, note 30]

_____. "Commentary on Resources from the Physical Sciences," in *Zygon: Journal of Religion & Science* Vol. 1 No. 1 (March 1966), pp 27-30. [Ch. 7, note 27]

Bargo, Michael Jr. *Choices and Decisions: A Guidebook for Constructing Values*. San Diego: University Associates, 1980. [Ch. 12, note 24]

Barrick, W. Boyd and Spencer, John R., eds. *In the Shelter of Elyon* Sheffield, England: JSOT Press, 1984. [Ch. 5, notes 25 and 54]

Barrington, Ernest James William, ed. *Hormones and Evolution*, Vol. 1 Princeton, N.J.: Academic Press, 1979. [Ch. 2, note 11]

Barrow, Robin. *Happiness and Schooling*. New York: St. Martin's Press 1980. [Ch. 12, notes 5 and 15]

Bartley, William Warren III. *The Retreat to Commitment*. Open Court Publishing Company; LaSalle, Illinois; 1984. [Ch. 7, note 32; Ch. 10 note 13]

_____ and Radnitzky, G., eds. *Evolutionary Epistemology, Theory of Rationality and the Sociology of Knowledge*. Open Court Publishing Company: LaSalle, Illinois; 1986. [Ch. 9, note 22]

Bates, Christine and Stevens, Darryl. "Illusion and Well-Being: Toward a New Agenda for Mental Health," *Journal of Counseling and Development* Vol. 67 No. 10 (June 1989), pp. 590-91. [Ch. 7, note 4]

Bayliss, Jeffrey R. and Halpin, Zuleyma Tang (1982). "Behavioural antecedents of sociality," Chapter 13 in *Learning, Development, and Culture*, ed. H.C. Plotkin. John Wiley & Sons, New York. [Ch. 3, note 13]

Becker, Gary S. *The Economic Approach to Human Behavior*. Chicago: University of Chicago Press, 1976. [Ch. 13, note 12]

_____. *A Treatise on The Family*. Cambridge, Massachusetts: Harvard University Press, 1991. [Ch. 13, note 12]

Bell, Daniel and Kriston, Irving, eds. *The Crisis in Economic Theory*. New York: Basic Books, Inc. 1982. [Ch. 13, note 13]

Bellah, Robert N.; Madsen, Richard; Sullivan, William M.; Swidler, Ann and Tipton, Steven M. *Habits of the Heart: Individualism and Commitment in American Life*. New York: Harper & Row Perennial Library, 1985. [Ch. 15, note 23]

Berger, Peter and Luckman, Thomas. "Sociology of Religion and Sociology of Knowledge," pp.410-418 in *Sociology and Religion: A Book of Readings*, ed. Norman Birnbaum and Gertrud Lenzer. [Ch. 16, note 3

Bernal, Martin. *Cadmean Letters*. Winom Lake: Eisenbarauns, 1990 [Ch. 6, note 3]

Berra, Yogi. See Ch. 9, note 25 for a piece of apocrypha attributed to him.

Bettelheim, Bruno. *The Informed Heart: Autonomy in a Mass Age.* New York: Free Press, 1960. [Ch. 12, note 22]

Bible, The Living. Wheaton, Illinois: Tyndale House Publishers, 1971. [Ch. 6, note 10]

Biblical References (KJV unless specified otherwise):

Genesis 2: 19-20 [Ch. 2, note 57]
Genesis 12:10 [Ch. 13, note 7]
Genesis 16:13 [Ch. 5, note 26]
Genesis 19:37 [Ch. 5, note 51]
Genesis 21:13 [Ch. 5, note 26]
Genesis 21:33 [Ch. 5, note 26]
Genesis 22 [Ch. 5, note 34]
Genesis 26:1 [Ch. 13, note 7]
Genesis 41:56-47:20 [Ch. 13, note 7]
Genesis 49 [Ch. 5, note 24]
Exodus 3:14 [Ch. 8, note 23]
Exodus 6:2-3 [Ch. 5, note 27]
Leviticus 25-27 [Ch. 16, note 20]
Numbers 21:29 [Ch. 5, note 50]
Joshua 24:2 [Ch. 5, note 26]
Judges 9:4 and 46 [Ch. 5, note 26]
Judges 11:24-30 [Ch. 5, notes 36 and 50]
Ruth 1:1 [Ch. 13, note 7]
II Samuel 21:1 [Ch. 13, note 7]
I Kings 18:2 [Ch. 13, note 7]
I Kings 11:1-8 and 33 [Ch. 5, note 50]
II Kings 3:27 [Ch. 5, note 36]
II Kings 6:25 [Ch. 13, note 7]
II Kings 16:10 and 21:1 [Ch. 5, note 26]
II Kings 23:13 [Ch. 5, note 50]
Proverbs 1:7, 4:7, 8:5 [Ch. 7, note 39]
Proverbs 13:24 [Ch. 11, note 6]
Ecclesiastes 12:13 [Ch. 12, note 10]
Jeremiah 48:7-46 [Ch. 5, note 50]
Matthew 5:18 [Ch. 11, note 14]
Matthew 6:33 [Ch. 12, note 4]
Matthew 7:7 [Ch. 15, note 15]
Mark 2:21-27 [Ch. 7, note 40]
St. John 14:16-17 [Ch. 6, note 18]
Acts 14:12 Revised Standard Version [Ch. 15, note 2]
Romans 3:23 [Ch. 10, note 15]
Romans 6:14 [Ch. 11, note 15]

Romans 6:23 [Ch. 10, note 19]
Romans 7:6, 8:2 and 21 [Ch. 15, note 19]
I Corinthians 3:19 [Ch. 7, note 39; Ch. 15, note 16]
I Corinthians 8:9 and 10:29 [Ch. 15, note 19]
II Corinthians 3:17 [Ch. 15, note 19]
Galatians 2:4, 5:1 and 5:13. [Ch. 15, note 19]
Ephesians 2:8 [Ch. 11, note 13]
II Titus 3:16 [Ch. 6, note 4]
James 4:17. [Ch. 10, note 20]

Bicchieri, Cristina. *Rationality and Coordination.* Cambridge University Press, 1993. [Ch. 14, note 18]

Bielicki, T. "Deviation-amplifying cybernetic systems and hominid evolution." *Materialy i Prace Antrhopologicyne* 77:57-60; 1969: 57-60. (Cited in Tunnell, below.) [Ch. 3, note 7]

Birket-Smith, Kaj. *Eskimos.* New York: Crown Publishers, Inc., 1971. [Ch. 5, note 15]

Birnbaum, Norman and Lenzer, Gertrud, eds. *Sociology and Religion: A Book of Readings.* Englewood Cliffs, New Jersey: Prentice-Hall 1969. [Ch. 8, note 33; Ch. 16, notes 2, 3, 8, 10, and 16]

Bix, James. *Interpreting Evolution: Darwin & Teilhard de Chardin.* Buffalo, New York: Prometheus Books, 1991. [Ch. 8, notes 29 and 35]

Black, James E. and Greenough, William T. "Developmental Approaches to the Memory Process." In *Learning and Memory*, ed. Joe L. Martinez, Jr. and Raymond P. Kesner. New York: Academic Press, 1986. [Ch. 3, note 28]

Bleeker, C. Jouco and Widengren, George, eds. *Historia Religioniam.* Leiden, the Netherlands: E.J. Brill Publishers, 1969 and 1971. [Ch. 8, note 2]

Bleibtreu, H., ed. *Evolutionary Anthropology.* Boston: Allyn and Bacon, 1969. [Ch. 2, note 58]

Bloom, Allan David. *The Closing of the American Mind.* New York: Simon and Schuster, 1987). [Ch. 15, note 12]

Boesch, Christophe. "Teaching Among Wild Chimpanzees;" *Animal Behavior* for March, 1991, pp 530-2. [Ch. 15, note 4]

Bonner, John Tyler. *The Evolution of Culture in Animals.* Princeton New Jersey: Princeton University Press, 1980. [Ch. 3, note 6; Ch. 9, notes 2 and 19]

Bonwick, James. *Egyptian Belief and Modern Thought.* Indian Hills, Colorado: The Falcon's Wing Press, 1956. [Ch. 5, note 10]

Booth, Wayne C. *Critical Understanding: The Powers and Limits of Pluralism.* Chicago: The University of Chicago Press, 1979. [Ch. 7, note 30]

Borgen, Peder. *Paul Preaches Circumcision and Pleases Men and Other Essays on Christian Origins*. Trondeim, Norway: Tapir Press, 1983. [Ch. 6, note 8]

Borger, Robert, ed. *Explanation in the Behavioral Sciences*. Cambridge at the University Press, 1970. [Ch. 5, notes 5 and 12]

Bosley, Richard. *On Good and Bad, whether happiness is the highest good.* New York: University Press of America, 1986. [Ch. 12, note 9]

Boulding, Kenneth E. *Evolutionary Economics*. Beverly Hills, California: Sage Publications, 1981. [Ch. 13, note 12]

_____. "Economics as a Not Very Biological Science," Chapter 9 (pp. 129-44) of *Biology and the Social Sciences: An Emerging Revolution*, ed. Thomas Wiegele. [Ch. 13, note 11]

Bowen, Howard R. *Investment in Learning: the Individual and Social Value of American Higher Education.* San Francisco: Jossey-Bass Publishers, 1978. [Ch. 15, note 12]

Bower, Bruce. "Gauging the Winds of War," in *Science News*, Vol. 139, No. 6 (February 9, 1991), pp. 88-9. [Ch. 14, note 10]

Bowlby, J. *Attachment and Loss*. Vol. 1: *Attachment*. London: Institute of Psycho-Analysis, 1982. [Ch. 4, note 3]

Bradie, Michael. *The Secret Chain: Evolution and Ethics*. Albany: State University of New York Press, 1994. [Preface, note 8]

_____. "What Does Evolutionary Biology Tell Us about Philosophy of Religion?" *Zygon: Journal of Religion & Science* Vol. 29 No. 1 (March 1994), pp. 45-53. [Preface, note 8].

Bradshaw, John L. and Rogers, Lesley J. *The Evolution of Lateral Asymmetries, Language, Tool Use and Intellect*. New York: Academic Press Inc., 1993. [Ch. 2, note 10]

Bramson, Robert and Susan. "What Kind of Thinker are You?" Vol. CII No. 10 (October 1985) of *Ladies' Home Journal*. [Ch 7, note 6]

Bratton, Fred Gladstone. *A History of the Bible*. Boston: Beacon Press, 1959. [Ch. 5, notes 20, 22 and 40]

Breasted, James Henry; *Development of Religion and Theology in Ancient Egypt*. New York: Charles Scribner's Sons, 1912. [Ch. 5, notes 9 and 40]

_____. *History of Egypt*. New York: Charles Scribner's Sons, 1910. [Ch. 5, note 9]

Briggs, Katharine M. *An Encyclopedia of Fairies, Hobgoblins, Brownies, Bogies, and Other Supernatural Creatures*. New York: Pantheon Books 1978. [Ch. 5, note 16]

Bronson, Bennet. "The Earliest Farming: Demography as Cause and Consequence," pp. 53-78 in *Population, Ecology, and Social Evolution*, ed. Steven Polgar. [Ch. 13, note 4]

Brown, Colin. *Miracles and the Critical Mind.* Grand Rapids, Michigan William B. Eerdmans Publishing Company, 1984. [Ch. 8, note 11]

Brown, Jerram. *The Evolution of Behavior*, 1st ed. New York: Norton 1975. [Ch. 2, note 44]

Brown, Sanborn C. "Commentary on Daniel Day Williams's 'Priest Prophet, and the Establishment'." *Zygon: Journal of Religion & Science* Vol. 2 No. 4 (December 1967), pp. 327-330. [Ch. 6, note 7]

Buddha, The Teaching of, published by Bukkyo Dendo Kyokai (Buddhist Promoting Foundation); Tokyo 1981. [Ch. 6, note 21]

Bunge, Mario. *The Mind-Body Problem.* New York: Pergamon Press 1980). [Ch. 10, note 9]

Burghardt, Gordon M. "Ethology and Operant Psychology," pp. 413-16 in *The Selection ofBehavior*, ed. Charles A. Catania and Steven Harnad. [Ch. 10, note 30]

Burns, John; Randall, David; Calhoun, Robert; and Abrahamson, Samuel *The Answer to Addiction: The Path to Recovery from Alcohol, Drug, Food, and Sexual Dependencies.* New York: Crossroad Publishing Company, 1990. [Ch. 11, note 8]

Busse, Richard P. "From Belief to Unbelief and Back to Belief," pp. 55-65 in *Zygon: Journal of Religion & Science* Vol. 29 No. 1 (March 1994). [Ch. 10, note 24]

Calvin, William H. *The Cerebral Symphony.* New York: Bantam Books 1990. [Ch. 2, note 19]

Campbell, Bernard. *Human Evolution.* Aldine Publishing Company New York, 1985. [Ch. 2, notes 35, 48, and 51; Ch. 3, notes 12, 17, and 31; Ch. 4, note 16]

Campbell, Donald T. *Methodology and Epistemology for Social Science.* Chicago: The University of Chicago Press, 1988. [Ch. 9, note 22]

Campbell, John H. "An Organizational Interpretation of Evolution," in *Evolution at a Crossroads: The New Biology and the New Philosophy of Science*, ed. David J. Depew and Bruce H. Weber. [Ch. 9, note 21]

Campbell, Joseph. *The Masks of God: Occidental Mythology.* New York: The Viking Press, 1964. [Ch. 5, notes 8 and 11; Ch. 16, note 5]

_____ and Moyers, William. *The Power of Myth.* Doubleday, New York: 1988. [Ch. 6, note 14]

Caplan, David; Lecours, André; and Smith, Alan. *Biological Perspectives on Language.* Cambridge, Mass: The MIT Press, 1984. [Ch. 2, note 56; Ch. 10, note 2]

Carson, Hampton L. "Patterns of Inheritance." In *Science as a Way of Knowing: III--Genetics*, published by the American Society of Zoologists, 1985. [Ch. 1, note 7]

Carter, Jimmy. *The Blood of Abraham.* Boston: Houghton Mifflin, 1985. [Ch. 11, note 22]

Catania, Charles A. and Harnad, Steven, eds. *The Selection of Behavior*. New York: Cambridge University Press, 1988. [Ch. 10, note 30]

Cavalli-Sforza, Luigi L. "Cultural Evolution." In *Science as a Way of Knowing: III—Genetics* (The American Society of Zoologists, 1985). [Ch. 2, note 48]

Cavanaugh, Michael. "The Precursors of the Eureka Moment as a Common Ground Between Science and Religion," pp. 191-204 in *Zygon: Journal of Religion & Science*, Vol. 29 No. 2 (June 1994). [Ch. 7, note 14; Ch. 9, note 23]

Cela-Conde, Camilo J. *On Genes, Gods and Tyrants: The Biological Causation of Morality*. Boston: D. Reidel Publishing Company, 1987. [Ch. 10, note 4]

Cerny, Jaroslav (1952). *Ancient Egyptian Religion*. Westport, Connecticut: Greenwood Press, 1952 [Ch. 5, note 9]

Cherry, Conrad. *Nature and Religious Imagination from Edwards to Bushnell*. Philadelphia: Fortress Press, 1980. [Ch. 5, note 1]

Chesnut, Glenn F. *Images of Christ: An Introduction to Christology*. Minneapolis: The Seabury Press, 1984. [Ch. 11, note 19]

Chomsky, Noam 1959, "Review of Skinner's verbal behavior." *Language* XXXV:26-58. [Ch. 3, note 19]

_____. *Language and Mind*. New York: Harcourt, Brace and World, 1967. [Ch. 3, note 19]

Cialdini, Robert B. *Influence: The Psychology of Persuasion*. New York: William Morrow and Co., 1984. [Ch. 9, note 20]

Clark, Wilfrid Edward LeGros. *Early Forerunners of Man: a morphological study of the evolutionary origin of the primates* (New York: AMS Press, 1979). [Ch. 1, note 4]

Clayton, Philip. *Explanation from Physics to Theology*. New Haven, Connecticut: Yale University Press, 1989. [Ch. 7, note 29]

Clutton-Brock, T.H. and Harvey, Paul H., eds. *Readings in Sociobiology*. (San Francisco: W.H. Freeman and Company, 1978). [Ch. 10, note 26]

Cobb, John B. Jr. *A Christian Natural Theology*. Philadelphia: Westminster Press, 1965. [Ch. 8, note 25]

Cody, Aelred. *A History of Old Testament Priesthood*. Rome: Pontifical Biblical Institute, 1969. [Ch. 6, note 7]

Connolly, S.J. *Priests and People in Pre-Famine Ireland*. New York: Gill and Macmillan, St. Martin's Press, 1982. [Ch. 5, note 17]

Cook, Norman D. *The Brain Code*. London: Methuen & Co., 1986. [Ch. 2, note 6]

Corben, Herbert C. *The Struggle to Understand: A History of Human Wonder and Discovery*. Buffalo, New York: Prometheus Books, 1991. [Ch. 7, note 22]

Cornfield, Gaalyahu, ed. *Adam to Daniel.* New York: The MacMillan Company, 1961. [Ch. 5, notes 27 and 38]

Cupitt, Don. *The Debate about Christ.* London: SCM Press, Ltd., 1979. [Ch. 8, note 18]

Damrosch, David. *Narrative Covenant: Transformations of Genre in the Growth of Biblical Literature.* New York: Harper & Row, 1987. [Ch. 5, note 52]

D'Antonio, Michael. *Heaven on Earth.* New York: Crown Publishers, Inc. 1992. [Ch. 11, note 8]

D'Aquili, Eugene. "The Summa Hefneriana: Myth, Megamyth, and Metamyth," pp. 371-381 in *Zygon: Journal of Religion & Science*, Vol. 29 No. 3 (September 1994). [Preface, note 14]

Darwin, Charles. *On the Origin of Species by Means of Natural Selection, or the Preservation of Favored Races in the Struggle for Life*, 1st ed. London: John Murray, 1859. (Facsimile of the first edition published by Harvard University Press, Cambridge, 1964.) [Ch. 1, note 1]

_____. *The Expression of the Emotions in Man and Animals.* Chicago: University of Chicago Press, 1965 <1872>. [Ch. 1, note 6]

Davidson, Julian M. and Richard J., eds. *The Pschobiology of Consciousness*, New York: Plenum Press, 1980. [Ch. 8, note 31]

Davies, Paul. *The Mind of God: The Scientific Basis for a Rational World.* Simon & Schuster, New York, 1992. [Ch. 8, note 7]

Dawkins, Richard. *The Selfish Gene.* New York: Oxford University Press, 1976. [Ch. 10, note 31; Ch. 11, note 24]

Dearman, Andrew, ed. *Studies in the Mesha Inscription and Moab.* Atlanta: Scholars Press, 1989. [Ch. 5, note 50]

DeBono, Edward. *Six Thinking Hats.* Boston: Little, Brown and Company; 1985. [Ch. 7, note 6]

DeMoor, Johannes C. *The Rise of Monotheism.* Leuven-Louvaine, Belgium: Leuven University Press, 1990. [Ch. 5, note 29]

de Góes, Cecilia and Martlew, Margaret. "Young Children's Approach to Literacy," Ch. 9 (pp. 217-236) in *The Psychology of Written Language*, ed. Margaret Martlew. [Ch. 6, note 3]

Depew, David J. and Weber, Bruce H., eds. *Evolution at a Crossroads: The New Biology and the New Philosophy of Science.* Cambridge, Massachusetts: The MIT Press, 1985. [Ch. 7, note 24; Ch. 9, notes 10, 13, and 21; Ch. 15, note 3]

de Révai, Mely Lechtich. Letter on unnumbered page in "Forum" section *National Geographic* Vol. 182, No. 5 (November 1992), reporting on the work of Dr. Russell Marker. [Ch. 13, note 2]

Desroche, Henri. "Socialism and the Sociology of Christianity," in *Sociology and Religion*, ed. Norman Birnbaum and Gertrud Lenzer. [Ch. 16, note 2]

Dewey, John. *Experience and Education.* New York: The MacMillan Company, 1938. [Ch. 7, note 28]

Deyo, Richard A.; Straube, Karen T.; and Disterhoff, John F. "Nimodipine Facilitates Associative Learning in Aging Rabbits." *Science* Vol. 241, p. 809 (February 10, 1989). [Ch. 2, note 24]

Diamond, Marian C. "Love Affair with the Brain." *Psychology Today* November 1984. [Ch. 2, note 17]

_____. "The Aging Brain: Some Enlightening and Optimistic Results" in *American Scientist* for January/February 1978 (Vol. 66 No. 1, pp. 66-71). [Ch. 4, note 20]

_____. Personal communication dated December 7, 1984. [Ch. 2, note 18]

Dobzhanzky, Theodosius Grigorievich. "Nothing in Biology Makes Sense Except in the Light of Evolution," included in *Evolution versus Creationism: The Public Education Controversy.* Ed. J. Peter Zetterberg; Oryx Press 1983. Phoenix. [Ch. 1, note 1]

_____ and Boesiger, Ernest. *Human Culture: A Moment in Evolution.* New York: Columbia University Press, 1983. [Ch. 1, note 17]

_____. *Mankind Evolving: The evolution of the human species.* New Haven: Yale University Press, 1962. [Ch. 2, note 2]

Dombrowski, Daniel A. "Must a Perfect Being Be Immutable?" at pp. 91-111 in *Hartshorne, Process Theology, and Theology*, ed. Robert Kane and Stephen H. Phillips. [Ch. 8, note 28]

Dorman, Susann. Personal communication dated June 12, 1993. [Ch. 15, note 23]

Driver, Tom F. *Patterns of Grace: Human Experience as Word of God.* San Francisco: Harper & Row, 1977. [Ch. 8, note 16]

Drummond, Robert J. and Stoddard, Ann H. "Learning Style and Personality Type," pp. 99-104 in Vol. 75 No. 1 (August 1992) of *Perceptual and Motor Skills*. [Ch. 7, note 6]

Durant, Will. *The Age of Faith.* New York: Simon and Schuster, 1950. [Ch. 7, note 21]

Durkheim, Emile. *The Elementary Forms of the Religious Life*, trans. by Joseph Ward Swain. Glencoe, Ill.: The Free Press, 1968 (original copright 1915 by George Allen & Unwin Ltd.). [Ch. 8, note 19]

_____. "The False Dichotomy of Real and Ideal," pp. 209-210 of *Sociology and Religion: A Book of Readings*, ed. Norman Birnbaum and Gertrud Lenzer. [Ch. 8, note 33]

Eaves, Lindon and Gross, Lora. "Exploring the Concept of Spirit as a Model for the God-World Relationship in the Age of Genetics" in *Zygon: Journal of Religion & Science* Vol. 27 No. 3, pp. 261-285 (September 1992). [Preface, note 15; Ch. 8, note 20; Ch. 9, note 1]

Educational Leadership. Entire October 1990 issue (Vol. 48 No. 2). [Ch. 7, note 6]

Eible-Eibesfeldt, I. *Ethology: The biology of behavior.* New York: Holt, Rinehart and Winston, 1970. [Ch. 3, note 9, Ch. 9, note 14]

Eller, Vernard. "The 'Mad' Morality: An Exposé," in *Christian Century*, pp. 1647-9 of the December 27, 1967 issue (Vol. 84). [Ch. 15, note 21]

Emde, Robert N. *Dialogues From Infancy*, by René A. Spitz. New York: International Universities Press, 1983. [Ch. 4, note 3]

Erikson, Erik. "Identity and the Life Crisis." No. 1 in *Psychological Issues* Vol. 1, No. 1. New York: International University Press, Inc., 1959. [Ch. 7, note 9]

_____. "Religion and the Crisis of Identity," p. 173-183 in *Sociology and Religion: A Book of Readings*, ed. Norman Birnbaum and Gertrud Lenzer. [Ch. 16, note 16]

Ernst, Cécil and Angst, Jules. *Birth Order: Its Influence on Personality*. New York: Springer-Verlag, 1983. [Ch. 7, note 7]

Evans, Richard I., ed. *Konrad Lorenz: The Man and His Ideas*. New York: Harcourt Brace Jovanovich, 1975. [Ch. 10, note 30]

Feinberg, Gerald. *Solid Clues*. New York; Simon & Schuster, 1985. [Ch. 7, note 36]

Ferm, Deane William. *Third World Liberation Theology*. Maryknoll, New York: Orbis Books, 1986. [Ch. 6, note 11]

Feyerabend, Paul. *Against Method*. London: Verso, 1988. [Preface, note 16]

Ferré, Frederick. "Unfazed by Mystery," pp. 363-70 in *Zygon: Journal of Religion & Science* Vol. 29 No. 3 (September 1994). [Preface, note 13]

Fine, Reuben. *The Meaning of Love in Human Experience*. New York: John Wiley & Sons, 1985. [Ch. 11, note 7]

Finn, William M. Jr. and Smith, Gerald Alonzo, eds. *The Morality of Scarcity: Limited Resources and Social Policy*. Baton Rouge, Louisiana: Louisiana State University Press, 1979. [Ch. 13, note 18]

Fishbein, Martin and Ajzen, Icek. *Belief, Attitude, Intention, and Behavior*. Reading, Massachusetts: Addison-Wesley, 1975. [Ch. 12, note 17]

Fishbein, William, ed. *Sleep, Dreams and Memory*. New York: Spectrum Publications, 1981. [Ch. 4, note 21]

Fiske, John. *The Idea of God as Affected by Modern Knowledge*. New York: Houghton, Mifflin and Company, 1890. [Ch. 8, note 4; Ch. 11, note 1]

Fitzmyer, Joseph A. *According to Paul*. New York: Paulist Press, 1993. [Ch. 15, note 19]

_____. *Pauline Theology*. Englewood Cliffs, New Jersey: Prentice Hall, Inc., 1967. [Ch. 15, note 19]

Fling, Thomas P. *Christian Philosophy*. Notre Dame, Indiana: University of Notre Dame Press, 1990. [Ch. 11, note 20]

Flowers, Betty Sue, ed. *The Power of Myth*. New York: Doubleday, 1988. [Ch. 6, note 14]

Fohrer, Georg. *Introduction to the Old Testament*. Nashville, Tennessee: Abington Press, 1968. [Ch. 5, note 31]

Forrester, Jay W. "Counter-Intuitive Behavior of Social Systems" in *Toward Global Equilibrium*, ed. Dennis L. Meadows and Donella H. Meadows. [Ch. 14, notes 11 and 13]

_____. "Churches at the Transition Between Growth and World Equilibrium," Chapter 13 (pp. 337-353) of *Toward Global Equilibrium*, ed. Dennis L. Meadows and Donella H. Meadows. [Ch. 16, note 14]

Fowler, Henry T. *A History of the Literature of Ancient Israel*. New York: The Macmillan Company, 1922. [Ch. 5, note 21]

Frankfort, H.A. et. al. *Before Philosophy: The Intellectual Adventure of Ancient Man*. Baltimore: Penguin Books, 1961. [Ch. 5, note 7]

Frazer, James George. "Magic and Religion," pp. 31-39 in *Sociology and Religion: A Book of Readings*, ed. Norman Birmbaum and Gertrud Lenzer. [Ch. 16, note 10]

Freedman, David Noel. "Who is Like Thee Among the Gods? The Religion of Early Israel," Ch. 16 (pp. 315-335) in *Ancient Israelite Religion*, ed. Patrick D. Miller Jr., Paul D. Hanson, and S. Dean McBride. [Ch. 5, note 24].

Frei, Hans W. *Types of Christian Theology*, ed. George Hunsinger and William C. Placher. New Haven, Connecticut: Yale University Press, 1992. [Preface, note 4]

Freud, Sigmund. *Civilization and its Discontents*. London: Hogarth Press, 1939. [Ch. 5, note 13]

_____. *Moses and Monotheism*. Vintage Books, New York: 1955. [Ch. 5, note 42]

Fromm, Erich. *You Shall Be as Gods*. New York: Holt, Rinehart and Winston, 1966. [Ch. 5, note 19]

Fry, Geoffrey K. *The Growth of Government*. London: Frank Cass and Company, Ltd., 1979. [Ch. 14, note 18]

Frye, Roland Mushat, ed. *Is God a Creationist? The Religious Case Against Creation Science*. Charles Scribner's Sons 1983. New York. [Ch. 1, note 2]

Gabel, Creighton. *Analysis of Prehistoric Economic Patterns*. New York: Holt, Rinehart and Winston 1967. [Ch. 6, note 1]

Galanter, Mark ed. *Cults and New Religious Movements*. Washington, D.C.: American Psychiatric Association, 1989. [Ch. 11, note 8]

Gardner, B.T. and A.R. 1969; "Teaching sign language to a chimpanzee." *Science* 165:664-672. [Ch. 2, note 50]

Gardner, Howard. *Multiple Intelligences*. New York: Basic Books, 1993. [Ch. 7, note 6]

_____ and Winner, Elfen. "The Child is Father to the Metaphor," in *Psychology Today* Vol. 12 No. 12 (May '79) pp. 81-91. [Ch. 8, note 32]

Gaur, Albertine. *A History of Writing.* New York: Charles Scribner's Sons, 1984. [Ch. 6, note 3]

Gaus, Gerald F. *Value and Justification.* New York: Cambridge University Press, 1990. [Ch. 12, notes 18, 19, 21, 22]

Gazzaniga, Michael S. "Gut Thinking;" pp. 68-71 in *Natural History*, Vol. 104 no.2 (February 1995). [Ch. 2, note 30]

Gell-Mann, Murray. *The Quark and The Jaguar.* New York: W.H. Freeman and Company, 1994. [Ch. 7, note 26]

Gever, John; Kaufmann, Robert; Skole, David; and Vörösmarty, Charles. *Beyond Oil: The Threat to Food and Fuel in the Coming Decades.* Cambridge, Massachusetts: Ballinger Publishing Company, a subsidiary of Harper & Row Publishers, Inc., 1986. [Ch. 13, note 14; Ch. 14, notes 14, 15, and 17]

Gilkey, Langdon. *Creationism on Trial: Evolution and God at Little Rock.* Minneapolis: Winston Press, 1985. [Ch. 1, note 2]

_____. "Nature, Reality, and the Sacred: A Meditation in Science," in *Zygon: Journal of Religion & Science* Vol. 24 No. 3 (September 1989), pp. 283-298. [Preface, note 15; Ch. 7, note 38]

Gilligan, Carol; Ward, Janie Victoria; and Taylor, Jill McLean, eds. *Mapping the Moral Domain.* [Cambridge, Massachusetts: Center for the Study of Gender, Education & Human Development (at Harvard University Graduate School of Education), 1988]. [Ch. 10, note 7]

Glassman, Robert B. "Free Will Has a Neural Substrate" pp. 67-81 in *Zygon: Journal of Religion & Science* Vol. 18 No. 1 (March 1968). [Ch. 9, note 19]

_____. Personal communication, May 17, 1995. [Ch. 9, note 9]

Godfrey, Laurie R., ed. *Scientists Confront Creationism.* W.W. Norton & Company, New York, 1983. [Ch. 1, note 2]

Gollin, Eugene S., ed. *Developmental Plasticity: Behavioral and Biological Aspects of Variation in Development.* New York: Academic Press, a subsidiary of Harcourt Brace Jovanivich; 1981. [Ch. 4, note 19; Ch. 15, note 6]

Goodall, Jane. *The Chimpanzees of Gombe: Patterns of Behavior.* Cambridge, Massachusetts: The Belknap Press of Harvard University Press, 1986. [Ch. 2, note 37; Ch. 9, note 7]

Görlitz, Dietmar and Wohlwill, Joachim F., eds. *Curiosity, Imagination, and Play.* London: Lawrence Erlbaum Associates, Publishers; 1987. [Ch. 2, note 38]

Gottlieb, Gilbert. "Conceptions of prenatal development: behavioral embryology," Chapter 17 in *Learning, Development, and Culture*, ed.

H.C. Plotkin. [Ch. 4, note 2]

Goulder, Michael, ed. *Incarnation and Myth: The Debate Continued.* Grand Rapids, Michigan: William B. Eerdman's Publishing Company, 1979. [Ch. 8, note 18]

Greenfield, Larry. Personal communication dated May 10, 1995. [Ch. 8, note 1]

Gregg, Richard. *Symbolic Inducement and Knowing: A Study in the Foundations of Rhetoric*. Columbia: South Carolina: University of South Carolina Press, 1984. [Ch. 4, note 9; Ch. 9, note 23]

Griffin, David Ray, ed. *The Re-enchantment of Science: Postmodern Proposals*. Albany: State University of New York Press, 1980. [Ch. 8, note 25]

____. "Charles Hartshorne's Postmodern Philosophy" in *Hartshorne, Process Philosophy, and Theology*, ed. Robert Kane and Stephen H. Phillips. [Ch. 8, note 26]

____ and Sherburne, Donald W., eds. *Process and Reality*, by Alfred North Whitehead. New York: The Free Press, 1978. [Ch. 8, note 24]

Gruber, Howard E. and Richard, Lucien. "Active Work and Creative Thought in University Classrooms," Chapter 7 (pp. 137-164 in *Promoting Cognitive Growth Over the Life Span*, ed. Milton Schwebel et. al. [Ch. 15, note 12]

Guerin, Philip J. Jr, Fay; Fay, Leo F.; Burden, Susan L.; and Kautto, Judith Gilbert. *The Evaluation and Treatment of Marital Conflict*. New York: Basic Books, Inc., 1987. [Ch. 7, note 33]

Guidano, Vittorio F. *Complexity of the Self.* New York and London: The Guilford Press, 1987. [Ch. 15, note 10]

Gustafson, James. *Ethics from a Theocentric Perspective, Volume 1.* Chicago: University of Chicago Press, 1981. [Ch. 6, note 20]

____. "Ethical Issues in the Human Future;" pp. 491-516 (Ch. 17) in *How Humans Adapt: A Biocultural Odyssey*, ed. Donald J. Ortner. [Ch. 16, note 18]

Hackett, Jo Ann. "Religious Traditions in Israelite Transjordan," Ch. 9 (pp. 125-136 in *Ancient Israelite Religion*, ed. Patrick Miller, Paul D. Hanson, and S. Dean McBride. [Ch. 5, note 35].

Hailman, Jack P. "Ethology Ignored Skinner to its Detriment," pp. 431-33 in *The Selection of Behavior*, ed. Charles A. Catania and Steven Harnad. [Ch. 10, note 30]

Haldane, J.B.S. *The Causes of Evolution.* London: Longmans, Green & Co., 1932. [Ch. 3, note 18]

Hall, Douglas John. *Thinking the Faith: Christian Theology in a North American Context*. Minneapolis, Minnesota: Augsburg Fortress Publishers, 1991. [Ch. 2, note 30]

Halperin, David A. "Group Processes in Cult Affiliation and Recruitment," Chapter 13 (pp. 223-234) in *Psychodynamic Perspectives o. Religion Sect and Cult*, ed. David A. Halperin. [Ch. 7, note 3]

_____, ed. *Psychodynamic Perspectives on Religion Sect and Cul.* Littleton, Massachusetts: John Wright PSG, Inc., 1983. [Ch. 7, note 3

Hamburger, Henry. *Games as Models of Social Phenomena.* San Francisco: W.H. Freeman and Company, 1979. [Ch. 3, note 21]

Hamilton, William. "The Evolution of Altruistic Behavior," pp. 31-33 o Section 1 ("Group Benefit or Individual Advantage?") of *Readings i. Sociobiology*, ed. T.H. Clutton-Brock and Paul H. Harvey (San Francisco: W.H. Freeman and Company, 1978). [Ch. 10, note 26]

Hapgood, Fred. *Why Males Exist: an inquiry into the evolution of sex* New York: Morow, 1979. [Ch. 1, note 4]

Hardin, Garrett, assembler. *Population, Evolution, and Birth Control*, 2 ed. San Francisco: W.H. Freeman Co., 1969. [Ch. 16, note 19]

Harris, Henry. "Rationality in Science," Chapter 3 (pp. 36-52) in *Scientifi Explanation*, ed. A.F. Heath. [Ch. 7, note 17]

Harris, Marvin. *Cows, Pigs, Wars and Witches: The Riddles of Culture* New York: Random House, 1974. [Ch. 10, note 7]

Harrison, Simon J. *Stealing People's Names: History and Politics in Sepik River Community.* Cambridge University Press, 1990. [Ch. 2 note 47]

Hartshorne, Charles. *Creative Synthesis and Philosophic Method.* Lanham Maryland: University Press of America, 1983. [Ch. 8, note 27]

Hawkes, Jacquetta, ed. *The World of the Past.* New York: Alfred A Knopf, 1963. [Ch. 5, note 9]

Haydon, A. Eustance. *Biography of the Gods.* New York: The Macmilla Company, 1941). [Ch. 8, note 2]

Heath, A.F., ed. *Scientific Explanation.* Oxford: Clarendon Press, 1981 [Ch. 7, notes 17 and 23]

Hebb, David O. "Epilogue" (pp. 220-223) to *The Mind-Body Problem* by Mario Bunge. New York: Pergamon Press, 1980. [Ch. 10, note 9

Hefner, Philip. "Entrusting the Life That Has Evolved," pp. 67-73 i *Zygon: Journal of Religion & Science* Vol. 29 No. 1 (March 1994) [Ch. 10, note 24]

_____. *The Human Factor.* Minneapolis: Fortress Press, 1993. [Preface note 11]

Heider, George C. *The Cult of Molek: A Reassessment.* Sheffield England: JSOT Press, 1985. [Ch. 5, note 26]

Heimer, Lennart and Robards, Martine J., eds. *Neuroanatomical Tract Tracing Methods.* New York: Plenum Press, 1981. [Ch. 2, note 18

Hendrickx, Herman. *The Miracle Stories of The Synoptic Gospels.* Sa Francisco: Harper & Row, 1987. [Ch. 8, note 11]

Hess, E.H. *Imprinting: Early Experience and the Developmental Psychology of Attachment*. New York: D. Van Nostrand Reinhold Company, 1973. [Ch. 7, note 1]

Hewitt, J.K., Fulker, D.W., and Broadhurst, A.L., "Genetic architecture of escape-avoidance conditioning in laboratory and wild populations of rats: A biometrical approach," *Behavior Genetics*, 1981, 11, 533-544. [Ch. 2, note 33]

Hewitt, Paul L. and Flett, Gordon L. "Perfectionism in the Self and Social Contexts: Conception, Assessment, and Association with Psychpathology," in *Journal of Personality and Social Psychology*, Vol. 60 No. 3 (March 1991), pp. 456-470. [Ch. 12, note 23]

Hick, John, ed. *The Myth of God Incarnate*. Philadelphia: The Westminster Press, 1977. [Ch. 8, note 18]

Hikins, James W. and Zagacki, Kenneth S. "Rhetoric, Philosophy, and Objectivism: An Attenuation of the Claims of the Rhetoric of Inquiry," in *Quarterly Journal of Speech* 74 (1988). [Ch. 7, note 31]

Hill, J.G. 1972. "On the evolutionary foundations of language." *American Anthropologist* 74:309. [Ch. 2, note 51]

Hirshleifer, Jack. "Economics from a Biological Viewpoint," Chapter 10 (pp. 145-184) of *Biology and the Social Sciences: An Emerging Revolution*, ed. Thomas Wiegele. [Ch. 13, note 11]

Honans, George C. "The Relevance of Psychology to the Explanation of Social Phenomena," in *Explanation in the Behavioral Sciences*, ed. Robert Borger. Cambridge at the University Press, 1970. [Ch. 5, note 12]

Horgan, John. "Report on the Correlates of War project," in *Scientific American* for April 1991, pp. 24-5. [Ch. 14, note 10]

Hornung, Erik. *Conceptions of God in Ancient Egypt*. Ithaca: Cornell University Press, 1982. [Ch. 5, note 11]

Horton, Tom. "The Endangered Species Act: Too tough, Too weak, or Too late?" in *Audubon*, Vol. 94, No. 2 (March/April 1992), pp. 68-74. [Ch. 14, note 8]

Howell, F.C. *Early Man*. New York: Time-Life Books, 1965. [Ch. 3, note 15]

Hrdy, B.S. 1974. "Male-male competition and infanticide among the langurs (Presbytis entellus) of Abu, Rajasthan." *Folia primatil.* 22:19-58. [Ch. 10, note 3]

Hunsinger, George and Placher, William C., eds. *Types of Christian Theology*, by Hans W. Frei. New Haven, Connecticut: Yale University Press, 1992. [Preface, note 4]

Hunt, Earl B. *Concept Learning*. New York: John Wiley and Sons, 1966. [Ch. 4, note 4]

Huyek, Margaret and Hoyer, William J. *Adult Development and Aging*. Belmont, California: Wadsworth Publishing Company, 1982. [Ch. 12, note 26]

Irons, William. "How Did Morality Evolve?" *Zygon: Journal of Religion & Science*, Vol 26 no. 1 (March 1991), pp. 49-89. [Ch. 3, note 22; Ch. 10, note 1]

_____. Personal communication dated April 25, 1995. [Ch. 3, note 20]

Jackson, Margaret Y. *The Struggle for Freedom: Phase I.* Chicago: Adams Press, 1976. [Ch. 6, note 5]

Janis, Irving L. *Groupthink: Psychological Studies of Policy Decisions and Fiascoes*, 2d ed. Boston: Houghton Mifflin Company, 1983. [Ch. 15, note 22]

Jarvie, I.C. "Understanding and Explanation in Sociology and Social Anthropology," in *Explanation in the Behavioral Sciences*, ed. Robert Borger. Cambridge at the University Press, 1970. [Ch. 5, note 5]

Joad, Cyril Edwin Mitchinson. *The Future of Life, a Theory of Vitalism*. London and New York: G.P. Putnam's Sons, 1928. [Ch. 2, note 15]

Johanson, Donald and Shreeve, James. *Lucy's Child: The Discovery of a Human Ancestor*. New York: Avon Books, 1990. [Ch. 3, note 11]

Johnson, Allen W. and Earle, Timothy. *The Evolution of Human Societies: From Foraging Group to Agrarian State*. Stanford, California: Stanford University Press, 1987. [Ch. 3, note 3; Ch. 13, notes 1, 2, and 4]

Johnston, D. Kay. "Adolescents' Solutions to Dilemmas in Fables: Two Moral Orientations--Two Problem Solving Strategies," pp. 49-72 in *Mapping the Moral Domain*, ed. Carol Gilligan, Janie Victoria Ward, and Jill McLean Taylor. [Ch. 10, note 7]

Johnston, Timothy D. "Learning and the evolution of developmental systems," Chapter 19 of *Learning, Development, and Culture*, ed. H.C. Plotkin. [Ch. 4, note 18]

Jones, Bernard E., ed. *Earnest Enquirers After Truth.* London: George Allen & Unwin Ltd., 1970. [Preface, note 3]

Jones, James W. "Personality and Epistemology: Cognitive Social Learning Theory as a Philosophy of Science." In *Zygon: Journal of Religion & Science* Vol. 24 No. 2 (March, 1989), pp. 23-38. [Ch. 5, note 14]

Jones, Linda Jarchow. "Wildflowers and Wonder: A Pastor's Wanderings in the Religion-Science Wilderness," in *Zygon*: Journal of Religion & Science, Vol. 29 No. 1 (March 1994), pp. 115-125. [Ch. 16, note 11]

Jordan, Michael. *Gods of the Earth.* New York: Bantam Press, 1992. [Ch. 8, note 2]

Jowett, B., analyst and translator. *The Dialogues of Plato*, Vol. 1. New York: Charles Scribner's Sons, 1907. [Ch. 6, note 15]

Joyce, James Avery, ed. *World Population Basic Documents*, Vol. 4. Dobbs Ferry, New York: Ocenia Publications, Inc., 1976. [Ch. 13, note 6]

Judson, Horace Freeland. *The Search for Solutions*. Baltimore: The Johns Hopkins University Press; 1987. [Ch. 4, note 12]

Kahneman, Daniel; Slovic, Paul; and Tversky, Amos; eds. *Judgment Under Uncertainty: Heuristics and Biases*. Cambridge, England: Cambridge University Press, 1982. [Ch. 9, note 19]

Kaltenbach, Jane. "Endocrine Aspects of Homeostasis." In *Science as a Way of Knowing," Vol. 5 (Form and Function)*. Published by the American Society of Zoologists, University of California, Riverside, 1988. [Ch. 2, note 25]

Kane, Robert and Phillips, Stephen H., eds. *Hartshorne, Process Philosophy, and Theology*. Albany: State University of New York, 1989. [Ch. 8, notes 25, 26, and 28]

Kauffman, Stuart A. "Self-Organization, Selective Adaptation, and Its Limits," pp. 169-207 in *Evolution at a Crossroads*, David J. Depew and Bruch H. Weber, eds. The MIT Press, Cambridge, Massachusetts, 1985. [Ch. 9, note 10]

Kaufman, Gordon. *The Theological Imagination*. Philadelphia: The Westminster Press, 1981. [Ch. 3, note 26; Ch. 5, note 1 and conclusion]

_____. *In Face of Mystery*. Cambridge: Harvard University Press, 1993. [Preface, note 10]

Kaufman, Kenn. "The Subject is Alex." *Audubon*, September-October 1991 (v. 93 n. 5 p. 52 (7). [Ch. 2, note 45]

Kaufman, Peter. *Redeeming Politics*. Princeton, New Jersey: Princeton University Press, 1990. [Ch. 10, note 17]

Kawai, M. 1965. "Newly acquired precultural behavior of the troop of Japanese monkeys on Koshima islet." *Primates* 6:1-30. Also see *Animals* 5:450-455. [Ch. 2, note 48; Ch. 3, note 23; Ch. 5, note 3]

Keesing, F.M. *Cultural Anthropology*, New York: Rinehart 1958. [Ch. 3, note 2]

Kendrick, John W. "Happiness is Personal Productivity Growth," in *Challenge* Vol. 30 No. 2 (May/June 1987), pp. 37-44. [Ch. 12, note 14]

Kennedy, John Studdert. *I Believe: Sermons on the Apostles' Creed*. New York: George H. Doran Company, 1921. [Ch. 8, note 17]

_____. *Lies!* London: Hodder and Stoughton Ltd., 1919. [Ch. 8, note 17]

Kennedy, Paul. *Preparing for the Twenty-First Century*. New York: Random House, 1993. [Ch. 14, note 3]

Klee, Robert L. "Micro-determinism and Concepts of Emergence," in *Philosophy of Science* 51:44-63 (March 1984). [Ch. 2, note 20]

Klein, Richard G. *The Human Career*. Chicago: University of Chicago Press, 1989. [Ch. 3, note 5]

Kliever, Lonnie D. *The Shattered Spectrum*. Atlanta: John Knox Press, 1981. [Ch. 8, note 16]

Kline, Meredity G. *By Oath Consigned.* Grand Rapids, Michigan: William B. Eerdmans Publishing Company, 1968. [Ch. 6, note 8]

Kohlberg, Lawrence, with Charles Levine and Alexandra Hewer. *Moral Stages: A Current Formulation and a Response to Critics*. New York: Karger, 1983. [Ch. 7, note 6; Ch. 10, note 10]

Konishi, Masakazu; Emlen, Stephen T.; Ricklefs, Robert E.; and Wingfield, John C. "Contributions of Bird Studies to Biology," *Science* October 27, 1989, pages 465-472. [Ch. 9, notes 16 and 18]

Kramer, Peter D. *Listening to Prozac*. New York: Penguin Books, 1993. [Ch. 1, note 10; Ch. 11, note 25]

Kroeber, Alfred Louis and Kluckhohn, Clyde. "Culture, a critical review of the concepts and definitions," *Papers Peabody Mus. Am. Archeol. Ethnol.* 47(1):1-223 [33-38], 1952. [Ch. 3, note 1]

Kuenne, Robert E. *General Equilibrium Economics, Part I.* New York: New York University Press, 1992. [Ch. 14, note 18]

Kuhn, Thomas. *The Structure of Scientific Revolutions*. Chicago: University of Chicago Press, 1962. [Ch. 7, note 23]

_____. *The Essential Tension.* Chicago: University of Chicago Press, 1977. [Ch. 7, note 23]

Kurtz, Paul; Alcock, James; Frazier, Kendrick; Karr, Barry; Klass, Philip J.; and Randi, James. "Testing Psi Claims in China: Visit of CSICOP Delegation," in *The Skeptical Inquirer* Vol. XII, No. 4 (Summer 1988). [Ch. 8, note 14]

Ladd, John. *The Structure of a Moral Code*. Cambridge, Massachusetts: Harvard University Press, 1957. [Ch. 11, note 10]

LaMar, Donna F. *Transcending Turmoil: Survivors of Dysfunctional Families*. New York: Plenum Press, 1992. [Ch. 7, note 2]

Langmuir, Irving. "Pathological Science," in *Physics Today* Vol. 42, No. 10 (October 1989), pp. 36-48. [Ch. 7, note 36]

Lasswell, Marcia E. and Lobsenz, Normal M. *No-Fault Marriage*. Garden City, New York: Doubleday, 1970. [Ch. 12, note 27]

Lawrence, Irene. *Linguistics and Theology: The Significance of Noam Chomsky for Theological Construction*. London: The American Theological Library Association, 1980. [Ch. 8, note 16]

Lenneberg, Eric. "A biological perspective on language." In *Evolutionary Anthropology*. H. Bleibtreu, ed. [Ch. 2, note 58]

Lenski, Gerhard. *The Religious Factor: A Sociological Study of Religion's Impact on Politics, Economics, and Family Life*. Garden City, New York: Doubleday, 1961. [Ch. 11, note 11]

Leslie, John, ed. *Physical Cosmology and Philosophy*. New York: Macmillan, 1990. [Ch. 8, note 21]

Levi, Isaac. *Hard Choices*. London: Cambridge University Press, 1986. [Ch. 12, note 28]

Lewis, Hunter. *A Question of Values*. San Francisco: Harper & Row, 1990. [Ch. 12, note 25]

Lewontin, Richard C. *Human Diversity*. New York: Scientific American Books, 1982. [Ch. 1, note 11]

_____. "Organism and environment," Chapter 9 of *Learning, Development, and Culture*, ed. H.C. Plotkin. [Ch. 1, note 18]

_____. Introduction to *Scientists Confront Creationism*, ed. Laurie R. Godfrey (New York: W.W. Norton & Company, 1983). [Ch. 1, note 2]

_____; Rose, Steven; and Kamin, Leon. *Not in Our Genes*. New York: Pantheon Books, 1984. [Ch. 1, note 12; Ch. 7, note 15, Ch. 10, note 32]

Lieberman, Philip. *The Biology and Evolution of Language*. Cambridge, Mass: Harvard University Press, 1984. [Ch. 2, note 52]

_____; Crelin, E.S.; and Klatt, D. 1972. "Phonetic ability and related anatomy of the newborn and adult human, Neanderthal man, and the chimpanzee." *American Anthropologist* 74:287-307. [Ch. 2, note 51]

Lindbeck, George A. *The Nature of Doctrine*. Philadelphia: The Westminster Press, 1984. [Ch. 6, note 6; Intro. Part 3, note 1]

Locke, John. *Essay Concerning Human Understanding*. Oxford: Clarendon Press, 1690. [Ch. 4, note 14]

Long, Charles H. *Alpha: The Myths of Creation*. Toronto: Ambassador Books, 1963. [Ch. 5, note 4]

Lopreato, Joseph. *Human Nature & Biocultural Evolution*. (Boston: Allen & Unwin, 1984. [Ch. 6, note 13]

Lorenz, Konrad. "A consideration of methods of identification of species-specific instinctive behavior patterns in birds" (1932). In *Studies in Animal and Human Behavior* Vol. I, pp. 57-100. [Ch. 9, note 3]

_____. "Companions as Factors in the Bird's Environment" (1935). In *Studies in Animal and Human Behavior* Vol. I, pp. 101-258. [Ch. 11, note 4]

_____. "The establishment of the instinct concept" (1937). In *Studies in Animal and Human Behavior* Vol. I, pp. 259-315. [Ch. 2, note 32]

_____. "Inductive and Teleological Psychology" (1942). In *Studies in Animal and Human Behavior* Vol. I, pp. 351-370. [Ch. 8, note 9]

_____. "Psychology and Phylogeny" (1954). In *Studies in Animal and Human Behavior* Vol. II, pp. 202-209. [Ch. 2, note 29]

_____. "Gestalt Perception as a Source of Scientific Knowledge" (1959). In *Studies in Animal and Human Behavior* Vol. II, pp. 281-322. [Ch. 4, note 10; Ch. 7, note 14; Ch. 15, note 13]

_____. "Do animals undergo subjective experience?" (1963). In *Studies in Animal and Human Behavior* Vol. II, pp. 323-337. [Ch. 2, note 43]

_____. *On Aggression*, translated by Marjorie Kerr Wilson. New York: Harcourt, Brace & World; 1963. [Ch. 3, note 22; Ch. 9, note 17; Ch. 14, note 19; Ch. 16, note 12]

_____. "Innate bases of learning." In *On the Biology of Learning*, ed. K. Pribram. New York: Harcourt, Brace and World, 1969. [Ch. 4, note 18]

_____. *Studies in Animal and Human Behavior* (2 volumes). Cambridge, Massachusetts: Harvard University Press, 1971. [Ch. 1, note 5; Ch. 2, notes 29, 31, and 43; Ch. 4, note 10; Ch. 7, notes 14 and 15; Ch. 11, note 4; Ch. 15, note 13]

_____. *The Year of the Greylag Goose*, translated by Robert Martin. New York: Harcourt Brace Jovanovich, 1978. [Ch. 9, note 17]

_____. "The Fashionable Fallacy of Dispensing with Description," in *Konrad Lorenz: The Man and His Ideas*, ed. Richard I. Evans. [Ch. 10, note 30.]

_____. *The Waning of Humanness*, translated by Robert Warren Kickert. Boston: Little, Brown and Company, 1987. [Ch. 15, notes 1 and 17]

Lynch, Edward A. *Religion and Politics in Latin America: Liberation Theology and Christian Democracy*. New York: Praegor, 1991. [Ch. 6, note 11]

Lynch, Gary. *Synapses, Circuits, and the Beginnings of Memory*. Cambridge, Mass: The MIT Press, 1986. [Ch. 2, note 13]

McClary, Robert A. and Moore, Robert Y. *Subcortical Mechanisms of Behavior*. New York: Basic Books, 1965. [Ch. 2, note 31]

McFague, Sallie (See also TeSelle, Sallie McFague). *Metaphorical Theology: Models of God in Religious Language*. Philadelphia: Fortress Press, 1982. [Ch. 5, note 6; Ch. 8, note 16]

_____. *Models of God: Theology for an Ecological Nuclear Age*. Philadelphia: Fortress Press, 1987. [Ch. 8, note 35]

McFarland, David. *Animal Behavior*. Menlo Park, California: The Benjamin/Cummings Publishing Company, Inc. 1985. [Ch. 2, note 42]

McGeer, P.C.: Eccles, J.C.; and McGeer, E.G. *Molecular Neurobiology of the Mammalian Brain*. New York: Plenum Press, 1978. [Ch. 9, note 12]

McGill, Thomas E., ed. *Readings in Animal Behavior*. New York: Holt, Rhinehart and Winston, 1965. [Ch. 2, note 42]

McGreal, Ian. *The Art of Making Choices*. Dallas: Southern Methodist University Press, 1953. [Ch. 12, note 28]

McGrew, W.C. *Chimpanzee Material Culture*. Cambridge: Cambridge University Press, 1992. [Ch. 5, note 2]

Mackey, James P., ed. *Religious Imagination*. Edinburgh: Edinburgh University Press, 1986. [Ch. 5, note 1]

McKinney, Frank; Cheng, Kimberly M.; and Bruggers, David J. "Sperm Competition in Apparently Monogamous Birds," Chapter 15 (pp. 523-545) in *Sperm Competition and the Evolution of Animal Mating Systems*, ed. Robert L. Smith. [Ch. 9, note 16]

Mackintosh, N.J. *Conditioning and Associative Learning*. New York: Oxford University Press, 1983. [Ch. 4, note 6]

McLean v. The Arkansas Board of Education, 529 F. Supp 1255 (1982). [Ch. 1, note 2]

MacLean, Paul D. "On the Origin and Progressive Evolution of the Triune Brain," pp. 291-316 in *Primate Brain Evolution*, ed. Este Armstrong and Dean Falk. [Ch. 2, note 4]

McLuhan, Marshall. *Understanding Media: The Extensions of Man*. New York: McGraw-Hill, 1964. [Ch. 16, note 4]

McMullin, Ernan. "Evolution and Special Creation," in Vol. 28 No. 3 (September 1993) of *Zygon: Journal of Religion & Science*. [Ch. 6, note 20]

Macquarrie, John. *Twentieth Century Religions Thought*. New York: Harper & Row, 1963. [Preface, note 2]

_____. *In Search of Humanity: A Theological & Philosophical Approach*. London: SCM Press Ltd., 1982. [Ch. 11, note 23]

Malinowski, Bronislow. *Magic, Science and Religion*. Glencoe, Illinois: The Free Press, 1954. [Ch. 16, notes 6 and 7]

Mandell, Arnold J. "Toward a Psychobiology of Transcendence: God in the Brain," Chapter 14 (pp. 379-464) of *The Pschobiology of Consciousness*, ed. Julian M. Davidson and Richard J. Davidson. [Ch. 8, note 31]

Manuel, Frank E. *The Eighteenth Century Confronts the Gods*. Cambridge, Massachusetts: Harvard University Press, 1959. [Ch. 5, note 57]

Martinez, Joe L. and Kesner, Raymond P., eds. *Learning and Memory: A Biological View*. New York: Academic Press, Inc. 1986. [Ch. 2, note 22; Ch. 3, note 28]

Martlew, Margaret, ed. *The Psychology of Written Language*. New York: John Wiley and Sons, 1983). [Ch. 6, note 3]

Massion, J., Paillord, J, Schultz, W. and Wiesendanger, M., eds. *Neural Coding of Motor Performance*. New York: Springer-Verlag, 1993. [Ch. 2, note 17]

Mathews, Shailer. *The Growth of the Idea of God*. The University of Chicago Press, 1931. [Ch. 8, note 3; Ch. 11, note 1]

Matras, Judah. *Introduction to Population*. Englewood Cliffs, New Jersey: Prentice-Hall, Inc., 1977. [Ch. 13, note 6]

Mattingly, Gerald L., "Moabitic Religion," in *Studies in the Mesha Inscription and Moab*, ed. Andrew Dearman. Atlanta: Scholars Press, 1989. [Ch. 5, note 50]

Maynard Smith, John. "The Concepts of Sociobiology." In *Morality as a Biological Phenomenon*, ed. Gunther S. Stent. Berkeley: University of California Press, 1980. [Ch. 3, note 32]

_____. *Evolution and The Theory of Games*. Cambridge University Press, 1982. [Ch. 3, note 21]

Meadows, Dennis L. and Donella H., eds. *Toward Global Equilibrium: Collected Papers*. Cambridge, Massachusetts: Wright-Allen Press, Inc., 1973. [Ch. 13, notes 2 and 8; Ch. 14, notes 11, 12, 13, and 16; Ch. 16, note 14]

Meadows, Donella; Richardson, John; and Bruckmann, Gerhart. *Groping in the Dark*. New York: John Wiley & Sons, 1982. [Ch. 14, note 13]

Meadows, Donella; Meadows, Dennis L.; and Randers, Jorgen. *Beyond the Limits of Growth*. Post Mills, Vermont: Chelsea Green Publishing, 1994. [Ch. 14, note 13]

Meek, T.J. *Hebrew Origins*. Toronto: University of Toronto Press, 1950. [Ch. 5, notes 33 and 48]

Meier, Deborah. "Reinventing Teaching," in Vol. 93 No. 4 (Summer 1992) of *Teachers College Record*. [Ch. 15, note 10]

Mérö, László. *Ways of Thinking*. London: World Scientific, 1990. [Ch. 7, note 6]

Midgley, Mary. "Toward a New Understanding of Human Nature: The Limits of Individualism," Ch. 18 (pp. 517-533) in *How Humans Adapt: A Biocultural Odyssey*, ed. Donald J. Ortner. [Preface, note 15; Ch. 3, note 17]

Mill, John Stuart. *Principles of Political Economy*. University of Toronto Press, 1965. [Ch. 13, note 17]

Miller, Patrick D.; Hanson, Paul D.; and McBride, S. Dean, eds. *Ancient Israelite Religion*. Philadelphia: Fortress Press, 1987. [Ch. 5, notes 24 and 35].

Miller, Perry; Calhoun, Robert L.; Pusey, Nathan M.; and Niebuhr, Reinhold. *Religion and Freedom of Thought*. New York: Doubleday and Company, 1954. [Ch. 16, note 17]

Moody, Paul Amos. *An Introduction to Evolution*, 3rd ed. New York: Harper & Row, 1970. [Ch. 1, note 1]

Moore, John. "Evolutionary Biology." In *Science as a Way of Knowing," Vol. 1 (Evolutionary Biology)*. Riverside, California: American Society of Zoologists, 1984. [Ch. 15, note 11]

_____. "Human Ecology." In *Science as a Way of Knowing," Vol. 2 (Human Ecology)*. Riverside, California: American Society of Zoologists, 1984. [Ch. 13, note 9]

_____. *Science As a Way of Knowing: The Foundations of Modern Biology*. Cambridge, Massachusetts: Harvard University Press, 1993. [Ch. 7, note 22]

_____. "Understanding Nature—Form and Function." In *Science as a Way of Knowing," Vol. 5 (Form and Function)*. Riverside, California: American Society of Zoologists, 1988. [Ch. 8, note 10]

Moravian, Alfred. *Silent Messages: Implicit Communication of Emotions and Attitudes*, 2d ed. Belmont, California: Wadsworth Publishing Company, 1981. [Ch. 2, note 53]

Morishima, Michio. *Capital and Credit*. Cambridge University Press, 1992. [Ch. 14, note 18]

Munz, Peter. "Philosophy and the Mirror of Rorty" pp. 345-398 in *Evolutionary Epistemology, Rationality, and the Sociology of Knowledge*, ed. W.W. Bartley and G. Radnitzky. [Ch. 7, note 37]

Murphy, Gardner. *Personality: A Biosocial Approach to Origins and Structure*. New York: Harper & Brothers, 1947. [Ch. 9, note 6]

Murphy, Jeffrie G. *Evolution, Morality, and the Meaning of Life*. Totowa, New Jersey: Rowman and Littlefield, 1982. [Ch. 11, note 9]

Murphy, Nancey. "Truth, Relativism, and Crossword Puzzles," *Zygon: Journal of Religion and Science* Vol. 24, No. 3, (September 1989), pp. 299-314. [Ch. 7, note 19]

_____. *Theology in the Age of Scientific Reasoning*. Ithaca, New York: Cornell University Press, 1990. [Ch. 7, note 38]

_____. Personal communication dated April 8, 1995. [Ch. 7, note 16]

Myers, J.B. and McCaulley, M.H. *Manual: A guide to the development of the Myers-Briggs Type Indicator*. Palo Alto, California: Consulting Psychologists Press, 1985. [Ch. 7, note 6]

Neill, A.S. *Summerhill School*. New York: St. Martin's Press, 1992. [Ch. 15, note 18]

Nelkin, Dorothy. *The Creation Controversy*. W. W. Norton & Company 1982, New York. [Ch. 1, note 2]

Neusner, Jacob. *Talmudic Thinking: Language, Logic, Law*. Columbia: University of South Carolina Press, 1992. [Ch. 6, note 9]

Newcombe, Nora S. and Baenninger, Maryann: "Biological Change and Cognitive Ability in Adolescence," in *Biology of Adolescent Behavior and Development*, ed. Gerald R. Adams, Raymond Montemayor, and Thomas P. Gullota. Newbury Park, California: Sage Publications, 1989. [Ch. 4, note 19]

Nicholson, Ernest. *God and His People: Covenant and Theology in the Old Testament*. Oxford: Clarendon Press, 1986. [Ch. 5, note 31]

Niebuhr, Reinhold. "The Commitment of the Self and the Freedom of the Mind," pp. 55-64 of *Religion and Freedom of Thought*, by Perry Miller, Robert L. Calhoun, Nathan M. Pusey, and Reinhold Niebuhr New York: Doubleday and Company, 1954. [Ch. 16, note 17]

Nitecki, Matthew H. and Doris V., eds. *Evolutionary Ethics*. Albany State University of New York Press, 1993. [Ch. 10, note 5]

Nussbaum, Martha and Sen, Amartya, eds. *The Quality of Life*. New York Oxford University Press, 1993. [Ch. 13, note 13]

O'Dea, Thomas and Aviad, Janet O'Dea. *The Sociology of Religion*, 2d ed. Englewood Cliffs, New Jersey: Prentice-Hall, Inc, 1983. [Ch. 16, notes 1 and 9]

Ortner, Donald J., ed. *How Humans Adapt: A Biocultural Odyssey*. Washington, D.C.: Smithsonian Institution Press, 1983. [Ch. 3, note 17; Intro. Part 4, note 1; Ch. 16, note 18]

Ottati, Douglas F. *Meaning and Method in Richard Niebuhr's Theology*. Lanham, Maryland: University Press of America, 1982. [Ch. 6, note 20]

Palombo, Stanley R. *Dreaming and Memory*. New York: Basic Books, 1977. [Ch. 4, note 21]

Pannenberg, Wolfhart. "Theological Appropriation of Scientific Understandings: Response to Hefner, Wicken, Eaves, and Tipler," in Vol. 24 No. 2 (June 1989) of *Zygon: Journal of Religion & Science* pp. 255-271. [Ch. 8, note 8]

Patai, Raphael. *The Hebrew Goddess*, 3d ed. Detroit: Wayne State University Press, 1990. [Ch. 5, note 24]

Patterson, Michael and Kesner, Raymond, eds. *Electrical Stimulation Techniques*. New York: Academic Press, 1981. [Ch. 2, note 8]

Peacocke, Arthur. *God and the New Biology*. London: Dent, 1986. [Preface, notes 6 and 9; Ch. 7, note 38]

_____. *The Sciences and Theology in the Twentieth Century*. Notre Dame, Indiana: University of Notre Dame Press, 1981. [Preface, note 7]

_____. *Creation and the World of Science*. New York: Oxford University Press, 1979. [Preface, note 6]

Peck, M. Scott. *The Road Less Travelled*. New York: Simon and Schuster, 1978. [Ch. 9, note 1]

_____. *People of the Lie: The Hope for Healing Human Evil*. New York: Simon and Schuster, 1985. [Ch. 10, note 21]

Perkinson, Henry. *Learning from our Mistakes: A Reinterpretation of Twentieth-Century Educational Theory*. Westport, Connecticut: Greenwood Press, 1984. [Ch. 4, note 10; Ch. 15, note 7]

Pfeiffer, J.E. *The Emergence of Man*. New York: Harper and Row, 1972. [Ch. 3, note 16] 'helps, Edmund S. *Structural Slumps*. Cambridge, Massachusetts: Harvard University Press, 1994. [Ch. 14, note 18]

'iaget, Jean. *Play, Dreams and Imitation in Childhood*. London: William Heinemann Ltd., 1951. [Ch. 4, note 10]

____. *The Principles of Genetic Epistemology*. New York: Basic Books, Inc. 1972; translated from the French by Wolfe Mays. [Ch. 15, note 5]

'lato. "Phaedo." In *The Dialogues of Plato*, Vol. 1, translated and analyzed by B. Jowett. New York: Charles Scribner's Sons, 1907. [Ch. 6, note 15; see also Ch. 15, note 1]

'lotkin, H.C., ed. *Learning, Development, and Culture*. New York: John Wiley & Sons, 1982. [Ch. 1, notes 13 and 18; Ch. 3, note 13; Ch. 4, notes 2, 8, 13, and 18]

____ and Odling-Smee, F.J. "Learning in the context of a hierarchy of knowledge gaining processes," chapter 20 of *Learning, Development, and Culture*, ed. H.C. Plotkin. [Ch. 4, notes 8 and 13]

'oirier, Frank E. "The St. Kitts Green Monkey" in *Folia Primatologica* 17:20-55 (1972). [Ch. 9, note 11]

'olgar, Steven, ed. *Population, Ecology, and Social Evolution*. The Hague: Mouton Publishers, 1975. [Ch. 13, note 4]

'olkinghorne, John C. *Reason and Reality: The Relationship Between Science and Theology*. London: SPCK, 1991. [Ch. 7, note 38]

____. *Science and Providence*. London: SPCK, 1989. [Ch. 8, note 21]

____. "The Nature of Physical Reality" pp. 221-236 in *Zygon: Journal of Religion & Science*, Vol. 26 No. 2 (June 1991). [Ch. 8, note 21]

'opper, Sir Karl. *A World of Propensities*. Bristol, England: Thoemmes, 1990. [Ch. 7, note 15, Ch. 15, note 13]

'ostel, Sandra. *Last Oasis: Facing Water Scarcity*. Washington, D.C.; Worldwatch Institute, 1992. [Ch. 14, note 5]

'ovinelli, Daniel J. and Godfey, Laurie R. "The Chimpanzee's Mind: How Noble in Reason? How Absent of Ethics?" pp. 277-324) in *Evolutionary Ethics*, ed. Matthew H. and Doris V. Nitecki. [Ch. 10, note 5],

'remack, A.J. and D.D. 1972; "Teaching language to an ape" *Scientific American* 227 (4):92-99) [Ch. 2, note 50]

'ribram, K., ed. *On the biology of learning*. New York: Harcourt, Brace and World, 1969. [Ch. 4, note 18]

'rideaux, Tom. *Cro-Magnon Man*. New York: Time-Life Books, 1978. [Ch. 3, note 25]

'ritchard, James B., ed. *Ancient Near Eastern Texts Relating to the Old Testament*, 3rd ed. Princeton: Princeton University Press, 1969. [Ch. 5, note 41]

'ryor, Karen. *Don't Shoot the Dog!* New York: Bantam Books, Inc., 1985. [Ch. 15, note 20]

Pugh, George E. *Biological Origin of Human Values*. New York: Basi Books, 1977. [Ch. 12, notes 16 and 20]
Purtill, Richard L. "Justice, Mercy, Supererogation, and Atonement," pp 37-50 in *Christian Philosophy*, ed. Thomas P. Fling. [Ch. 11, note 20
Putnam, Hilary. "Philosophers and Human Understanding," Chapter 6 c *Scientific Explanation*, edited by A.F. Heath. [Ch. 7, note 23]
Rand, Ayn. *The Virtue of Selfishness*. New York: New American Library 1965. [Ch. 13, note 16]
Randers, Jorgen and Meadows, Donella H. "The Carrying Capacity o Our Global Environment: A Look at the Ethical Alternatives," pp. 315 335 in *Toward Global Equilibrium*, ed. Dennis and Donella Meadows [Ch. 13, note 8; Ch. 14, note 16]
Rapaport, David. "A Historical Survey of Psychoanalytic Ego Psychol ogy," pp. 5-17 of *Psychology Issues* Vol. 1 no. 1. New York International University Press, 1959. [Ch. 7, note 9]
Raschke, Carl A. *The Alchemy of the Word: Language and the End c Theology*. Missoula, Montana: Scholars' Press, 1979. [Ch. 8, note 30
Raven, Charles E. *Natural Religion and Christian Theology*. London Cambridge at the University Press, 1953. [Preface, note 3]
Reimer, Joseph; Paolitto, Diana Pritchard; and Hersh, Richard H. *Promot ing Moral Growth*. Prospect Heights, Illinois: Waveland Press, Inc 1983. [Ch. 15, note 17]
Rensch, Bernhard. *Biophilosophy*. New York: Columbia Universit Press, 1971, translated by C.A.M. Sym. [Ch. 2, note 14]
_____. *From man to demigod*. New York: Columbia University Press 1972. [Ch. 2, note 49]
Rescher, Nicholas. *The Coherence Theory of Truth*. Washington, D.C. University Press of America, 1982. [Ch. 7, note 19]
Reynolds, Vernon. *The Biology of Religion*. London and New York Longman, 1983. [Ch. 7, note 38]
Richardons, Peter. *Paul's Ethic of Freedom*. Philadelphia: The Westminste Press, 1979. [Ch. 15, note 19]
Robertson vs. Seattle Audubon Society et. al, 118 L. Ed. 2nd. 73 (decide March 15, 1992. [Ch. 14, note 7]
Rock, Irvin and Palmer, Stephen. "The Legacy of Gestalt Psychology, *Scientific American* v. 263 No. 6 (December 1990), pp. 84-90. [Ch 7, note 10]
Rolston, Holmes III. "Religion in an Age of Science; Metaphysics in a Era of History." Pp. 65-87 in Vol. 27 No. 1 (March 1992) of *Zygon Journal of Religion & Science*. [Ch. 8, note 22]
_____. "Rights and Responsibilities on the Home Planet, pp. 425-439 i Vol. 28 No. 4 (December 1993) of *Zygon: Journal of Religion & Science*. [Ch. 14, note 6]

_____. "Science Education and Moral Education," pp. 347-355 of Vol. 23 No. 2 (September 1988) of *Zygon: Journal of Religion & Science*. [Ch. 15, note 8]

Rorty, Richard. *Objectivity, Relativism and Truth*. New York: Cambridge University Press, 1991. [Preface, note 16]

Ruimschotel, Dick. *Explanation, Causation, and Psychological Theories*. Amsterdam: Swets and Zeitinger, 1989. [Ch. 5, note 12]

Rue, Loyal D. *Amythia*. Tuscaloosa, Alabama: The University of Alabama Press, 1989. [Preface, note 12; Ch. 5, note 20; Ch. 8, notes 4, 19, 22, and 34; Ch. 10, note 11; Intro. Part 3, note 2; Ch. 16, note 21]

_____. *By the Grace of Guile: The Role of Deception in Natural History and Human Affairs*. New York: Oxford University Press, 1994. [Preface, notes 3 and 17; Ch. 8, note 9; Ch. 9, note 7]

Ruse, Michael. "Genesis Revisited: Can We Do Better than God?" Pp. 297-316 in Vol. 19 No. 3 (September 1984) of *Zygon: Journal of Religion & Science*. [Ch. 1, note 14]

_____. *Sociobiology: Sense or Nonsense?* (Dordrecht, The Netherlands: Reidel, 1985). [Ch. 10, notes 23 and 33]

_____. "Evolutionary and Christian Ethics: Are They in Harmony," pp. 5-24 of *Zygon: Journal of Religion & Science* Vol. 29 No. 1 (March 1994). [Ch. 10, note 24]

_____. "From Belief to Unbelief--and Halfway Back," pp. 25-35 of *Zygon: Journal of Religion & Science* Vol. 29 No. 1 (March 1994). [Ch. 10, note 24]

Sackett, G.P. 1966. "Monkeys reared in isolation with pictures as visual input." *Science* 154:1468-1473. [Ch. 2, note 54]

Sahlin, Marshall. *The Use and Misuse of Biology: An Anthropological Critique of Sociobiology*. Ann Arbor: University of Michigan Press, 1976. [Ch. 10, notes 29 and 32]

Sarna, Nahum M. *Exploring Exodus*. New York: Schocken Books, 1986. [Ch. 5, note 39]

Scherlemann, Robert P., ed. *Naming God*. New York: Paragon House, 1985. [Ch. 5, note 6]

Schmitz-Moormann, Karl. "The Future of Teilhardian Theology," in *Zygon: Journal of Religion and Science* Vol. 30, No. 1 (March 1995), pp. 117-129. [Ch. 8, note 29]

Schoen, Edward L. *Religious Explanations: A Model from the Sciences*. Durham, North Carolina: Duke University Press, 1985. [Ch. 7, note 29]

Schubert, Glendon and Masters, Roger D., eds. *Primate Politics*. Carbondale, Illinois: Southern Illinois University Press, 1991. [Ch. 14, note 1]

Schwartzkopf, Steven H. "Design of a Controlled Ecological Life Suppor System," in *Bioscience*, Vol. 42, No. 7 (July/August 1992), pp. 526-35 [Ch. 13, note 10]

Schwebel, Milton; Maher, Charles A.; and Fagley, Namcy S., eds *Promoting Cognitive Growth Over the Life Span*. Hillsdale, New Jersey: Lawrence Erlbaum Associates, 1990. [Ch. 15, note 12]

Scott, John. *Hunger*. Time, Inc., 1966. [Ch. 13, note 8]

Segal, Robert A. *Joseph Campbell: An Introduction*. New York: Penguir Books, 1987. [Ch. 6, note 14]

_____. "Paralleling Religion and Science: The Project of Robin Horton," pp. 177-198 in *Annals of Scholarship* Vol. 10, No. 2 (Spring 1993) [Ch. 6, note 16]

Seiler, Th. B. and Wannenmacher, W., eds. *Concept Development and the Development of Word Meaning*. Berlin and New York: Springer-Verlag, 1983. [Ch. 4, note 5]

Seitelberger, Franz. "Neurobiological Aspects of Intelligence." In *Concepts and Approaches in Evolutionary Epistemology*, ed. F.M. Wuketits Dordrecht and Boston: D. Reidel Publishing Company, 1984. [Ch. 2 notes 16 and 27]

Sellers, Jane B. *The Death of Gods in Ancient Egypt*. New York: Penguir Books, 1992. [Ch. 5, note 28]

Senner, Wayne M. *The Origin of Writing*. Lincoln: University o Nebraska Press, 1989. [Ch. 6, note 2]

Shakespeare, William. *Henry IV*, Part I, Act III, Scene i, line 53, Henley ed. [Ch. 9, note 24]

Shantzis, Steven B. and Behrens, William W. III. "Population Contro Mechanisms in a Primitive Agricultural Society," Chapter 9 of *Towar Global Equilibrium: Collected Papers*, ed. Dennis L. Meadows anc Donella H. Meadows. [Ch. 13, note 2; Ch. 14, note 12]

Shapiro, Henry L. "Impact of Culture on Genetic Mechanisms." In *The Nature and Transmission of the Genetic and Cultural Characteristic of Human Populations*. Milbank Memorial Fund, 1957. [Ch. 3, note 27]

Shapley, Harlow, ed. *Science Ponders Religion*. New York: Appleton Century-Crofts, Inc., 1960. [Ch. 7, note 38]

Sharp, Daryl. *Personality Types: Jung's Model of Typology*. Toronto Inner City Books, 1987. [Ch. 7, note 6]

Sherry, David F. and Galef, Bennett G. Jr. "Social Learning Withou Imitation: More about milk-bottle opening by birds." *Animal Behavio* November 1990, p. 987. [Ch. 2, note 46]

Shirley, Hunter B. *Mapping the Mind*. Chicago: Nelsen-Hall, 1983. [Ch 2, note 7]

Shoven, John B. and Whalley, John. *Applying Global Equilibrium.* Cambridge University Press, 1992. [Ch. 14, note 18]

Simmell, Edward C.; Hahn, Martin E.; and Walters, James K; eds. *Aggressive Behavior: Genetic and Neural Approaches.* Hillsdale, New Jersey: Lawrence Erlbaum Associates, 1983. [Ch. 1, note 10]

Simpson, George Gaylord. *Horses; The Story of the horse family in the modern world and through sixty million years of history.* Garden City, N.Y.: Doubleday, 1961. [Ch. 2, note 5]

Sinnet, Edmund W. *The Biology of Spirit*, New York 1955, The Viking Press. [Ch. 9, note 8]

Skinner, B.F. *Verbal Behavior.* New York: Appleton-Century-Crofts, 1957. [Ch. 4, note 8]

Smart, Ninian. *The Religious Experience of Mankind.* New York: Charles Scribners Sons, 1984 (3d ed.). [Ch. 5, note 37]

Smith, Morton. *Palestinian Parties and Politics That Shaped the Old Testament.* London: SCM Press Ltd., 1987. [Ch. 5, note 53]

Smith, Quentin. "The Anthropic Coincidences, Evil and the Disconfirmation of Theism." In *Religious Studies* Vol. 28 No. 3 (September 1992), pp. 347-50. [Ch. 8, note 21]

Smith, Robert L., ed. *Sperm Competition and the Evolution of Animal Mating Systems.* New York: Academic Press, Inc., 1984. [Ch. 9, note 16]

Snelling, N.J., ed. *The Chronology of the Geological Record.* London: Blackwell Scientific Publications, 1985. [Ch. 1, note 3]

Sperry, Roger W. "A modified Concept of Consciousness," *Psychological Review*, 76:532-536, 1969. [Ch. 9, note 21]

_____. "Paradigms of Belief, Theory and Metatheory," in *Zygon: Journal of Religion and Science* Vol. 27, No. 3 (September 1992), pp. 245-259. [Ch. 9, note 21]

_____. "Search for Beliefs to Live by Consistent with Science" in *Zygon: Journal of Religion & Science*, Vol. 26, No. 2 (June, 1991). [Ch. 8, note 15]

Spitz, René Arpad and Wolf, Katherine M. "The smiling response: A contribution to the ontogenesis of social relations." *Genetic Psychological Monographs* 34:57-125 (1946). [Ch. 4, note 3]

Spitz, René Arpad. *Dialogues From Infancy*, ed. Robert N. Emde. New York: International Universities Press, 1983. [Ch. 4, note 3]

Springer, Sally P. and Deutsch, Georg. *Left Brain, Right Brain.* New York: W.H. Freeman and Company, 1993, Fourth ed. [Ch. 2, note 10]

Stanley, Christopher D. *Paul and the Language of Scripture.* Cambridge, England: Cambridge University Press, 1992. [Ch. 6, note 9]

Stanesby, Derek. *Science, Reason & Religion.* Dover, N.H.: Croom Helm, 1985. [Ch. 7, note 38]

Star, Susan Leigh. *Regions of the Mind.* Stanford, California: Stanfor University Press, 1989. [Ch. 2, note 9]

Stark, Rodney and Bainbridge, W.S. *A Theory of Religion.* New York Peter Lang Publishers, 1987. [Ch. 5, note 13]

Stent, Gunther S. "Hermeneutics and the Analysis of Complex Biologica Systems" pp. 209-225 in *Evolution at a Crossroads: The New Biolog and the New Philosophy of Science*, ed. David J. Depew and Bruce H Weber. [Ch. 7, note 24; Ch. 9, note 13; Ch. 15, note 3]

_____, ed. *Morality as a Biological Phenomenon.* Berkeley: Universit of California Press, 1980. [Ch. 3, note 32; Ch. 10, notes 1, and 19]

Strickberger, Monrow W. "The Structure and Organization of Geneti Material," pp. 769-780 of *Science as a Way of Knowing: III--Genetics* (The American Society of Zoologists, 1985). [Ch. 1, note 9]

Strong, James. *The Exhaustive Concordance of the Bible.* Nashville Tennessee: Abington Press, 1974 ed. [Ch. 6, note 4]

_____. "Greek Dictionary of the New Testament," included in his *Th Exhaustive Concordance of The Bible.* [Ch. 11, note 17]

Swinburne, Richard. "The argument from the fine-tuning of the universe in *Physical Cosmology and Philosophy* ed. John Leslie (New York Macmillan, 1990). [Ch. 8, note 21]

Taylor, Vincent. *The Historical Evidence for the Virgin Birth.* Oxford a the Clarendon Press, 1920. [Ch. 8, notes 12 and 13]

Teilhard de Chardin, Pierre. *The Phenomenon of Man.* New York: Harpe & Brothers Publishers, 1959. [Ch. 8, note 29]

TeSelle, Sallie McFague (see also McFague, Sallie). *Speaking in Parables A Study of Metaphor in Theology.* Philadelphia: Fortress Press, 1975 [Ch. 8, note 16]

Thagard, Paul. *Conceptual Revolutions.* Princeton, New Jersey: Princeto University Press, 1992. [Ch. 7, note 22]

Theissen, Gerd. *Biblical Faith.* London: SCM Press Ltd., 1984, citing A van Selms, "Temporary Henotheism," in *Symbolae Biblicae e Mesopotamicae*, FS F.M.T de Liagre Böhl, Leiden 1973 (pp. 341-8) [Ch. 5, note 49]

_____. *The Miracle Stories of the Early Christian Tradition.* Edinburgh T & T Clark, 1983. [Ch. 8, note 11]

Thomas Aquinas, Saint. *Summa Contra Gentiles, Book One: Go* translated by Anton C. Pegis. London: University of Notre Dam Press, 1975 <1259-1264>. [Ch. 6, note 19]

Tigay, Jeffrey H. *You Shall Have No Other Gods.* Atlanta: Scholars Press 1986. [Ch. 5, note 23]

Tiger, L. and Fox, R. *The Imperial Animal.* New York: Holt, Rinehart an Winston, 1971. [Ch. 3, note 8]

Tillich, Paul. *The Courage to Be.* New Haven: Yale University Press, 1952. [Ch. 11, note 3]

Tipler, Frank J. "The Omega Point as *Eschaton*: Answers to Pannenberg's Questions for Scientists" in Vol. 24 No. 2 (June 1989) of *Zygon: Journal of Religion & Science* pp. 217-253. [Ch. 8, note 8]

Torrance, Thomas F. *Christian Theology and Scientific Culture.* New York: Oxford University Press, 1981. [Ch. 7, note 38]

Triggle, D.J. and C.R. *Chemical Pharmacology of the Synapse.* New York: Academic Press, 1976. [Ch. 4, note 22]

Trivelpiece, Susan G. and Wayne Z. "Anarctica's Well-bred Penguins," Vol. 98 (December 1989) of *Natural History*, pp. 28-37. [Ch. 9, note 15]

Trivers, Robert. *Social Evolution.* Menlo Park, California: The Benjamin/ Cummings Publishing Company, Inc., 1985. [Ch. 9, note 7]

Troeltsch, Ernst. "Religion, Economics, and Society," pp. 197-204 in *Sociology and Religion: A Book of Readings*, ed. Norman Birnbaum and Gertrud Lenzer. [Ch. 16, note 8]

Tunnell, Gary G. *Culture and Biology: Basic Concepts in Anthropology*, pp. 15-16. Minneapolis: Burgess Publishing Company, 1973. [Ch. 2, note 40; Ch. 3, notes 4, 7, 26, and 29]

Turner, Victor. *The Anthropology of Performance.* New York: PAJ Publications, 1987. [Ch. 7, note 8]

____, and Bruner, Edward M., eds. *The Anthropology of Experience.* Chicago: University of Illinois Press, 1986). [Ch. 7, note 8]

Untermann, Isaac. *The Talmud.* New York: Record Press, 1952. [Ch. 6, note 9]

van Selms, A. "Temporary Henotheism," in *Symbolae Biblicae et Mesopotamicae*, FS F.M.T de Liagre Böhl, Leiden 1973, (pp. 341-8). [Ch. 5, note 49]

Veenhoven, Ruut. *Conditions of Happiness.* Boston: D. Reidel Publishing Company, 1984. [Ch. 12, notes 6, 7, and 8]

Verene, Donald Phillip. *Vico's Science of Imagination.* Ithaca: Cornell University Press, 1981. [Ch. 5, note 56]

von Frisch, K. and Lindauer, M. "The 'Language' and Orientation of the Honey Bee," Chapter 32 (pp. 347-358) of *Readings in Animal Behavior*, ed. Thomas E. McGill (New York: Holt, Rhinehart and Winston, 1965). [Ch. 2, note 42]

von Schilcher, Florian and Tennant, Neil. *Philosophy, Evolution, and Human Nature.* London: Routledge & Kegan Paul, 1984. [Ch. 2, note 55]

Vygotsky, Lev S. "The Prehistory of Written Language," Ch. 11 (pp. 279-292) of *The Psychology of Written Language*, ed. Margaret Martlew. [Ch. 6, note 3]

Warren, Jennifer. “Nevadans are Gambling with their Health.” *Lo Angeles Times* for Sunday, August 2, 1992; page A-1. [Ch. 16, note 13

Webber, Thomas L. *Deep Like the Rivers: Education in the Slave Quarte Community 1831-1865.* New York: W.W. Norton & Company Inc. 1978. [Ch. 6, note 5]

Webster's Third New International Dictionary, unabridged. G. & C Merriam Co., Chicago 1976. [Ch. 2, note 39]

Weingartner, Rudolph H. *Undergraduate Education: Goals and Means* Phoenix, Arizona: Oryx Press, 1993. [Ch. 15, note 11]

Weintraub, Boris. “Report of an experiment in learning by an octopus,” i the “Geographica” section (page not numbered) of Vol. 182, No. (December 1992) of *National Geographic.* [Ch. 15, note 4]

Weitz, Shirley. *Sex Roles.* New York: Oxford University Press, 1977. [Ch 4, note 17]

Wenke, Robert J. *Patterns in Prehistory.* Cambridge: Oxford Universit Press, 1984. [Ch. 3, note 16]

Wheeler, Daniel D. and Janis, Irving L. *A Practical Guide for Makin Decisions.* New York: The Free Press, 1980. [Ch. 12, note 28]

White, Andrew D. *A History of the Warfare of Science with Theology i Christendom.* New York: D. Appleton and Company, 1936. [Preface note 3]

White, L., with Dillingham, B. 1973. *The Concept of Culture.* Minneapo lis: Burgess Publishing Co. [Ch. 3, note 8]

White, Lynn Jr. “The Historical Roots of our Ecologic Crisis.” *Scienc* Vol. 155 (March 10, 1967) pp. 1203-7. [Ch. 16, note 18]

Whitehead, Alfred North. *Process and Reality.* New York: The Fre Press, 1978 <The MacMillan Company, 1929>, eds Donald Ra Griffin and Donald W. Sherburne. [Ch. 8, note 24]

Whyte, L.L. *The Unconscious Before Freud.* New York: St. Martin' Press, 1978. [Ch. 7, notes 12 and 13]

Wiegele, Thomas C., ed. *Biology and the Social Sciences: An Emergin Revolution.* Boulder, Colorado: Westview Press, 1982. [Ch. 13, not 11]

_____. *Biopolitics.* Boulder: Westview Press, 1979. [Ch. 14, note 2]

Wiggington, Eliot. *Sometimes a Shining Moment: The Foxfire Experi ence.* Garden City, New York: Anchor Press/Doubleday, 1985. [Ch 15, note 17]

Wiles, Peter and Rath, Guy, ed. *Economics in Disarray.* New York: Bas Blackwell, Inc., 1984. [Ch. 13, note 13]

Williams, Daniel Day. “Priest, Prophet, and the Establishment.” *Zygon Journal of Religion & Science* Vol. 2 No. 4 (December 1967), pp. 309 326. [Ch. 6, note 7]

Williams, G.C. *Adaptation and Natural Selection: A Critique of Current Evolutionary Thought*. Princeton University Press, 1966. [Ch. 2, note 41]

Wilson, David B, ed. *Did the Devil Make Darwin Do It?* The Iowa State University Press, Ames 1983. [Ch. 1, note 2]

Wilson, Edward O. "Sociobiology and the Idea of Progress," pp. 197-216 in *Biological Foundations and Human Nature*, ed. Miriam Balaban. [Ch. 3, note 14]

____. *Sociobiology: The New Synthesis*. Cambridge, Massachusetts: Belknap Press of Harvard University Press, 1975. [Ch. 10, note 22]

____. "The Relation of Science to Theology," pp. 425-434 in Vol. 15 No. 4 (December 1980) of *Zygon: Journal of Religion & Science*. [Ch. 10, note 22]

____. *On Human Nature*. Cambridge: Cambridge University Press, 1978). [Ch. 10, note 27]

____. "Biology and the Social Sciences," pp. 245-263 of Vol. 25, No. 3 (September 1990) of *Zygon: Journal of Religion & Science*. [Ch. 10, note 34; Ch. 13, note 15]

Wilson, J.A., ed. *Ancient Near Eastern Texts Relating to the Old Testament*, 3rd ed. Princeton, New Jersey: Princeton University Press, 1969. [Ch. 5, note 41]

Wilson, Robert R. *Sociological Approaches to the Old Testament*. Philadelphia: Fortress Press, 1984. [Ch. 5, note 32]

Winslow, James T.; Hastings, Nick; Carter, C. Sue; Harbaugh, Carroll R.; and Insel, Thomas R. "A Role for central vasopressin in pair bonding in monogamous prairie voles." *Nature* Vol. 365, No. 6446 (October 7, 1993) pp. 545-548. [Ch. 4, note 16]

Wooley, Sir Leonard. *Abraham*. London: Faber and Faber Limited, 1936. [Ch. 5, notes 18 and 30]

Wuketits, Franz M., ed. *Concepts and Approaches in Evolutionary Epistemology*. Dordrecht and Boston: D. Reidel Publishing Company, 1984. [Ch. 2, notes 16 and 19; Ch. 4, note 15]

____. "Concepts and Approaches in Evolutionary Epistemology," in *Concepts and Approaches in Evolutionary Epistemology*, Franz Wuketits, ed. [Ch. 4, note 15]

Yankner, Bruce A. and Mesulam, M. Marcel. "Beta-amyloid and the pathogenesis of Alzheimer's disease." *New England Journal of Medicine*, December 26, 1991 (v. 325 n. 26), p 1849 (9). [Ch. 2, note 23]

Yarrow, Peter; Stookey, Noel Paul; and Travers, Mary (performing as Peter, Paul, and Mary). "The Great Mandala (The Wheel of Life)," from Album 1700. Burbank, California: Warner Bros. Records, Inc., *Album # 1700*. [Ch. 14, note 9]

York, Derek and Farquhar, Ronald M. *The Earth's Age and Geochronology* (New York: Pergamon Press, 1973). [Ch. 1, note 3]

Zetterberg, J. Peter, ed. *Evolution versus Creationism: The Public Education Controversy*. Phoenix: Oryx Press, 1983. [Ch. 1, note 1]

Index

(For authors see also Bibliography)

-C-

-F-

-G-

-N-

-Q-

-R-

-U-

-V-

-W-

-Y-

-Z-

About the Author

Michael Cavanaugh was reared with two brothers and a sister by his mother and father, the latter a Baptist preacher. He began to study science/religion issues in early adulthood, and when it became financially feasible to do so he laid aside his law practice, assumed the life of an independent scholar, and started exploring those issues in depth. He has been happily married to Carolyn McGinnis Cavanaugh, a mental health counselor, for almost thirty years.

Readers might be surprised to learn that the author has always been an active churchman. He was president of a religious group in college, and has held most local lay positions including Deacon, Board of Trustees Chairman, and Sunday School Superintendent. For ten years he taught University Juniors and Single Adults in Sunday School, and he currently participates in a lively "Courage Class." He sees no inconsistency between participating in church and the ideas in this book; indeed those ideas compel his participation.

Readers may also wonder that a lawyer is writing about biology and theology. Although widely read in both (and in related areas like anthropology, sociology, philosophy, history, ethology, paleontology, genetics, neurology, and psychology, as well as in specialized fields like evolutionary epistemology, biophilosophy, sociobiology, and biblical criticism), the author could not possibly have in-depth knowledge about any of those fields, much less all of them. Therefore it seems reasonable to ask what expertise he brings to the task of writing a book like this.

The answer resides in the nature of the legal profession, which trained him to wade through volumes of diverse materials. More important, his interests and talents there have always been in 1) gathering evidence, 2) deriving concepts from that evidence, and 3) presenting the evidence and concepts in an order designed to help listeners understand his case. Indeed, the legal profession has a distinguished record for producing philosophers and theologians, precisely because its emphasis on evidence carries over into every venue, including not only the courthouse but also the schoolhouse and churchhouse. Thus there is reason to hope the author's reading and thinking may have combined with his professional expertise to produce a practical and meaningful synthesis.

Ultimately, however, it is the judgment of readers, and not the author's background, that will determine whether *Biotheology* is persuasive. It also remains to be judged whether this synthesis can accommodate the many new concepts and facts that will emerge in coming years.